COMPUTER SIMULATION OF SPACE PLASMAS

Advances in Earth and Planetary Sciences

Computer Simulation of Space Plasmas

Edited by

H. Matsumoto

Radio Atmospheric Science Center, Kyoto University, Kyoto, Japan

and

T. Sato

Institute for Fusion Theory, Hiroshima University, Hiroshima, Japan

Terra Scientific Publishing Company

Tokyo, Japan

D. Reidel Publishing Company

A MEMBER OF THE
KLUWER ACADEMIC PUBLISHERS GROUP

Dordrecht / Boston / Lancaster

Library of Congress Cataloging in Publication Data

CIP

Main entry under title:

Computer simulation of space plasmas.

(Advances in earth and planetary sciences)
Includes bibliography and index.
1. Space plasmas -- Data processing. 2. Space plasmas -- Mathematical models. I. Sato, Tetsuya. II. Matsumoto, H. (Hiroshi), 1942- . III. Series.
QC809.P5C62 1985 523.1'12 84-27639

ISBN-13: 978-94-010-8850-3 e-ISBN-13: 978-94-009-5321-5
DOI: 10.1007/978-94-009-5321-5

Published by Terra Scientific Publishing Company (TERRAPUB),
307 Shibuyadai-haim, 4-17 Sakuragaoka-cho, Shibuya-ku, Tokyo 150, Japan,
in co-publication with D. Reidel Publishing Company, Dordrecht, Holland

Sold and distributed in the U.S.A. and Canada
by Kluwer Academic Publishers,
190 Old Derby Street, Hingham, MA 02043, U.S.A.,
in Japan by Terra Scientific Publishing Company (TERRAPUB),
307 Shibuyadai-haim, 4-17 Sakuragaoka-cho, Shibuya-ku, Tokyo 150, Japan

In all other countries, sold and distributed
by Kluwer Academic Publishers Group,
P.O. Box 322, 3300 AH Dordrecht, Holland

PREFACE

Computer simulation is now widely recognized as a powerful tool and useful method at the current stage of research in space plasma physics. The expected role of computer simulation is to bridge the existing gap between theories and experiments/observations and to give a profound physical insight into highly tangled and nonlinearly coupled space plasma phenomena. One of the goals of space plasma physics in 1980's and 1990's is to elucidate the quantitative causal relationships of global and local energy flows in space plasma environment and establish the space plasma physics via cooperative studies among three important elements of observations, theories and computer simulations.

Based on such recognition, Dr. M. Ashour-Abdalla (UCLA/USA), Dr. R. Gendrin (CNET/FRANCE) and both of us met together at the 20th General Assembly of URSI at Washington D. C. in 1981 to discuss what we should do and what we could do, reaching a conclusion that it is time to establish an International School of Space Simulations (ISSS). The objectives of the ISSS thus organized are firstly to educate and stimulate graduate students and young scientists, secondly to exchange information on updated simulation techniques and thirdly to have mutual discussions among observational, theoretical and simulational scientists in the field of space physics. The first ISSS were organized by Prof. P. Coleman, Prof. T. Obayashi, Dr. H. Okuda in addition to the above four members. The first ISSS was held at Kansai Seminar House in Kyoto from Nov. 1 to Nov.12, 1982. More than one hundred scientists and students attended the ISSS from various countries of the world. The ISSS ended in great success with an understanding that this sort of tutorial and academic activities are useful and should be continued. The ISSS is thus planned to be held at every one-and-half or two years in Japan, USA and Europe in turns.

The present volume is edited not as simple proceedings of selected lectures at the First ISSS but rather as a form of textbook of computer simulations of space plasmas for students and young scientists. Therefore only the lectures dealing with simulation techniques and philosophies are selected and other papers on theories and observations given at the First ISSS are not included. The subject index is limited only to the terms related to techniques and concepts used in computer simulations.

Finally we would like to thank Mrs. Keiko Miwa, Mr. Toru Yamada and students of RASC/Kyoto Univ. for their assistance of editorial works and retyping all the materials in this book. Without their efficient helps this book would not have been in the current form.

Kyoto and Hiroshima
September, 1984

H. Matsumoto
T. Sato

CONTENTS

NUMERICALLY-SIMULATED FORMATION AND PROPAGATION OF INTERPLANETARY SHOCKS :179

S. T. Wu

ANOMALOUS TRANSPORT BY KELVIN-HELMHOLTZ INSTABILITIES :203

Akira Miura

PARTICLE BEHAVIOR IN THE MAGNETOSPHERE :227

R. A. Wolf and R. W. Spiro

HYBRID SIMULATION TECHNIQUES APPLIED TO THE EARTH'S BOW SHOCK :255

D. Winske and M. M. Leroy

PART I

PARTICLE SIMULATIONS

INTRODUCTION TO PARTICLE SIMULATION MODELS AND THEIR APPLICATION TO ELECTROSTATIC PLASMA WAVES IN SPACE

H. Okuda

Plasma Physics Laboratory, Princeton University
Princeton, New Jersey 08544, U.S.A.

ABSTRACT

In the first part of this article, a review of plasma simulation models using finite-size particles is given with an emphasis on the electostatic models. Physical properties of such simulation models as well as numerical tecniques will be studied in some detail. Modification to plasma kinetic theory, collisions and fluctuations, Vlasov-Landau equations will be derived along with an introduction to a spatial grid, charge-force interpolation and finite-difference schemes.

In the second part of this article, a study of electrostatic ion cyclotron waves driven by a field-aligned current will be studied by numerical simulations. Recent observations of large amplitude electrostatic ion cyclotron waves ($e\phi/T_e \simeq 1$) and ion conics have renewed wide interest in the physics of auroral field lines. We show by numerical simulations that large amplitude ion cyclotron waves can cause an intense ion heating across magnetic field resulting in the formation of ion conics. In addition, two-dimensional simulations of the electrostatic ion cyclotron waves indicate a possibility of energy condensation into a d.c. structure across magnetic field. Such a d.c. structure is associated with large density modulations, $n_{max}/n_{min} \sim 1$, across magnetic field and may be responsible for the generation of auroral arc elements.

H. Matsumoto and T. Sato, Computer Simulation of Space Plasmas,

PART I: Introduction to Particle Simulation Models in Plasma Physics

1. INTRODUCTION

The subject of the first part of this lecture is numerical plasma simulations using particles in which various models are introduced in order to solve nonlinear Vlasov equation. Physical properties of such simulation models and numerical techniques will be studied in some detail. Some applications to space plasma physics are given in the second part. Recent development in high speed, large scale computers and the advancement of our numerical technique have made it possible to simulate an environment which has direct impact on realistic plasma experiments. Numerical simulation is particularly useful when nonlinear effects are important for which only a limited number of analytic methods are available. Furthermore, one often finds, with a help from numerical simulations, that the results obtained by means of numerical simulations are not what analytic theory predicts or what you have guessed beforehand. One is guided therefore, to develop a different analytic model or a new physical interpretation leading to a discovery of new plasma phenomena.

Among many simulation models, emphasis will be placed on the particle simulation models in which a large number of particles are followed in time in their self-consistent and external electromagnetic fields. Knowledge of such particle orbits or characteristics is equivalent to solving the nonlinear Vlasov equation. Solving the Vlasov equation directly is in general much more difficult than particle simulations, particularly in multi-dimensions. Vlasov equation, which has an infinite degree of freedom, represents a dissipationless plasma in which great care must be taken for numerical smoothing. Paricle distributions can be distorted to an arbitrary complex degree so that numerical smoothing must be introduced in order to avoid numerical divergence. In particle simulation, such smoothing is done most naturally through particle collisions.

In Sec.2, we shall introduce particle simulation models using finite-size particles. Modification to plasma kinetic theory including dispersion relations to linear oscillations, collisions and fluctuations are given. In Sec.3, numerical methods using finite-size particles in a spatial grid are introduced along with finite-difference method. Several specific models useful for low frequency plasma waves and large plasma volume are also discussed.

2. ELECTROSTATIC PARTICLE SIMULATION MODELS

2.1 Kinetic Equations

We shall briefly review several fundamental kinetic equations for a plasma and point out their relation to particle simulation models. One of the fundamental kinetic equations is the Klimontovich equation for N-particle distribution functions for electrons and ions. Defining the distribution function for s-species

$$f_s(\underset{\sim}{x},\underset{\sim}{v},t) = \sum_{j\varepsilon s} \delta(\underset{\sim}{x} - \underset{\sim}{x}_j(t))\, \delta(\underset{\sim}{v} - \underset{\sim}{v}_j(t)), \tag{1}$$

the Klimontovich equation states

$$\frac{df_s}{dt} = \frac{\partial f_s}{\partial t} + \underset{\sim}{v} \cdot \frac{\partial f_s}{\partial \underset{\sim}{x}} + \frac{q_s}{m_s} (\underset{\sim}{E} + \frac{1}{c} \underset{\sim}{v} \times \underset{\sim}{B}) \cdot \frac{\partial f_s}{\partial \underset{\sim}{v}} = 0 \quad . \tag{2}$$

The electromagnetic fields $\underset{\sim}{E}$ and $\underset{\sim}{B}$ are determined from Maxwell's equations

$$\nabla \times \underset{\sim}{E} = - \frac{1}{c} \frac{\partial \underset{\sim}{B}}{\partial t} \tag{3}$$

$$\nabla \times \underset{\sim}{B} = \frac{4\pi}{c} \underset{\sim}{J} + \frac{1}{c} \frac{\partial \underset{\sim}{E}}{\partial t} \tag{4}$$

$$\nabla \cdot \underset{\sim}{E} = 4\pi \rho \tag{5}$$

$$\nabla \cdot \underset{\sim}{B} = 0 \tag{6}$$

where

$$\underset{\sim}{J}(\underset{\sim}{x}, t) = \int d\underset{\sim}{v} \sum_s q_s f_s(\underset{\sim}{x}, \underset{\sim}{v}, t) \underset{\sim}{v} \tag{7}$$

and

$$\rho(\underset{\sim}{x}, t) = \int d\underset{\sim}{v} \sum_s q_s f_s(\underset{\sim}{x}, \underset{\sim}{v}, t) . \tag{8}$$

These equations must be supplemented by the initial and boundary conditions for the uniqueness of their solution. Generally speaking, there is no way of solving the Klimontovich equation analytically since, for example, we have no knowledge of the initial conditions for each plasma particle. The Klimontovich equation which is highly singular is not suitable even for numerical computation. On the other hand, the characteristics of the Klimontovich equation in phase space can be followed with relative ease using a high speed computer. The characteristics of the Klimontovich equation is nothing but the trajectory of each particle in phase space and is defined by

$$\frac{d\underset{\sim}{v}_j}{dt} = \frac{q_j}{m_j} [\underset{\sim}{E}(\underset{\sim}{x}_j) + \frac{1}{c} \underset{\sim}{v}_j \times \underset{\sim}{B}(\underset{\sim}{x}_j)] \tag{9}$$

$$\frac{d\underset{\sim}{x}_j}{dt} = \underset{\sim}{v}_j \tag{10}$$

from which the charge and current densities are defined by

$$\rho_\delta(\underset{\sim}{x}, t) = \sum_s q_j \, \delta(\underset{\sim}{x} - \underset{\sim}{x}_j(t)) \tag{11}$$

and

$$\underset{\sim}{J}_\delta(\underset{\sim}{x}, t) = \Sigma \, q_j \, \underset{\sim}{v}_j \, \delta(\underset{\sim}{x} - \underset{\sim}{x}_j(t)) \quad . \tag{12}$$

This way of solving the Klimontovich equation is best suited for numerical simulations by using large, fast digital computers. Differential equations must first be transformed to finite difference equations before they are programmed into a computer. It often takes only a few microsecond to integrate or "push" the equation of motion for one particle for one time step Δt. Thus more than a few thousands to a few millions of particles may be used to simulate a plasma within a reasonable computing time using presently available computers. We shall note here that even a million of simulation particles is far too few to represent a real plasma. Each simulation particle must therefore be regarded as representing a large number of real plasma particles. In this sense a simulation particle may be called a superparticle. The consequence of using such a small number of particles is the enhancement of statistical fluctuations and collisional effects both of which tend to mask collective plasma behavior. In addition, δ-function representation of charge and current densities given by Eqs. (11) and (12) must be smoothed out in general. We will discuss in detail how such smoothing is done in numerical simulations.

When the Klimontovich equation is averaged over the ensembles of the initial conditions for N-particles, one obtains the Vlasov equation after neglecting the two-body collisons,

$$\frac{\partial f_s}{\partial t} + \underset{\sim}{v} \cdot \frac{\partial f_s}{\partial \underset{\sim}{x}} + \frac{q_s}{m_s} \left(\underset{\sim}{E} + \frac{1}{c} \underset{\sim}{v} \times \underset{\sim}{B} \right) \cdot \frac{\partial f_s}{\partial \underset{\sim}{v}} = 0 \ . \tag{13}$$

This equation is formally the same to the Klimontovich equation and its numerical solution requires special cares as the distribution functions evolve into fine structures in phase space.

2.2 Finite-size Particle Model

After discussing plasma kinetic eguations, we will describe plasma simulation models using finite-size particles in which only the electrostatic Coulomb interactions are retained. Electromagnetic and relativistic simulations are discussed elsewhere in the book. Historically, the first plasma simulation model developed was the one-dimensional sheet model consisting of zero-thickness charged sheets interacting with each other under the influence of electrostatic

Coulomb forces (Buneman 1959, Dawson 1962). While such a sheet model is useful in one-dimension, much of the computing is exhausted in calculating particle crossing which takes place at a distance much shorter than the Debye length. In plasma simulations, we are generally interested in the collective phenomena whose wavelengths are much longer than the Debye length. Furthermore, it is not obvious how to extend the sheet model into two and three dimensions where the Coulomb force between two particles becomes arbitrarily large as the separation of two particles becomes smaller. Since the force changes very rapidly at close encounter, one must use a small time step for preserving acceptible numerical accuracy. Again we note the informations at wavelengths shorter than the Debye length are not necessary. These considerations lead us to the use of finite-size particle models in plasma simulations in the late 1960's (Hockney 1966; Morse and Nielson, 1969; Birdsall and Fuss, 1969; Kruer <u>et al.</u>, 1973). Much more information on plasma simulations using particles can be found in Methods in Computatioinal Physics, Vols. 9 (1970) and Vol. 16 (1976), Hockney and Eastwood (1981), Potter (1973) and Birdsall and Langdon (in press).

Let us now consider a simulation model in which singular distributions of charges and potentials are smoothed out without modifying plasma properties at long wavelengths. Singularities in charge and force may be naturally removed by considering particles with finite extent, instead of classical zero-size particles. Thus we consider particles whose charge and hence current distributions are given by

$$\rho_j(\underset{\sim}{x}) = q_j \, S(\underset{\sim}{x} - \underset{\sim}{x}_j) \tag{14}$$

$$\underset{\sim}{J}_j(\underset{\sim}{x}) = q_j \, \underset{\sim}{v}_j \, S(\underset{\sim}{x} - \underset{\sim}{x}_j) \tag{15}$$

where $S(\underset{\sim}{x})$ is the shape factor which determines the distribution of charge of a finite-size particle. $S(\underset{\sim}{x})$ may be positive and we assume it is normalized so that

$$\int S(\underset{\sim}{x}) \, d\underset{\sim}{x} = 1 \quad . \tag{16}$$

While the choice of $S(\underset{\sim}{x})$ is arbitrary, one choice may be a gaussian shape,

$$S(\underset{\sim}{x}) = \frac{1}{(2\pi)^{3n/2} a^n} \exp\left[- \frac{x^2}{2a^2} \right] \tag{17}$$

where n is the dimensionality (n=1,2,3). Here a is the size of the particle.

One immediate consequence of the use of such extended particles is the elimination of singularities at short distance while at long distance, little modification takes place. In fact when two particles of finite extent approach closer overlapping with each other, the force between the two becomes smaller and smaller. Thus we would expect short range collisional interaction will be greatly reduced. We must carefully estimate just how much modification to classical plasma physis has been brought in by using such a model. We show in the following that long wavelength, collective plasma processes are little modified while short range collisional effects are greatly reduced.

2.3 Modification to Plasma Kinetic Theory Using Finite-size Particles

Let us consider what kind of modifications are necessary for a plasma of finite-size particles. For this purpose, we shall assume that the finite-size particles have no internal degree of freedom and furthermore they pass through with each other. The force on a particle centered at $(\underset{\sim}{x},\underset{\sim}{v})$ may be given by

$$\underset{\sim}{F}(\underset{\sim}{x},\ \underset{\sim}{v},\ t) = q \int dx'\ S(\underset{\sim}{x}' - \underset{\sim}{x})\ [\ \underset{\sim}{E}(\underset{\sim}{x}',\ t) + \frac{1}{c}\underset{\sim}{v} \times \underset{\sim}{B}(\underset{\sim}{x}',\ t)\ ]\ . \tag{18}$$

Charge and current densities for finite-size particles are given by

$$\rho(\underset{\sim}{x},\ t) = \int d\underset{\sim}{x}'\ S(\underset{\sim}{x} - \underset{\sim}{x}')\ \rho_\delta(\underset{\sim}{x}',\ t) \tag{19}$$

$$\underset{\sim}{J}(\underset{\sim}{x},\ t) = \int d\underset{\sim}{x}'\ S(\underset{\sim}{x} - \underset{\sim}{x}')\ \underset{\sim}{J}_\delta(\underset{\sim}{x}',\ t) \tag{20}$$

where ρ_δ and $\underset{\sim}{J}_\delta$ are charge and current densities for centers of finite-size particles defined by Eqs. (11) and (12). Since the above equations are of the form of convolution integral, they take a particularly simple form when transformed into Fourier space.

$$\underset{\sim}{F}(\underset{\sim}{k},\ \underset{\sim}{v},\ t) = q\ S(-\underset{\sim}{k})\ [\ \underset{\sim}{E}(\underset{\sim}{k},\ t) + \frac{1}{c}\ \underset{\sim}{v} \times \underset{\sim}{B}(\underset{\sim}{k},\ t)\ ] \tag{21}$$

$$\rho(\underset{\sim}{k},\ t) = S(\underset{\sim}{k})\ \rho_\delta(\underset{\sim}{k},\ t) \tag{22}$$

and

$$\underset{\sim}{J}(\underset{\sim}{k},\ t) = S(\underset{\sim}{k})\ \underset{\sim}{J}_\delta(\underset{\sim}{k},\ t) \tag{23}$$

where

$$S(\underset{\sim}{k}) \equiv \int d\underset{\sim}{x}\ S(\underset{\sim}{x})\ \exp(-i\underset{\sim}{k}\cdot\underset{\sim}{x}) \quad . \tag{24}$$

Note $S(\underset{\sim}{k})=1$ for point particles. Also $S(k=0)=1$. For gaussian particles where $S(x)=\exp(-x^2/2a^2)/(2\pi)^{1/2}a$, $S(k)=\exp(-k^2a^2/2)$. "a" will give the "size" of particles in real space and a^{-1} will be a measure of the spread in k-space. Generally speaking, $S(k)\sim1$ for $ka<1$ and $S(k)\sim0$ for $ka>1$. For square particles, $S(x)=1/\Delta x$ for $-\Delta x/2<x<\Delta x/2$ and zero otherwise so that $S(k)=\sin(k\Delta x/2)/(k\Delta x/2)$ which becomes zero for particle wavelengths where $\sin(k\Delta x/2)=0$ is satisfied. In the following we shall assume the shape factor is symmetric, $S(\underset{\sim}{k})=S(-\underset{\sim}{k})$, which is satisfied for most of the simulation models.

2.4 Vlasov-Maxwell Equations

It is straightforward to write down a set of Vlasov-Maxwell equations for finite-size particles. Assuming $f(\underset{\sim}{x},\underset{\sim}{v},t)$ is the distribution function for the center of finite-size particles, Vlasov equation is given by

$$\frac{\partial f}{\partial t} + \underset{\sim}{v} \cdot \frac{\partial f}{\partial \underset{\sim}{x}} + \frac{\underset{\sim}{F}}{m} \cdot \frac{\partial f}{\partial \underset{\sim}{v}} = 0 \tag{25}$$

where $\underset{\sim}{F}(\underset{\sim}{x},\underset{\sim}{v},t)$ is defined by Eq. (18). Charge and current densities are given by

$$\rho(\underset{\sim}{x}, t) = q \int d\underset{\sim}{x}' \, d\underset{\sim}{v} \, S(\underset{\sim}{x} - \underset{\sim}{x}') \, f(\underset{\sim}{x}', \underset{\sim}{v}, t) \tag{26}$$

and

$$\underset{\sim}{J}(\underset{\sim}{x}, t) = q \int d\underset{\sim}{x}' \, d\underset{\sim}{v} \, S(\underset{\sim}{x} - \underset{\sim}{x}') \, f(\underset{\sim}{x}', \underset{\sim}{v}, t) \, \underset{\sim}{v} \, .$$

Much of the plasma properties using finite-size particles may be studied from linear dispersion relation for small amplitude oscillations. Writing

$$f(\underset{\sim}{x}, \underset{\sim}{v}, t) = f_0(\underset{\sim}{v}) + f_1(\underset{\sim}{x}, \underset{\sim}{v}, t) \tag{27}$$

where $f_0(\underset{\sim}{v})$ is the zero-order equilibrium distribution and f_1 is a small perturbation, the Vlasov equation may be linearized in the absence of a d.c. magnetic field to find

$$\frac{\partial f_1}{\partial t} + v \frac{\partial f_1}{\partial x} + \frac{F_1}{m} \frac{\partial f_0}{\partial v} = 0 \tag{28}$$

where

$$F_1 = q \int dx' \, S(x' - x) \, E(x') \tag{29}$$

and

$$\nabla \cdot \underset{\sim}{E} = 4\pi q \int dx' \, dv \, S(x' - x) \, f_1(x', v, t) \quad . \tag{30}$$

Fourier analyzing in space and time, f_1, $E \sim \exp(ikx - i\omega t)$, it is straightforward to write

$$(- i\omega + ikv) \, f_1 = - \frac{q}{m} S(x) \, E(k) \, f_0'$$

and

$$ikE(k) = 4\pi \, q \, S(k) \int dv \, f_1(k, v, t)$$

so that the familiar plasma dielectric constant $\varepsilon(k,\omega)$ is given by

$$\varepsilon(k, \omega) = 1 - \frac{\omega_{pe}^2}{k} S^2(k) \int \frac{f_0'}{\omega - kv} dv \quad . \tag{31}$$

Note the only modification to the conventional dielectric constant for zero-size particles is the modification of ω_{pe}^2 to $\omega_{pe}^2 S^2(k)$. This modification results from the reduction of the force $\underset{\sim}{F}$ between two finite-size particles. Since $S(k) \sim 1$ for $ka < 1$, little modification takes place at long wavelength while for $ka \gtrsim 1$, $S(k) \sim 0$ so that the interactions in plasmas are greatly reduced. If a is chosen appropriately, one can eliminate the phenomena at wavelengths shorter than a. One of such choices is to assume a equal to the Debye length λ_D since modes with $k\lambda_D>1$ are often unimportant in a plasma. This does not necessarily mean "a" cannot be much larger than λ_D and in fact in some applications $a=(10\sim20)\ \lambda_D$ are chosen (Okuda et al., 1979). Detailed discussions on the modification of the dispersion relation for finite-size particle plasmas is found in Langdon and Birdsall (1970).

2.5 Collisions and Fokker-Planck Equations

We have seen that the use of finite-size particles reduced the interactions at wavelengths shorter than the size of a particle "a". We therfore expect that the collisions in a plasma of finite-size particles can be reduced greatly if a is chosen of the order of the Debye length λ_D. We also expect the reduction in particle collisions in a plasma of finite-size particles in three dimensions is more than in two dimensions. This is because the Coulomb force in three dimensions varies as r^{-2} while it goes as r^{-1} in two dimensions so that the effects of smoothing by the use of finite-size particles at short distance are more enhanced in three dimensions.

In order to study the reduction factor, let us first calculate the force and the potential between two bare particles of finite extent. Consider a situation where one particle is located at the origin while the other is located at $\underset{\sim}{x}$. The electric field or the potential due to the particle at the origin is given by

$$\nabla \cdot \underset{\sim}{E} = 4\pi\ q_1\ S(\underset{\sim}{x}) \tag{32}$$

and the force on the second particle at x is given by

$$\underset{\sim}{F} = q_2 \int \underset{\sim}{E}(x')\ S(\underset{\sim}{x} - \underset{\sim}{x}')\ dx' . \tag{33}$$

Fourier transforming Eqs. (32) and (33) and introducing the potential V for the force $\underset{\sim}{F}$

$$\underset{\sim}{F} = -\nabla V \tag{34}$$

we find

$$V(\underset{\sim}{k}) = \frac{4\pi q_1 q_2 S^2}{k^2} \tag{35}$$

and

$$V(\underset{\sim}{x}) = \frac{1}{(2\pi)^n} \int d\underset{\sim}{k}\ V(\underset{\sim}{k})\ \exp(i\underset{\sim}{k}\cdot\underset{\sim}{x}) \tag{36}$$

where n is the dimensionality. Note that

$$V(\underset{\sim}{x}) \sim \frac{q_1 q_2}{a} \tag{37}$$

for x<a so that the potential does not diverge even when the two particles overlap on top with each other so long as "a" is finite.

When "a" is taken to be λ_D, one expects that the collisions must be greatly reduced as the particle interaction at a distance shorter than "a" is greatly reduced. More detailed calculations on the force and the potential is found in Okuda and Birdsall (1970).

It is straightforward to derive the Fokker-Planck equation for a plasma of finite-size particles (Langdon and Birdsall 1970, Okuda and Birdsall 1970). The collision operator is given by

$$\left.\frac{\partial f}{\partial t}\right|_{coll} = \frac{\omega_{pe}}{n\lambda_D^3} I\left(\frac{a}{\lambda_D}\right)\frac{V_t^3}{16\pi}\frac{\partial}{\partial \underset{\sim}{v}}\cdot\int d\underset{\sim}{v}'\frac{g^2\underset{\sim}{I}-\underset{\sim}{g}\underset{\sim}{g}}{g^3}\cdot\left(\frac{\partial}{\partial \underset{\sim}{v}}-\frac{\partial}{\partial \underset{\sim}{v}'}\right)\cdot f(\underset{\sim}{v},t)f(\underset{\sim}{v}',t) \quad (38)$$

where

$$g = \underset{\sim}{v} - \underset{\sim}{v}' \quad (39)$$

and

$$I = \int_0^\infty dk^2 \frac{S^4}{k^2(1+S^2/k^2\lambda_D^2)^2} \quad . \quad (40)$$

$\underset{\sim}{I}$ is a unit tensor. Note that all the necessary modification is contained in I given by Eq.(40) so that the Fokker-Planck equation formally remains the same. This is because the velocity space is unmodified by the use of finite-size particles. For S=1 and no shielding, $I=2\ln n\lambda_D^3$ and one recovers the familiar estimate of collisions

$$\left.\frac{\partial f}{\partial t}\right|_{coll} \sim \omega_{pe}\frac{\ln n\lambda_D^3}{n\lambda_D^3} \quad . \quad (41)$$

For a plasma of finite-size particles, I is greatly resuced due to the appearance of S^4 when $a \simeq \lambda_D$. Numerical integration in Eq.(40) reveals $I\sim0.1$ for $a=\lambda_D$ and $I\sim10^{-3}$ for $a=10\lambda_D$ for a gaussian particle. (Langdon and Birdsall, 1970; Okuda and Birdsall, 1970). Note that the corresponding values of I for zero-size particles is $I=2\ln n\lambda_D^3 \sim 10$ so that a large reduction is achieved by adapting finite-size particles.

3. NUMERICAL METHODS USING FINITE-SIZE PARTICLES

3.1 Electrostaic Simulation Models

Let us consider finite-size particles with the gaussion form factor in one dimension so that $S(x)=\exp(-x^2/2a^2)/\sqrt{2\pi}\,a$ and $S(k)=\exp(-k^2a^2/2)$. We have shown that the short wavelength modes, $ka \gtrsim 1$, are heavily suppressed while leaving the long wavelength modes unmodified. This would reduce the particle collisions as we have shown already. It is clear from Eqs. (9) and (10) that the first step in plasma simulation is to calculate the force on a particle. This can be done in several ways. Note, however, that the number of long wavelength modes (collective modes) in a plasma is in general much smaller than the number of particles. In fact the ratio of the number of collective modes to the number of particles is equal to the plasma parameter. We should therefore calculate the force on a particle from electric and magnetic fields rather than directly summing up the Coulomb force between two interacting particles. Such a pair-wise calculation would be prohibitively time consuming and should be avoided unless accurate information is required at short wavelengths.

Let us consider an electrostatic simulation model again. The electric potential $\Phi(\underset{\sim}{x},t)$ is given by Poisson's equation

$$\nabla^2 \Phi = -4\pi \sum_j q_j S(\underset{\sim}{x} - \underset{\sim}{x}_j) \quad . \tag{42}$$

Assuming periodic boundary conditions for Φ for simplicity, Eq.(42) is Fourier transformed into $\underset{\sim}{k}$-space.

$$\Phi(\underset{\sim}{k}) = \frac{4\pi}{k^2} S(\underset{\sim}{k}) \sum_j q_j \exp(-i\underset{\sim}{k}\cdot\underset{\sim}{x}_j) \tag{43}$$

where the Fourier transform of the form factor $S(\underset{\sim}{k})$ is assumed to be the same for all the particles.

The force on a particle at $\underset{\sim}{x}$, is given by

$$\underset{\sim}{F}(\underset{\sim}{x}) = q_i \int S(\underset{\sim}{x} - \underset{\sim}{x}') \underset{\sim}{E}(\underset{\sim}{x}') d\underset{\sim}{x}' \tag{44}$$

so that its Fourier transform is given by

$$\underset{\sim}{F}(\underset{\sim}{k}) = q_i S(\underset{\sim}{k}) \underset{\sim}{E}(\underset{\sim}{k}) \quad . \tag{45}$$

The force on a particle at $\underset{\sim}{x}=\underset{\sim}{x}_i$ is found by inverting Eq.(45) into x-space.

$$\underset{\sim}{F}(\underset{\sim}{x}_j) = q_i \sum_{\underset{\sim}{k}} S(\underset{\sim}{k}) \underset{\sim}{E}(\underset{\sim}{k}) \exp(i\underset{\sim}{k}\cdot\underset{\sim}{x}_j) \tag{46}$$

where

$$\underset{\sim}{E}(\underset{\sim}{k}) = i\underset{\sim}{k}\,\Phi(\underset{\sim}{k}) \quad . \tag{47}$$

It is clear from Eqs. (43) and (46), that the most time consuming part of the field calculations hinges on the evaluation of the phaser, $\exp(\pm i\underset{\sim}{k}\cdot\underset{\sim}{x}_j)$, since MN such operations are required at each time step. Here M is the number of the Fourier modes and N is the number of simulation particles. We will review several methods evaluating the phaser.

3.2 Spectral Method

In this method, only several Fourier modes are retained in the simulations and Eqs. (43) and (46) are calculated as they appear. Assuming the form of the electric field

$$E(x) = \sum_{k_{min}}^{k_{max}} E(k) \exp(ikx)$$

in one-dimension, Eq. (46) reduces to

$$F(x_i) = 2q_i \sum_{k_{min}}^{k_{max}} \frac{S^2(k)}{k} \sum_j q_j \sin\{k(x_j - x_i)\}$$
$$= 2q_i \sum_{k_{min}}^{k_{max}} \frac{S^2(k)}{k} [\sin(kx_i)\sum_{j=1}^{N} q_j \cos(kx_j) - \cos(kx_i)\sum_{j=1}^{N} q_j \sin(kx_j)] \quad . \tag{48}$$

Note that $\Sigma \cos kx_j$ and $\Sigma \sin kx_j$ is independent of i and therefore can be summed at once. Since there are M Fourier modes to be retained, the total number of operations will be of the order of MN, much less than N^2 required for the pairwise calculation. Since the calculations of sine and cosine are time-consuming, one may make use of formulae of trigonometry such as $\sin 2k = 2\sin k \cos k$, $\cos 2k = \cos^2 k - \sin^2 k$ and so on. Such a code has been developed and used to simulate a long plasma device but with only a few longest wavelength modes play important roles for the development of plasma turbulence (Cheng and Okuda, 1977).

3.3 Multipole Expansion Method

While the spectral method is accurate and faster than calculating the Coulomb force pairwise, it is not fast enough for large calculations, particularly in multi-dimensions, where the number of particles and the Fourier modes increases rapidly with the dimension. We should like to find a method which involves of the order of (M + M) operations rather than MN. This is done by introducing a spatial grid as shown below.

Consider again Poisson's equation in $\underset{\sim}{k}$-space,

$$k^2 \Phi(\underset{\sim}{k}) = 4\pi S(\underset{\sim}{k}) \sum_j q_j \exp(-i\underset{\sim}{k}\cdot\underset{\sim}{x}_j) \quad . \tag{49}$$

Since we have already given up the exact representation of particle location by adapting the use of finite-size particles, it appears reasonable to replace the phasor $\exp(-i\underset{\sim}{k}\cdot\underset{\sim}{x}_j)$ by using an approximate particle location. This is done by using a spatial grid. Let us now consider a spatial grid and write the particle location

$$\underset{\sim}{x}_j = \underset{\sim}{n}_j \Delta + \delta\underset{\sim}{x}_j \tag{50}$$

where $\underset{\sim}{n}_j$ is a set of integers representing the nearest grid point of the j-th particle and $\delta\underset{\sim}{x}_j = \underset{\sim}{x}_j - \underset{\sim}{n}_j\Delta$ is the displacement. Δ is the size of a spatial grid and need not be the same in each direction although we assume it is the same here for simplicity. Now the phase factor in Poisson's equation may be expanded about the nearest grid point. Poisson's equation is now written as

$$k^2\Phi(k) = 4\pi S(k) \sum_j q_j \exp(-i\underset{\sim}{k}\cdot\underset{\sim}{n}_j\Delta)[1 - i\underset{\sim}{k}\cdot\delta\underset{\sim}{x}_j - \frac{(\underset{\sim}{k}\cdot\delta\underset{\sim}{x}_j)^2}{2!} + \cdots\cdots] \quad . \tag{51}$$

Summation over the particles can be replaced by the summation over the grid point so long as the particles have the same nearest grid point. For example,

$$\sum_{j=1}^{N} q_j \exp(-i\underset{\sim}{k}\cdot\underset{\sim}{n}_j\Delta) = \sum_{n=1}^{L}\Big(\sum_{j\varepsilon n} q_j\Big)\exp(-i\underset{\sim}{k}\cdot\underset{\sim}{n}_j\Delta) \equiv q\sum_{n=1}^{L} g_n^0 \exp(-i\underset{\sim}{k}\cdot\underset{\sim}{n}\Delta) \quad (52)$$

where $g_n^0 = \sum_{j\varepsilon n}(\pm 1)$ is the net charge (monopole) at the n-th grid point. Similarly,

$$\sum_{j=1}^{N}(-i\underset{\sim}{k}\cdot\delta\underset{\sim}{x}_j)q_j\exp(-i\underset{\sim}{k}\cdot\underset{\sim}{n}_j\Delta) = q\sum_{n=1}^{L} g_n^1(-i\underset{\sim}{k})\exp(-i\underset{\sim}{k}\cdot\underset{\sim}{n}\Delta) \quad (53)$$

where $g_n^1 = \sum_{j\varepsilon n}(\pm)\delta\underset{\sim}{x}_j$ is the net displacement (dipole) of the charges at the n-th grid point. One can continue the expansion to higher orders if necessary. Noting that the wavenumber is given by $k=2\pi m/L$ for a periodic system, Eqs. (52) and (53) correspond to discrete Fourier transforms defined by

$$\hat{g}_m \equiv \sum_{n=1}^{L} g_n \exp(-i\ 2\pi mn/L) \quad (54)$$

whose inverse Fourier transform is given by

$$g_n \equiv \frac{1}{L}\sum_{m=1}^{L} \hat{g}_m \exp(i\ 2\pi mn/L)\ . \quad (55)$$

The force on a particle at $\underset{\sim}{x}_j$ is calculated similarly as

$$\begin{aligned} \underset{\sim}{F}(\underset{\sim}{x}_j) &= q_j \sum_{k} \underset{\sim}{E}(\underset{\sim}{k})\ S(\underset{\sim}{k})\ \exp(i\underset{\sim}{k}\cdot\underset{\sim}{x}_j) \\ &= q_j \sum_{k} \underset{\sim}{E}(\underset{\sim}{k})\ S(\underset{\sim}{k})\ \exp(i\underset{\sim}{k}\cdot\underset{\sim}{n}_j\Delta)\ \exp(i\underset{\sim}{k}\cdot\delta\underset{\sim}{x}_j) \\ &= q_j \sum_{k} \underset{\sim}{E}(\underset{\sim}{k})\ S(\underset{\sim}{k})\ \exp(i\underset{\sim}{k}\cdot\underset{\sim}{n}_j\Delta)\ (1 + i\underset{\sim}{k}\cdot\delta\underset{\sim}{x}_j + \cdots\cdots) \\ &= \underset{\sim}{F}(\underset{\sim}{n}_j\Delta) + \underset{\sim}{\nabla}\underset{\sim}{F}(\underset{\sim}{n}_j\Delta)\cdot\delta\underset{\sim}{x}_j + \cdots \end{aligned} \quad (56)$$

where the first term of Eq.(56) is the force on a particle at the nearest grid point and the second term is the correction due to dipole moment associated with the gradient of the field. The number of operations associated with the multipole expansion is only proportional to the number of particles for the charge density and force calculations (Kruer et al., 1973). In addition, the number of operations associated with the discrete Fourier transform scales as M ln M where M is the number of Fourier modes. It is shown that the higher order terms associated with the expansions given by Eqs.(51) and (56) converge rapidly so that under normal circumstances one can truncate at the dipole term (Chen and Okuda, 1975; Okuda and Cheng, 1978).

One of the drawbacks of the multipole expansion method is that it requires several arrays for storing field quantities. For example, for two-dimensional electrostatic simulations, three two-dimensional arrays are necessary for charge density up to the dipole moments. For the electric field calculations, we must store E_x, E_y, $\partial E_x/\partial x$, $\partial E_x/\partial y$, and $\partial E_y/\partial y$. Note for electrostatic simulations $\partial E_y/\partial x=\partial E_x/\partial y$. For three-dimensional simulations or electromagnetic simulations, the number of arrays required can often exceed the capacity of available computers. It is important to keep the number of the field arrays to the minimum in order to be able to perform large scale, multi-dimensional simulations. This is because the field arrays or grid quantities are randomly accessed for the calculation of charge density and force on a particle so that it is desirable to store the field arrays in the central memory of a computer rather than in an auxiliary memory such as disks and tapes. It is often necessary , on the other hand, to store particle data (positions and velocities) on disks and tapes in order to be able to use large number of particles. In this case, only a small fraction of particles are brought into the central memory of a computer for the calculations of charge density and force and they are sent back to the outside memory after such calculations. We now consider another simulations method where interpolations of charge and force on the grid points are used. Such a method is often called PIC (particle-in-cell) (Morse and Nielson, 1969) and CIC (cloud-in-cell) (Birdsall and Fuss, 1969) methods.

3.4 Interpolations and Weighting

The lowest order interpolation for charge or force is equivalent to the monopole approximation in which the entire amount of charge is assigned to the nearest grid point (NGP). Similarly the force on a particle is approximated by that at the nearest grid point. Schematically NGP weight function may be sketched as shown in Figure 1(a) where for particle location x with $-\Delta/2<x<\Delta/2$, the entire charge or force is assigned at the nearest grid point. Therefore the weight function w_0 is given by

$$w_0(x) = \begin{cases} 1/\Delta & \text{for } -\Delta/2<x<\Delta/2 \\ 0 & \text{otherwise} \end{cases} \quad . \tag{57}$$

In this case the particles appear to have a shape of a slab in the NGP interpolation with the width of Δ.

In the next approximation, charge on a particle is linearly interpolated between two nearest grid points as shown in Figure 1(b) so that its weight function is given by

$$w_1(x) = \begin{cases} (\ x + \Delta)/\Delta^2 & -\Delta<x<0 \\ (-x + \Delta)/\Delta^2 & 0<x<\Delta \\ 0 & \text{otherwise} \end{cases} \quad . \tag{58}$$

It is interesting to observe that the Fourier tranform of the weight function is given by

$$w_n(k) = [\ \sin(k\Delta x/2)/(k\Delta x/2)\ ]^{n+1} \tag{59}$$

so that $w_n(k=0)=1$ for any n. For large n, $w_n(k)$ is unity near $k\Delta x{\sim}0$ and nearly zero otherwise. For n=3, the weight function is given by

$$w_3(x) = \begin{cases} (\,3/4 - x^2\,)/\Delta^3 & |x| < \Delta/2 \\ [\,1/4 - |x\pm\Delta| + (x\pm\Delta)^2\,]/\Delta^3 & |x\pm\Delta| < \Delta/2 \\ 0 & \text{otherwise} \end{cases} \tag{60}$$

as shown in Figure 1(c) and this interpolation is sometimes called the quadratic spline. It is clear that the higher order interpolations involve operations including more grid points in x-space so that its

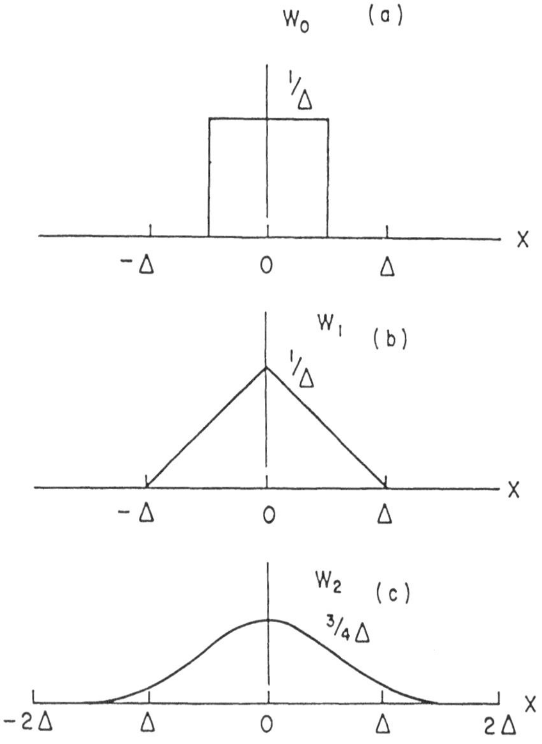

Figure 1. Weighting functions of charge density at the nearest grid point for the nearest grid point (a), linear interpolation (b) and quadratic interpolation (c).

Fourier transform is more and more narrowly peaked for $k\Delta x \sim 0$ which reduces numerical aliases (Langdon, 1971, Okuda, <u>et al.</u>, 1979). In general, the interpolation for the charge density takes the form of

$$\rho_g^n = \sum_j q_j \, w(x_n - x_j) \tag{61}$$

where ρ_g^n is the charge density at the n-th grid point. w is the weight function and x_n is the n-th grid point. Similarly, the force on a particle is given by

$$F(x_j) = q_j \sum_n E_n \, W(x_n - x_j) \; . \tag{62}$$

Since the charge density and electric field are defined only on a set of discrete grid points, a set of Fourier modes

$$k_p = k \pm p k_g \qquad (p = 0,1,2,3,\cdots\cdots)$$

will give the same variation in x where $k_g=2\pi/\Delta$ is the grid wave number. Detailed analysis of such argorithms reveals a presence of numerical instabilities associated aliases ($p\neq 0$) when the Debye length is much smaller than the grid size (Langdon, 1971; Chen and Okuda, 1975). Such numerical instabilities may be quenched by employing a higher order interpolations which reduces the amplitude of the aliases (Okuda and Cheng, 1978; Okuda et al., 1979).

3.5 Finite Difference Equation

Let us now consider the left-hand-side of the equation of motion as given by Eqs. (9) and (10). The problem here is how to approximate the differential equation by the finite-difference equation. The difference equation must be simple so that it can be integrated fast and yet at the same time it must be reasonably accurate after many iterations.

We consider again the electrostatic model in which the force on a particle is determined only from the particle location.

$$\begin{aligned} m_j \frac{d\underset{\sim}{v}_j}{dt} &= q_j \, \underset{\sim}{E}[\underset{\sim}{x}_j(t)] \\ \frac{d\underset{\sim}{x}_j}{dt} &= \underset{\sim}{v}_j(t) \qquad . \end{aligned} \tag{63}$$

The standard method for this case is the leap-frog scheme in which particle positions and velocities are defined at two different sets of time steps shifted by $\Delta t/2$. It is given schematically by

$$\begin{aligned} \frac{\underset{\sim}{v}^{t} - \underset{\sim}{v}^{t-\Delta t}}{\Delta t} &= \frac{q_j}{m_j} E^{t-\Delta t/2} \\ \frac{\underset{\sim}{x}^{t+\Delta t/2} - \underset{\sim}{x}^{t-\Delta t/2}}{\Delta t} &= \underset{\sim}{v}^{t} \qquad . \end{aligned} \tag{64}$$

It is clear that the leap-frog scheme is time-centered and is reversible in time regardless of the size of the time step Δt. The accuracy of the leap-frog scheme may be estimated by Taylor expanding $\underset{\sim}{v}^t$ and $\underset{\sim}{v}^{t-\Delta t}$ at around $\underset{\sim}{v}^{t-\Delta t/2}$. We find $\underset{\sim}{v}^t - \underset{\sim}{v}^{t-\Delta t} = \underset{\sim}{v}'^{\,t-\Delta t/2}\Delta t - v^{(,,,)t-t/2}\Delta t^3/24$ so that the error per time step will be of the order of $(\omega\Delta t)^3/24$ giving $10^{-3}\sim 10^{-4}$ for $\omega\Delta t = 0.1\sim 0.2$ which are commonly used. Note the leap-frog scheme is particularly simple and reguires storage for $\underset{\sim}{v}$ and $\underset{\sim}{x}$ only at one time step since the new $\underset{\sim}{v}$ and $\underset{\sim}{x}$ replace those in the past step.

In order to consider stability, let us examine a harmonic oscillator in one-dimension where the equation motion takes the form of

$$\frac{d^2x}{dt^2} = -\omega_0^2 x \quad . \tag{65}$$

The general solution to Eq. (65) is given by

$$x(t) = c_1 \cos\omega_0 t + c_2 \sin\omega_0 t \tag{66}$$

where c_1 and c_2 are determined from the initial position and velocity of the particle. The corresponding leap-frog scheme is given by

$$\frac{x^{t+\Delta t} - 2x^t + x^{t-\Delta t}}{\Delta t^2} = -\omega_0^2 x^t \quad . \tag{67}$$

The stability is found by assuming $x^t \sim \exp(i\omega t)$ to find

$$\sin(\omega\Delta t/2) = \pm\, \omega_0 \Delta t/2 \quad . \tag{68}$$

It is clear that for a large Δt satisfying $\omega_0\Delta t>2$, there is no real solution to ω and the leap-frog scheme is unstable (Birdsall and Langdon, in press). Since the numerical error is of the order of $(\Delta t)^3$ in the leap-frog scheme, one can reduce Δt in order to improve the accuracy of the computation for the same length of integration.

Another important application of the leap-frog scheme is the equation of motion in a static uniform magnetic field where Eq. (63) is replaced by the Lorentz force,

$$\frac{d\underset{\sim}{v}}{dt} = \frac{q}{m}\left(\underset{\sim}{E} + \frac{1}{c}\underset{\sim}{v} \times \underset{\sim}{B}_0 \right) \quad . \tag{69}$$

The time-centered, leap-frog scheme may be written as (Buneman 1967)

$$\frac{\underset{\sim}{v}^{t+t\Delta t/2} - \underset{\sim}{v}^{t-\Delta t/2}}{\Delta t} = \frac{q}{m}\left(\underset{\sim}{E}^t + \frac{1}{c}\frac{\underset{\sim}{v}^{t+\Delta t/2} + \underset{\sim}{v}^{t-\Delta t/2}}{2} \times \underset{\sim}{B}_0 \right) \tag{70}$$

where the velocity v^t at the right-hand-side is replaced by the average of $\underset{\sim}{v}^{t+\Delta t/2}$ and $\underset{\sim}{v}^{t-\Delta t/2}$. It is straightforward to prove that Eq.(70) conserves particle energy in the absence of the electric field,

$$(\underset{\sim}{v}^{t+\Delta t/2})^2 = (\underset{\sim}{v}^{t-\Delta t/2})^2$$

regardless of the size of Δt. This is proven by multiplying $(\underset{\sim}{v}^{t+\Delta t/2} - \underset{\sim}{v}^{t-\Delta t/2})$ to both sides of Eq. (70). This is reasonable since the effect of $\underset{\sim}{B}_0$ is to rotate particle velocity around B_0. Since Eq.(70) is an algebraic equation with respect to $\underset{\sim}{v}$, one can solve for $\underset{\sim}{v}^{t+\Delta t/2}$ to find

$$\underset{\sim}{v}^{t+\Delta t/2} = \frac{1-(\omega_c\Delta t/2)^2}{1+(\omega_c\Delta t/2)^2}\,\underset{\sim}{v}^{t-\Delta t/2} - \hat{z}\times\underset{\sim}{v}^{t-\Delta t/2}\,\frac{\omega_c\Delta t}{1+(\omega_c\Delta t/2)^2}$$
$$+ \frac{q\Delta t}{m}\,\frac{\underset{\sim}{E}}{1+(\omega_c\Delta t/2)^2} + \frac{q^2\Delta t^2}{2m^2c}\,\frac{\underset{\sim}{E}\times\underset{\sim}{B}_o}{1+(\omega_c\Delta t/2)^2} \quad . \tag{71}$$

Boris has shown that Eq. (71) can be obtained by adding the electric field acceleration of the particle velocity by $\Delta t/2$ time step, and a full rotation by the magnetic field accompanied by another $\Delta t/2$ step electric acceleration (Boris 1970). This can be seen by rewriting Eq. (70) as

$$\underset{\sim}{v}^{t+\Delta t/2}-\underset{\sim}{v}^{t+\Delta t/2}\times\underset{\sim}{\omega}_c\frac{\Delta t}{2} = \underset{\sim}{v}^{t-\Delta t/2}+\underset{\sim}{v}^{t-\Delta t/2}\times\underset{\sim}{\omega}_c\frac{\Delta t}{2} + \frac{q}{m}\,\underset{\sim}{E}^t\Delta t \tag{72}$$

where $\underset{\sim}{\omega}_c=q\underset{\sim}{B}_o/mc$ (Dawson, 1983). Eq.(72) is written as

$$\underset{\approx}{R}(-\theta\Delta t/2)\ \underset{\sim}{v}^{t+\Delta t/2} = \underset{\approx}{R}(\theta\Delta t/2)\ \underset{\sim}{v}^{t-\Delta t/2} + \frac{q}{m}\,\underset{\sim}{E}^t\ \Delta t$$

where $\underset{\approx}{R}$ is a rotation matrix about the magnetic field for an angle $\theta\Delta t/2$. Taking $\underset{\sim}{B}_o$ in the z direction, $\underset{\approx}{R}$ is found

$$\underset{\approx}{R} = \begin{vmatrix} \cos(\theta\Delta t/2) & -\sin(\theta\Delta t/2) & 0 \\ \sin(\theta\Delta t/2) & \cos(\theta\Delta t/2) & 0 \\ 0 & 0 & 1 \end{vmatrix}$$

with $\tan(\theta\Delta t/2)=\omega_c\Delta t/2$ (Buneman 1967). Solving for $\underset{\sim}{v}^{t+\Delta t}$ gives

$$\underset{\sim}{v}^{t+\Delta t/2} = \underset{\approx}{R}(\theta\Delta t)\ \underset{\sim}{v}^{t-\Delta t/2} + \frac{q}{m}\,\underset{\approx}{R}(\theta\Delta t/2)\cdot\underset{\approx}{E}^t\Delta t \quad . \tag{73}$$

3.6 Guiding Center Model and Predictor-Corrector Method

One of the dynamical systems in which the leap-frog scheme cannot be applied is the guiding center model. In this model, fast gyration around magnetic field lines is averaged out so that only the slow drift motion of guiding centers is followed in time. Clearly the frequency of the physical process involved must be smaller than the gyro-frequency, Ω, and at the same time the wavelength of the variation of physical quantities must be longer than the gyroradius.

Let us consider, for example, a two-dimensional guiding center model in which a plasma is embedded in a uniform strong magnetic field in z-direction. Such a plasma may be regarded as two-dimensional since the motion along magnetic field quickly smooths out any variations along magnetic field while the motion across it is much slower. Particle motion may be described by its guiding center drift due to electric field given by $c\underset{\sim}{E}\times\underset{\sim}{B}/B^2$. The equation of motion would be

$$\frac{dx}{dt} = \frac{c}{B} E_y(\underset{\sim}{x})$$

$$\frac{dy}{dt} = -\frac{c}{B} E_x(\underset{\sim}{x})$$

where only the first order time derivatives appear so that the leap-frog scheme cannot be applied here. One can try a leap frog scheme by stepping two time steps at once as shown below.

$$x^{n+1} = x^{n-1} + 2\Delta t \frac{c}{B} E_y^n$$

$$y^{n+1} = y^{n-1} - 2\Delta t \frac{c}{B} E_x^n \quad .$$

One of the problems using such a scheme is that "even" and "odd" time steps are coupled only through the electric field, so that particle positions at even and odd steps are found to diverge as the iterations are repeated in time. It is therefore necessary to average the even and odd steps as frequently as possible. Such a scheme is called a predictor-corrector method. In the predictor step, the same equation is used.

$$x^{n+1*} = x^{n-1} + 2\Delta t \frac{c}{B} E_y^n$$

$$y^{n+1*} = y^{n-1} - 2\Delta t \frac{c}{B} E_x^n \qquad (75)$$

where (x^{n+1*}, y^{n+1*}) is the predicted value. In the corrector step,

$$x^{n+1} = x^n + \Delta t \frac{c}{B} (E_y^n + E_y^{n+1*})/2$$

$$y^{n+1} = y^n - \Delta t \frac{c}{B} (E_x^n + E_x^{n+1*})/2 \qquad (76)$$

are used to find (x^{n+1}, y^{n+1}) where E^* is the electric field determined from the predicted positions, (x^*, y^*). Such a scheme gives satisfactory results for the simulations of the guiding center model in which low frequency particle motion across magnetic field has been studied with respect to turbulent diffusion associated with microinstabilities (Lee and Okuda, 1978).

The example given here is one of the simplest equations of motion. Particle drifts such as due to magnetic field gradient, magnetic field curvature and external force may be added to the $c\underset{\sim}{E}\times\underset{\sim}{B}/B^2$ drift in a similar manner.

3.7 Quasi-neutral Particle Simulation Model

The guiding center model describes the slow motion of a plasma across magnetic field assuming a plasma quickly becomes uniform along magnetic field by the rapid themal motion along the field lines. When such a rapid motion along magnetic field must be considered, it is necessary to use a small time step determined from the electron motion along magnetic field lines. When considering low frequency ion waves such as ion acoustic, ion cyclotron and drift waves, there are situations in which detailed electron motion in phase space may not be essential. Of course, there are certainly cases where precise electron dynamics is crucial especially when considering microinstabilities driven by the electron Landau damping.

Here we consider a model for stuying ion waves in which detailed electron motion is not necessary. In particular we would like to drop electron inertia since it is responsible to generate high frequency electron plasma oscillations. In the presence of low frequency waves where the phase speed of the wave along magnetic field, $\omega/k_{\parallel}$, is much smaller than the electron thermal speed, v_e, electron response may be approximated by the Boltzmann distribution

$$n_e = n_o \exp(e\phi/T_e) \quad . \tag{77}$$

Poisson's equation for the system is given by

$$\nabla^2\phi = - 4\pi \; e \; [\; n_i - n_o \exp(e\phi/T_e) \;] \quad . \tag{78}$$

so that for $e\phi/T_e \ll 1$, we may expand $\exp(e\phi/T_e) = 1 + e\phi/T_e$ obtaining

$$\nabla^2\phi = -4\pi \; en_i + 4\pi \; en_o(1+e\phi/T_e) \quad .$$

For a uniform plasma, n_o and Te are constants so that Poisson's equation is Fourier transformed in k-space after substituting $n_i = n_o + \delta n_i$, finding

$$\phi_k = \frac{4\pi e\delta n_{ik}}{k^2 + k_{De}^2}$$

in which the electrons appear as the Debye shielding cloud of the potential around the ion charge density. It is not necessary therefore to calculate electron density in this model. It should be noted that the simulation model using the technique just described is not only much faster than the conventional electron-ion model but also it is much quieter because of the absence of high frequency electron plasma waves (Okuda et al., 1978).

when a plasma is inhomogeneous in one-direction,

$$n_e(x) = n_o(x) \; [\; 1 + e\phi(x)/T_e \;]$$

so that, with $n_o(x) = \bar{n}_o + \delta n_o(x)$, Poisson's equation takes the form of

$$\nabla^2\phi - \frac{4\pi e^2 \bar{n}_o}{T_e} \phi = - 4\pi e \; \delta n_i + \frac{4\pi e^2 \delta n_o}{T_e} \phi$$

This equation may be solved by iteration about $\bar{n}_o$ in which, for the first iteration, $\delta n_o \phi$ term is neglected to find ϕ. In the next iteration $\delta n_o \phi$ term is back to obtain a new potential.

So far, we have treated the electrons in the adiabatic limit in which no resonant wave-particle interactions are taken into account. There are, however, occasions where small non-adiabatic electrons may play an important role in controlling microinstabilities in a collisionless plasma. In order to recover the resonant wave-particle interactions, we must follow electrons in velocity space so that electrons are no longer considered a fluid. Unfortunately, however, as soon as the electrons recover their discreteness and inertia, high frequency plasma oscillations appear and the advantages gained from the adiabatic electron model vanish. There are several techniques to cure the situation, however, they are of limited use and there is no model which works for general purpose.

One of the models is to recognize that the resonant electrons with the waves are low evergy electrons satisfying $\omega/k_{\parallel} \ll v_e$ so that one may divide the total electron velocity distribution into resonant and non-resonant particles. For non-resonant particles, Boltzmann distribution should be applied while the resonant particles whose velocity is much less than the thermal speed are treated as discrete particles and their orbits in phase space are followed in time. Since the low energy electrons move slowly, it is possible to use a large time step, larger than ω_{pe}^{-1}.

Poisson's equation may be written as

$$(k^2 + k_{De}^2) \Phi_k = 4\pi e (n_i - n_e{}^r)$$

where $n_e{}^r$ is the density from the resonant electrons. Such a model has been used to study, for example, ion acoustic instaility in one-dimension. One of the questions naturally arises is the choice of the boundary between resonant and non-resonant particles. Obviously some of the resonant electrons do diffuse in velocity space as a result of quasilinear diffusion so that they may appear in the non-resonant region in velocity space. If that happens, those particles start emitting high frequency plasma waves so that such particles must be removed from the computation. There are other kind of models in which, for example, electrons are treated as a massless fluid with macroscopic density, velocity, temperature and so forth (Sgro and Nielson, 1976).

3.8 Initial Conditions

Before starting a simulation, one must specify initial conditions in phase-space. This is ordinarily done by specifying particle distribution functions at t=0. This may appear simple so long as we know what we want to simulate. The choice of initial conditions, that is to assign the initial position and velocity to each simulation particle is not as simple as it appears. This is because the initial condition is often given by a particle distribution function in phase space, $f(\underset{\sim}{x},\underset{\sim}{v},t=0)$, while only a finite number of particles or degree of freedom is available to represent the initial distribution.

Consider, for example, a Maxwellian plasma in a uniform plasma in one-dimension. In order to represent a uniform distribution in space, one might load particles exactly uniformly along the x axis with equal spacing, or one might load particles randomly using a uniform random number generator. Certainly there are many other ways of choosing the initial positions. There is no way of deciding which one is the best in general. It all depends on the problem under consideration. If one is interested in, for example, a coherent wave in a plasma, exact uniform spacing of particle positions might be better. If, on the other hand, one is interested in turbulence where many modes are present, then uses of random numbers might give a better choice. Similar statement can be made for a velocity space distribution.

Among many different ways of initial loading, the quiet start in which the phase space is loaded as uniformly as possible has interesting characteristics (Byers and Grewal, 1970). For a spatially uniform Maxwellian, for example, quiet start can be obtained by loading particles uniformly in x and repeating the velocities used to represent a Maxwellian velocity distribution at one location in x for all the values of different x. In this way, uniform phase space is maintained and there would be no random noise associated with the initial loading. Such a loading scheme can be used for any distribution which is uniform in the coordinate space. If a plasma is intrinsically unstable with respect to plasma instbilities, initial noise associated with machine round-offs will grow to large amplitude until limited by nonlinear effects.

Clearly the advantage of using a quiet start is to suppress the unwanted noise coming from the initial loading such as the one associated with the use of random numbers and therefore the quiet start is particularly useful when studying weak instabilities or small amplitude phenomena. When studying strong instabilities or large amplitude phenomena, differences arising from different initial loadings are expected small.

There are several short-comings associated with the quiet start, however. One obvious drawback is that the method cannot be applied in general to a nonuniform plasma. The other drawback, which is more fundamental, is the presence of multi-beam instability in the presence of beams associated with the quiet start. Quiet start may be considered to consist of an ensemble of cold beams whose distribution is given by

$$f = \sum_i n_i \; \delta(\underset{\sim}{v} - \underset{\sim}{v}_{bi}) \quad . \tag{79}$$

Instabilities arising from the presence of multi-beams have been studied in detail (Dawson, 1960) and it is shown that the growth rate is proportional to $k\delta v$ where δv is the spacing of the beam. By choose the sufficient number of beams, the growth rate may be reduced to a much smaller value compared with the growth rate of the physical instability one is looking for. If that is the case, beam instability associated with the presence of multi-beams may be harmless.

Another interesting trick in plasma simulations useful when a small region in phase-space play an important role is particle weighting. Such an example is the well-known bump-on-tail instability in which a weak electron beam is injected to a background thermal plasma. Here the beam density, n, can be as small as 1% of the background density N, so that most of the particles may be used to represent the bulk distribution where little physics takes place. Most of the nonlinear effects such as particle trapping occur on beam particles where there are only a small fraction of simulation particles representing them.

It is possible to choose n~N so that statistical fluctuations associated with the small number of beam particles do not mask the physics one is trying to study. If n~N is chosen, then one must assign smaller charge and mass to beam particles while keeping q/m unchanged. It is clear that so long as q/m stays the same, Vlasov equation or equation of motion remains the same. The only change in the simulation code is in the charge density claculation in which beam particles carry smaller amounts of charge. It may be pointed out that, while Vlasov equation remains the same, collisional effects associated with the two-body correlation will be modified even if q/m is kept unchanged.

PART II: Application of Particle Simulation to Electrostatic Ion Cyclotron Waves on Auroral Field Lines

1. INTRODUCTION

It is well known that the electrostatic ion cyclotron (EIC) waves may be destabilized by drifting electrons (current) through the stationary ions (Drummond and Rosenbluth, 1962; Kindel and Kennel, 1971; Okuda et al., 1981). EIC waves are particularly important in an isothermal plasma ($T_e \simeq T_i$) where the ion acoustic waves may be stable due to Landau damping. The threshold drift speed is a fraction of electron thermal speed so that EIC waves may easily be excited either by a current or an ion beam.

Laboratory experiments as well as space craft measurements have been reported recently on the observation of EIC waves in tokamaks (TFR Group, 1978), linear devices (Yamada and Hendel, 1978), and space plasmas (Kintner et al., 1979; Yau, et al., 1983). Large amplitude density fluctuations, heating of ions across magnetic field and anomalous cross-field particle diffusion have been observed in these measurements.

There are a number of theoretical considerations on the linear as well as nonlinear behavior of the current-driven EIC instabilities. Linear theory predicts that the unstable modes satisfy

$$\begin{aligned} \omega &\simeq n\,\Omega_i \\ k_\perp \rho_i &\simeq n \end{aligned} \tag{80}$$

where ρ_i is the ion thermal gyroradius and n is an integer representing a cyclotron harmonic.

when the electron drift speed is above the threshold, EIC waves grow to large amplitude until limited by nonlinear effects. In the absence of electron source, plateau formation on the electron distribution due to quasilinear diffusion gives rise to the nonlinear saturation of the EIC waves resulting in a modest ion heating (Drummond and Rosenbluth, 1962). There are situations, however, in which a flux of fresh electrons constantly replenishes the distribution function of electrons so that complete stabilization due to plateau formation cannot take place. Plasma heating experiments by injection of electorn beams (Yamada and Hendel 1978), ion beams (Eubank et al., 1979) and field-aligned auroral currents where the ionosphere acts as a resrvior of fresh electrons are such examples. For these cases, the duration of beams is much longer than the characteristic time scale of the EIC waves so that a presence of beam source plays an important role on the nonlinear behavior of the EIC waves, ion heating and plasma transport.

Here we would like to present results from numerical simulations on EIC waves with and without a presence of source. Numerical results obtained from the initial value simualtions without a plasma source are given in Sec.2. In Sec.3, we present simulation results in which a plasma is subject to electron beam injection at one end of the system. In Sec.4, the simulations are extendead to two dimensions where large amplitude density striations across magnetic field appear. These density striations may be responsible for the generation of auroral are elements.

2. RESULTS OF SIMULATIONS WITHOUT A SOURCE

In this Section, we shall present results obtained from one-and two-dimensinal numerical simulations on the EIC instabilities. The simulation model used is an electrostatic model in which full dynamics is retained for the ion motion while guiding center drift approximation is used for the electrons. This approximation for the electrons is valid for low frequency, $\omega \ll \Omega_e$, and long wavelength, $k_\perp \rho_e \ll 1$, oscillations where electrons are treated as guiding center particles. Electron motion along magnetic field is solved exactly. In essence, a set of equations for this model are

$$m_i \frac{d\underset{\sim}{v}_i}{dt} = e \left(\underset{\sim}{E}(\underset{\sim}{x}_i) + \frac{1}{c} \underset{\sim}{v}_i \times \underset{\sim}{B} \right)$$

$$m_e \frac{dv_{\parallel e}}{dt} = - e E_{\parallel}(\underset{\sim}{x}_e)$$

$$\underset{\sim}{v}_{\perp e} = \frac{c\underset{\sim}{E}(\underset{\sim}{x}_e) \times \underset{\sim}{B}}{B^2}$$

$$\frac{d\underset{\sim}{x}_i}{dt} = \underset{\sim}{v}_i$$

$$\nabla^2 \phi = - 4\pi e (n_i - n_e)$$

$$\underset{\sim}{E} = -\nabla\phi$$

which have been solved on a spatial grid using a finite difference integration in time (Lee and Okuda, 1978).

The initial conditions for particles are a stationary Maxwellian for the ions and a drifting Maxwellian for the electrons along magnetic field. Initially, spatial distribution is uniformly loaded using random numbers. The parameters of the simluations are $m_i/m_e=1837, \omega_{pe}/\Omega_e=0.2$ and $T_e=T_i$ initially.

Let us first study results obtained from a one-dimensional

simulation in which a uniform magnetic field is taken in the y-z plane with $B_y/B_z = 0.1$. Perturbations are allowed only in the y-direction (one-dimensional model). $v_{de}/v_{te}=1.4$, $L=1024\ \Delta$ where L is the system length and Δ is the mesh size and $\lambda_e=\Delta$. The allowed wavelengths on the grid points in this system is given by

$$k_m = \frac{2\pi m}{L} \quad (m = 0, \pm 1, \pm 2, \cdots, \pm\frac{L}{2})$$

so that $k_\perp \rho_i$ varies 0.05 to 10 with the spacing $\delta(k_\perp \rho_i)=0.05$. 70,000 ions and electrons are used which gives about 70 particles per grid for each species.

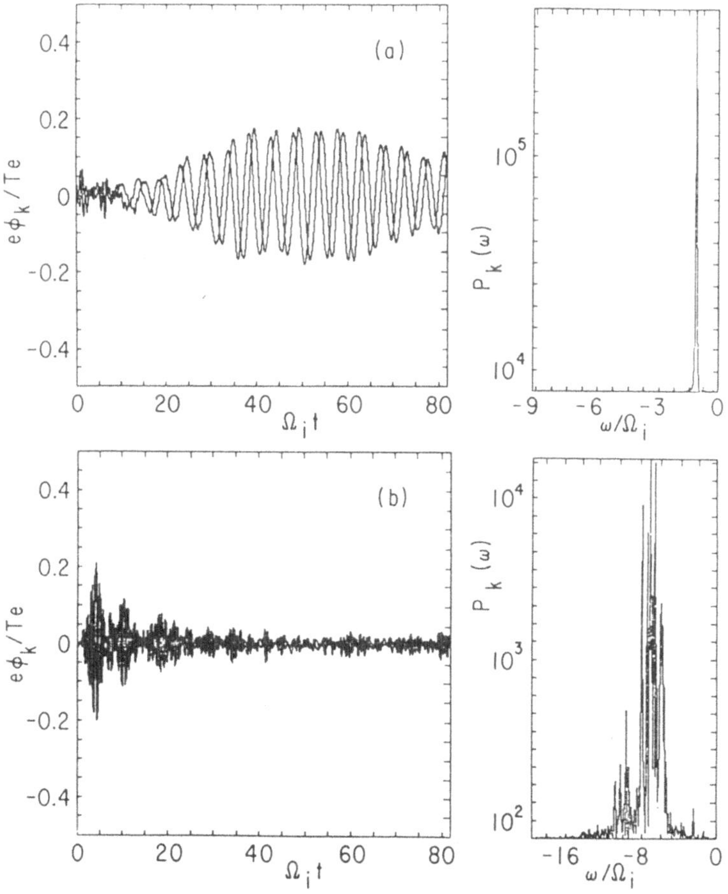

Figure 2. Time history of the real and imaginary parts of the potential, $e\phi_k(t)/T_e$, and its frequency for the 3rd (a) and the 23rd harmonics (b).

Figures 2 (a) and (b) indicate the time history of the real and imaginary parts of the electrostatic potential, $e\Phi_k(t)/T_e$, for the third mode, (a), and the 23rd mode and their frequency spectrum, (b). For the third mode, $k_\perp\rho_i=0.16$, only the fundamental cyclotron harmonic is excited while for the 23rd mode, $k_\perp\rho_i=1.2$, several harmonics are clearly seen. The amplitude for these modes reaches $e\Phi_k/T_e=0.2$ before saturation.

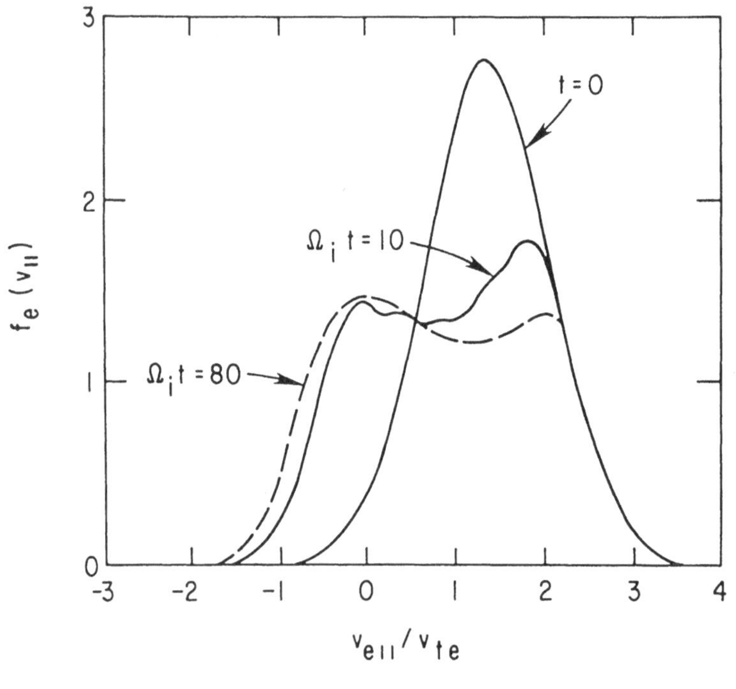

Figure 3. Electron velocity distribution along magnetic field at different times. Note the development of a plateau.

Nonlinear saturation takes place when the electron velocity distribution develops a plateau as shown in Figure 3. At $\Omega_i t=10$, a plateau is formed for the region of velocity space, $v_\parallel < v_{de}$, which then creates a positive slope $\partial f_e/\partial v_\parallel > 0$ for $v_\parallel > v_{de}$ as seen in Figure 3. This new region of the positive slope is flattened later as the long wavelength modes correspoding to larger phase velocity are destabilized.

Heating of ions perpendicular to magnetic field and the loss of electron kinetic evergy along magnetic field is shown in Figure 4. The relative change for both ions and electrons remains small as discussed in Sec. 3 because the plateau formation on the electron distribution releases only a fraction of electron kinetic energy.

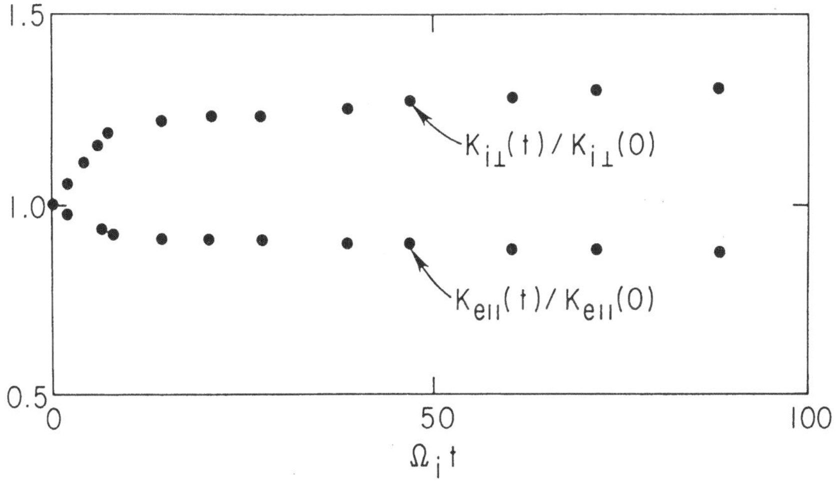

Figure 4. Heating of ion perpendicular energy and cooling of electron kinetic energy along magnetic field.

Simulations are extended to 2-dimensions in which spatial variations are allowed in the x-y plane with the external magnetic field oriented as before. Results of 2-dimensional simulations is expected to reveal processes not allowed in one-dimensional results such as mode-coupling among modes at different angles of propagation and cross-field particle diffusion across magnetic field.

Figure 5 shows a plot of test particle positions at an instant of time, $\Omega_i t=30$, for electrons (top) and ions (bottom) which were initially located at a narrow strip located at x=20. The spread in the x-direction suggests the presence of cross-field particle diffusion due to the ion cyclotron instabilities. The measured diffusion coefficient, $D_x = \langle(\Delta x_e)^2\rangle/t \simeq 5\times10^{-3}\Delta^2\omega_{pe}$ suggesting a presence of large anomaly for the transport coefficients associated with ion cyclotron waves.

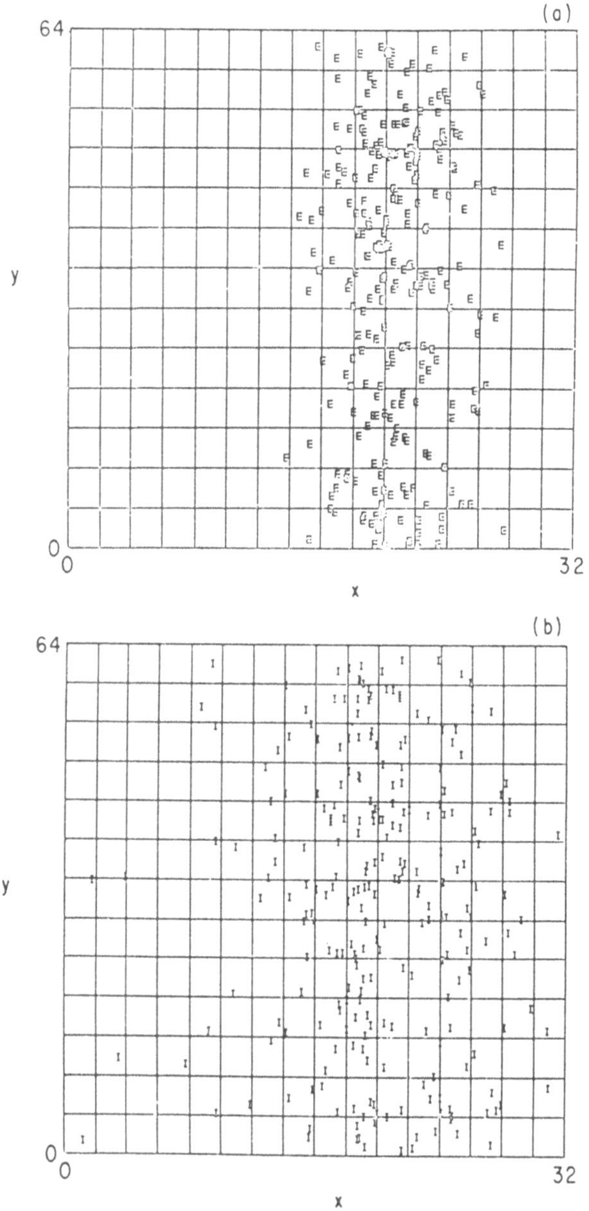

Figure 5. Plot of the test particle positions at $\Omega_i t=30$ for electrons (top) and ions (bottom) which were initially located at x=20.

3. RESULTS OD SIMULATIONS WITH A SOURCE

We have seen earlier in Sec.2 that the ion cyclotron instabilities saturate at a low level giving rise to a modest heating of ions in the absence of electron source. Here we present results of simulations obtained using a model in which a plasma is subject to a source which constantly injects electrons into a plasma system with the initial drifting Maxwellian velocity distribution as show in Figure 6. Such a model may represent an electron beam injection experiment

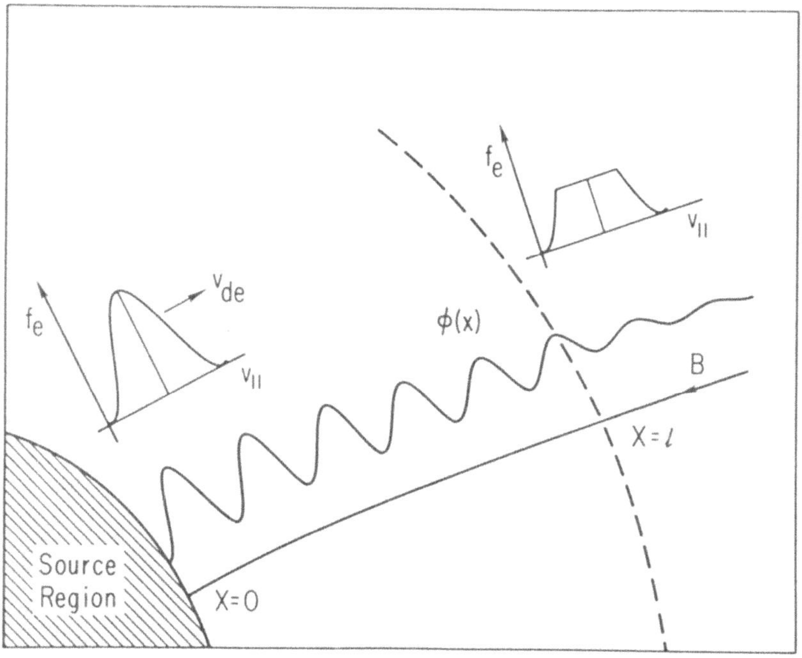

Figure 6. Theoretical model for auroral field lines in which the ionosphere is the source of cold, drifting Maxwellian electrons.

in a plasma or an auroral current along magnetic field driven by the ionosphere-magnetosphere coupling (Okuda and Ashour-Abdalla, 1983). It is clear that the ion heating in this case is expected to be much stronger in the presence of an electron source due to the inhibition of plateau formation on the electron distribution function.

Figure 7 shows the profiles of ion density normalized by the average density for the entire system length plotted at three different times. The parameters of the simulations are the same to the one-dimensional simulations reported in Sec.2 with L=1024 Δ,

$v_{de}=1.4v_{te}$, $m_i/m_e=1837$, $k_\parallel/k_\perp=0.1$ and $\Omega_e/\omega_{pe}=5$. The dashed vertical lines denotes the boundary lines for different bins for diagnostic purposes. At time $\Omega_i t=110$ (upper panel), the larger density perturbations, $\delta n/n \simeq 0.25$ are mostly confined to the first two bins. At later times shown in the middle and lower panels, the ion density perturbations are seen to propagate and extend further right, suggesting the unstable EIC waves can propagate with the beam. From Figure 7,one can estimate the propagation speed along magnetic field to be 0.007 v_{de}. This speed is much smaller than the electron drift speed and is in good agreement with the self-induced propagation speed calculated theoretically (Okuda and Ashour-Abdalla, 1983)

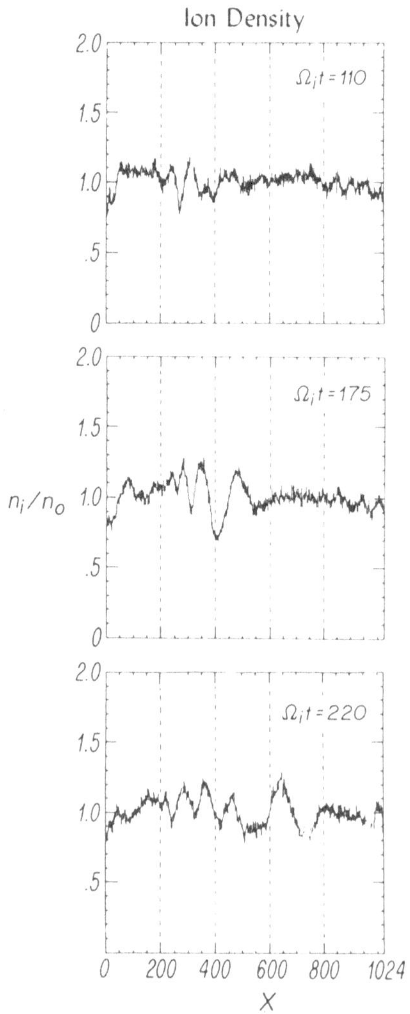

Figure 7. Ion density profile at three different steps. Note a large density perturbation propagates along magnetic field.

The fact that the density modulation is associated with the ion cyclotron waves is confirmed by measuring the frequency at several different locations along field lines. This is shown in Figure 8 where the time history of electron density at two different locations, $x=320\Delta$ and $x=416\Delta$ are shown in the upper panel (a), while in the lower panel, (b), frequency spectra of the density fluctuations are shown. The frequency analysis confirms the presence of coherent peaks above Ω_i. In addition, there are much smaller, but clearly coherent peaks near $\omega=k_{\parallel}c_s$ indicating oblique ion acoustic waves propagating nearly perpendicular to magnetic field.

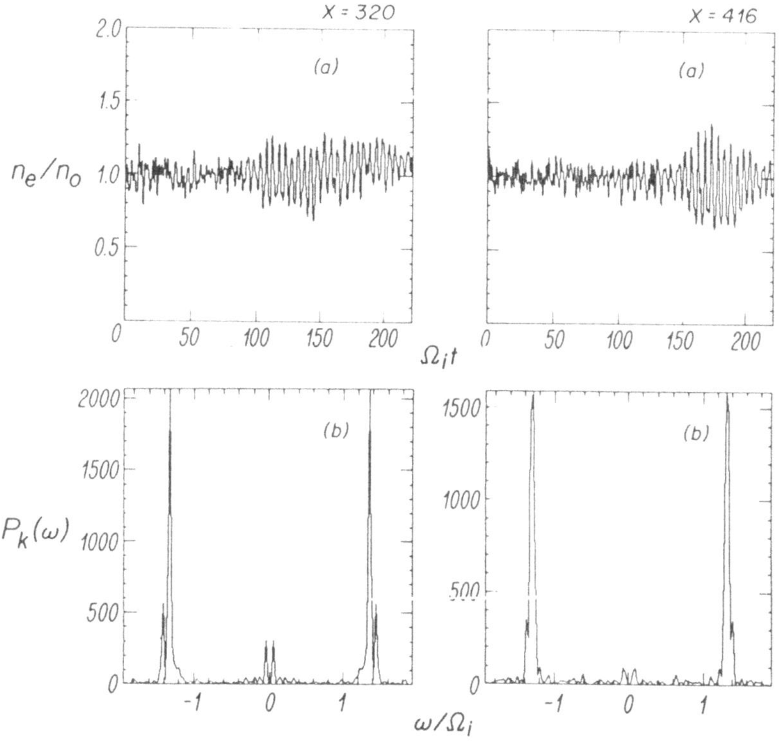

Figure 8. Temporal behavior of the electron density perturbation at $x=320\Delta$ and $x=416\Delta$ and its frequency spectrum.

The effects of ion cyclotron waves on ions are shown in Figure 9 where the perpendicular ion velocity distributions are shown for three bins at $\Omega_i t=220$. The large amplitude density perturbations associated with ion cyclotron waves have penetrated up to the third bin by this time. We therefore expect that the temperature in the first and second bins is more or less heated up to the maximum value determined from the marginal stability analysis (Okuda and Ashour-Abdalla, 1983).

It is very interesting to realize the presence of high energy tail in Figure 9 where the total distribution is separated into bulk and tail parts. The initial Maxwellian distribution is also shown for comparison in the figure. The heated distribution has a high energy tail extending almost as much as 100 times of the initial thermal energy. The temperature of the bulk distribution is about 5-10 times of the initial Maxwellian while that of the tail distribution 50-100 times of the initial temperature.

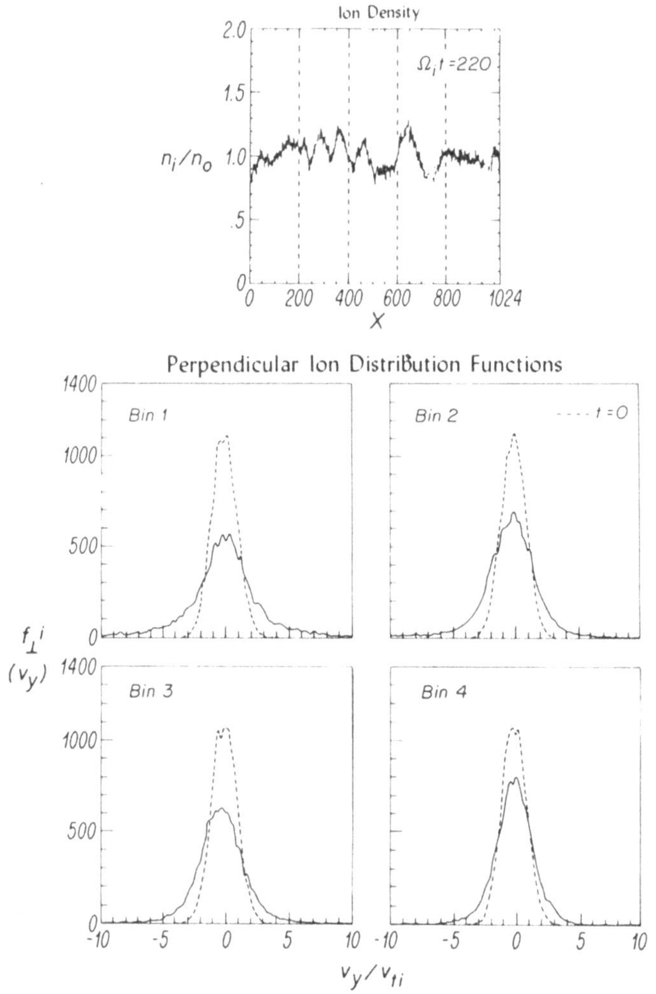

Figure 9. Ion density profile and the ion perpendicular velocity distribution at four locations at $\Omega_i t=220$.

Figure 10 shows similar plots for the electron distribution at $\Omega_i t=175$ for three different bins. It is seen that the electron distribution generally has a plateau at far right such as in bin 4, whereas the distribution tends to have a positive slope for $v_{\parallel}<v_{de}$ in bins 1 and 2. This suggests that the electron distribution is determined by a delicate balance between wave-induced diffusion and input from the source. Note the plateau formation on the electrons is a subtle process which does not require much energy at all.

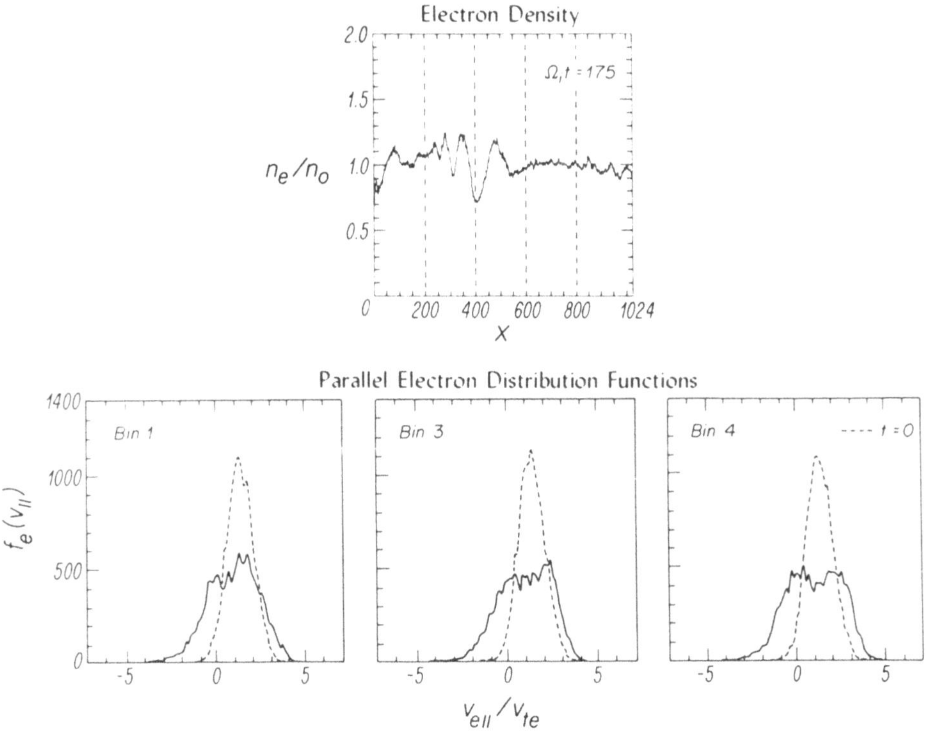

Figure 10. Electron density profile and the electron parallel velocity distribution at three different locations at $\Omega_i t=175$.

4. FORMATION OF AURORAL ARC ELEMENTS

Observations of ion cyclotron waves (Kintner et al., 1979; Yau et al., 1983) and of conic ion distributions on high-latitude field lines (Sharp et al., 1977; Klumpar, 1979; Ungstrup et al., 1979; Gorney et al., 1983) motivated a series of extensive theoretical studies of ion cyclotron turbulence on auroral field lines (Lysak et al., 1981; Papadopoulos et al., 1980; Singh et al., 1981; Dusenbery and Lyons,

1981; Ashour-Abdalla et al., 1981; Okuda and Ashour-Abdalla, 1983). These theoretical studies focused on ion heating due to the ion cyclotron turbulence and showed that both O+ and H+ ions can be accelerated perpendicular to field lines resulting in enhanced ion fluxes near 90^0 pitch angle.

In this section we focus on the nonlinear mode-coupling effects of the electrostatic ion cyclotron (EIC) turbulence across the magnetic field. The simulation model is assumed two-dimensional in which the ion cyclotron waves are allowed to propagate in the x-y plane (i.e., $k_z=0$) in an external uniform magnetic field which is in the y-z plane satisfying $B_z \gg B_y$. The ratio $B_y/B_z=0.1$ is chosen in our simulation model. The electron perpendicular motion is approximated by the guiding center drift in the simulation while the electron parallel motion and the ion dynamics are treated fully without approximation (Lee and Okuda, 1978). The model is initialized by a drifting Maxwellian electron distribution streaming along the external field and a stationary Maxwellian ion distribution. Note that the x axis is parallel to the north-south direction; the y axis is parallel to the east-west direction. The tilting of the magnetic field in the y-z plane (with $B_y/B_z=0.1$) is as approximation of the field (B_y) due to the field-aligned currents (Iijima and Potemra, 1976). The assumed velocity profile is an approximation of the condition of the electron flux being accelarated through the inverted-V potential drop.

Note that our choice of the magnetic field orientation excludes the current-driven instabilities propagating along field lines such as Buneman and ion-acoustic instabilities. Earlier two-dimensional simulations suggested that such instabilities saturate at low level and do not effect the EIC instability appreciably (Pritchett et al., 1981). Furthermore the perpendicular electric fields associated with such instabilities propagating along field lines are small and therefore cannot generate density striations across field lines.

The simulation is performed on a 128×64 grid in the x-y plane with 73,728 electrons and ions. The electron streaming speed is $V_{do}=2v_{te}$ where v_{te} is the electron thermal speed. The ratio of the electron gyrofrequency to plasma frequency is $\omega_{ce}/\omega_{pe}=5$ and the time step of integration is $\omega_{pe}\Delta t=5$. The ion to electron mass ratio is $m_i/m_e=400$. The initial ion gyroradius is $\rho_i=4$ grid size. The electron streaming speed is maintained by the recycling schemes (Okuda and Ashour-Abdalla, 1981) with a recycling rate $\delta=5\times10^{-3}/\Delta t$. This recycling scheme is a technique to maintain a flow of the drifting Maxwellian electron distribution along the inverted-V field lines. An electron streaming speed of ~2 v_{te} or greater can be expected to prevail along a large fraction of the acceleration region, and certainly below the acceleration region. It should be noted that the simulation model as described above is strictly valid only in regions where the electrons have been accelerated through at least a fraction of the total potential drop and not too far below the acceleration region where the cold ambient electrons are still a minor constituent.

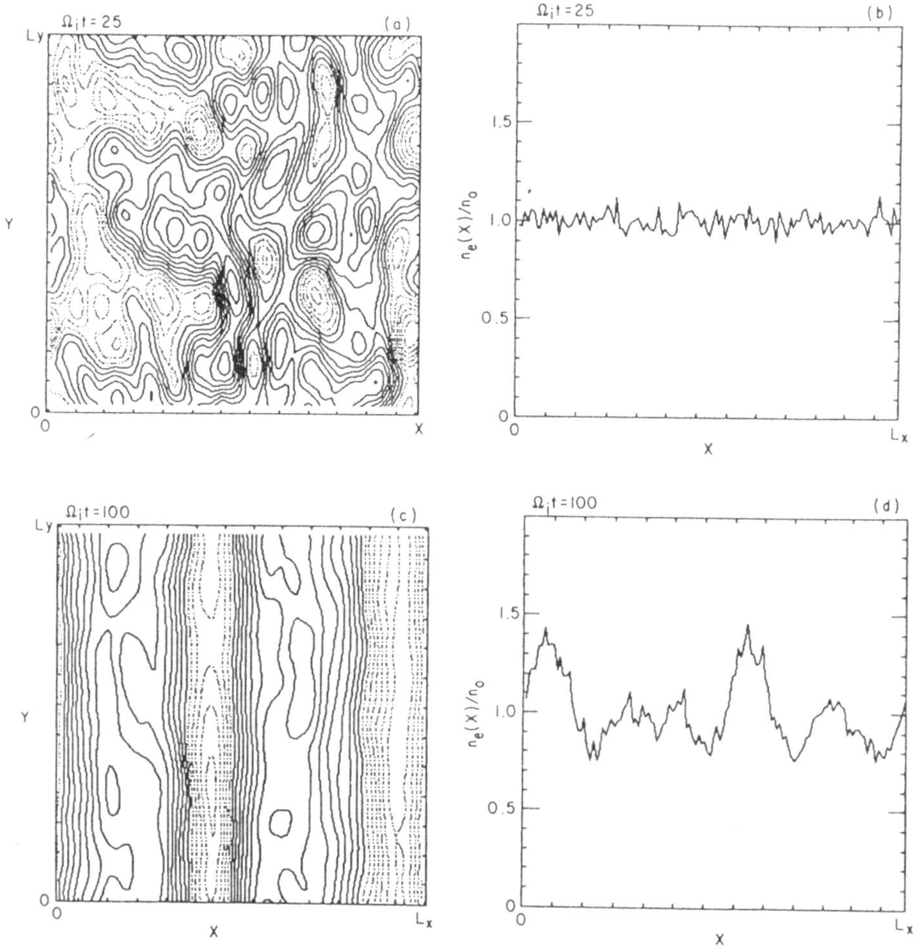

Figure 11. The number density profile $n_e(x)$ and the equipotential contours at $\Omega_i t=25$ and 100 in a two-dimensional simulation. Note the development of density striations across magnetic field.

Figure 11 shows the equipotential contours and the electron density $n_e(x)$ averaged over y associated with the ion cyclotron waves. During early stages of the simulation ($\Omega_i T=25$) the equipotential contours due to the ion cyclotron turbulence exhibit little preferred alignment with little density striations. Note the presence of small scale contours whose sizes are on the order of the initial ion gyroradius. As the simulation develops well into the nonlinear stage ($\Omega_i T=100$), the equipotential contours and the constant density contours start to striate along the y axis (east-west direction). The amplitude of the density modulation grows in time and saturates at a maximum variation of $n_{max}/n_{min} \simeq 2$. Note that the ion gyroradius is several times larger than the initial value due to the perpendicular ion heating by the ion cyclotron turbulence (Okuda and Ashour-Abdalla, 1981).

Variations in the electron streaming speed in our simulations are small compared with variations in the electron number density so that current density striation is produced mainly by the number density striation. It follows that the electron energy flux is also striated in accordance with the electron number density in our simulations. Thus, the intensity of the striations in the field-aligned current and the electron energy flux are both on the order of the number density striation $(n_{max}-n_{min})/n_{min} \simeq 100\%$. It is important to note that fine scale variations of field-aligned current density up to ~50% of the background current density within an inverted-V scale precipitation region have been measured by sounding rocket experiments conducted by Evans et al. (1977). However, association of these fine scale current density enhancements with auroral arc elements have yet to be established observationally. To recapitulate we emphasize that the striations in number density, field-aligned current and electron energy flux due to ion cyclotron turbulence are the main results of our simulations. The proposed connection between the striations in our simulations and the auroral arc elements is based on plausible arguments rather than observational facts.

The physical mechanism leading to the east-west striations in our simulation can be identified with a nonlinear mode-coupling process between electrostatic ion cyclotron waves. Frequency analysis of wave modes in the simulation reveals that the striation corresponds to a condensation of the wave energy into a zero-frequency d.c. mode (Okuda and Dawson, 1973). Such a condensation phenomenon has been demonstrated in unstable drift waves (Cheng and Okuda, 1978; Sagdeev et al., 1978a) and Alfven waves (Sadgeev et al., 1978b). On this basis, the striation in our simulation can be interpreted as resulting from nonlinear mode coupling between two ion cyclotron waves having the same $k_{\parallel}$ but different $k_{\perp}$(Cheng and Okuda, 1978; Sagdeev et al., 1978a,b). Since the frequencies of ion cyclotron waves are very close to the ion gyrofrequency when the ion perpendicular temperature is much greater than the ion parallel temperature (Okuda and Ashour-Abdalla, 1981), the nonlinear beating of two ion cyclotron waves with the same $k_{\parallel}$ can produce a zero-frequency mode with $k_{\parallel}=0$. Once such modes are generated across the magnetic field, they produce quasi-steady state density striations which persist on a time scale much longer than the ion gyroperiod owing to the slow dissipation associated with the ion viscosity (Okuda and Dawson, 1973). Only a slow diffusive process can destroy such d.c. structures. We are aware of the possibility that the condensation of ion cyclotron turbulence into a zero-frequency mode is just one example of a class of nonlinear phenomena which generate quasi-stationary macrostructures from microinstabilities.

5. DISCUSSIONS

We have given an example to show that how to apply particle simulations to a study of ion heating and electron beam propagation in a magnetic field. Only the electrostatic ion cyclotron instabilities have been considered as they are the most relevant instabilities on auroral field lines. It is clear that different boundary conditions will result in very different results of ion heating and beam propagation.

ACKNOWLEDGEMENTS

This work was supported by the National Science Foundation grant ATM83-11102 and the United States Department of Energy Contract No. DE-AC02-76-CHO-3073.

REFERENCES

Akasofu, S.-I., J. Atmos. Terr. Phys., 21, 287, 1961.
Ashour-Abdalla, M., H. Okuda and C. Z. Cheng, Geophys. Res. Lett., 8, 795, 1981.
Birdsall, C. K., and D. Fuss, J. Comput. Phys., 3, 494, 1969.
Birdsall, C. K., and A. B. Landon, Plasma Physics via Computer Simulation (McGraw-Hill, New York, in press).
Boris, J. P., "Relativistic Plasma Simulation" in Proceedings of the Fourth Conference on Numerical Simulation of Plasma, November, 1970, page 3.
Buneman, O., Phys. Rev., 115, 503, 1959.
Buneman, O., J. Comput. Phys., 1, 517, 1967.
Byers, J. A., and M. Greual, Phys. Fluids, 13, 1819, 1970.
Chen, L., and H. Okuda, J. Comput. Phys., 19, 339, 1975.
Cheng, C. Z. and H. Okuda, Nucl. Fusion, 185, 587 1978.
Cheng, C. Z., and H. Okuda, J. Comput. Phys., 25, 133, 1977.
Davis, T. N., Space Sci. Rev., 22, 77, 1978.
Dawson, J. M., Phys. Fluids, 5, 445 1962.
Dawson, J. M., Review of Modern Phys., 55, 403 1983.
Dawson, J. M., Phys. Rev., 118, 391, 1960.
Drummond, W. E., and M. N. Rosenbluth, Phys. Fluids, 5, 1507 1962.
Dusenbery, P. B. and L. R. Lyons, J. Geophys. Res., 86, 7627, 1981.
Eubank, H., R. Goldston, V. Arunasalam, M. Bitter, K. Bol, D. Boyd, N. Bretz, J. P. Bussac, S. Cohen, P. Colestock, S. Davis, D. Dimock, H. Dylla, P. Efthimion, L. Grisham, R. Hawryluk, K. Hill, E. Hinnov, J. Hosea, H. Hsuan, D. Johnson, G. Martin, S. Medley, E. Meservey, N. Sauthoff, G. Schilling, J. Schinell, G. Schmidt, F. Stauffer, L. Stewart, W. Stodiek, R. Stooksberry, J. Strachan, S. Suckewer, H. Takahashi, G. Tait, M. Ulrickson, S. von Goeler, M. Yamada, C. Tsai, W. Stirling, W. Dagenhart, W. Gardner, M. Menon, and H. Haselton, Phys. Rev. Lett, 43, 270, 1979.
Evans, D. S., N. C. Nayard, J. Troim, T. Jacobsen, and A. Egeland, J. Geophys. Res., 82, 2235, 1977.
Frank, L. A. and K. L. Ackerson, J. Geophys. Res., 76, 3612, 1971.
Gorney, D. J., A. Clarke, D. Croley, J. Fennell, J. Luhmann, and P.Mizera, J. Geophys. Res., 86, 83, 1981.
Hockney, R. W., Phys. Fluids, 9, 1826 1966.
Hockney, R. W., and J. W. Eastwood, "Computer Simulation Using Particles" (McGraw-Hill, New York), 1981.
Iijima, T. and T. A. Potemra, J. Geophys. Res., 81, 3999, 1976.
Kan, J. R., Space Sci. Rev., 31, 71, 1982.
Kan, J. R., and H. Okuda, J. Geophys. Res., 88, 6339, 1983.
Kim, J. S. and R. A. Volkman, J. Geophys. Res., 68, 3187, 1963.
Kindel, J. M. and C. F. Kennel, J. Geophys. Res., 76, 3055, 1971.
Kintner, P. M., M. C. Kelley, R. D. Sharp, A. G. Ghielmetti, M. Temerin, C. Cattell, P. F. Mizera, and J. F. Fennell, J. Geophys. Res., 84, 7201, 1979.
Klumpar, D. M., J. Geophys. Res., 84, 4229, 1979.
Kruer, W. L., J. M. Dawson, and B. Rosen, J. Comput. Phys., 13, 114, 1973.

Langdon, A. B., and C. K. Birdsall, Phys. Fluids, 13, 2115 1970.
Langdon, A. B., "Effects of the Spatial Grid in Simulated Plasmas, J. Comput. Phys., 1971.
Lee, W. W., and H. Okuda, J. Comput. Phys., 26, 139, 1978.
Lysak, R. L., Electron and Ion Heating by strong Electrostatic Turbulence in Physics of Auroral Arc Fromation, ed. S. I. Akasofu and J. R. Kan, AGU (Washington), 1981.
Maggs, J. E. and T. N. Davis, Planet. Space Sci., 16, 205, 1968.
Meng, C. -I., Space Sci. Rev., 22, 223, 1978.
Miura, A., H. Okuda and M. Ashour-Abdalla, Geophys. Res. Lett., 10, 353, 1983.
Mizera, P. F. and J. F. Fennell, Geophys. Res. Lett., 4, 311, 1977.
Morse, R. L., and C. W. Nielson, Phys. Fluids 12, 2418, 1969.
Nishikawa, K., H. Okuda, and A. Hasegawa, Geophys. Res. Lett., 10, 553, 1983.
Okuda, H., and C. K. Birdsall, Phys. Fluids, 13, 2123, 1970.
Okuda, H. and J. M. Dawson, Phys. Fluids, 16, 408, 1973.
Okuda, H., and C. Z. Cheng, Comput. Phys. Comm., 14, 169, 1978.
Okuda, H., J. M. Dawson, A. T. Lin, and C. C. Lin, Phys. Fluids, 21, 476, 1978.
Okuda, H., W. W. Lee, and C. Z. Cheng, Comput. Phys. Comm., 17, 233 1979.
Okuda, H., C. Z. Cheng, and W. W. Lee, Phys. Rev. Lett, 46, 427, 1980.
Okuda, H., W. W. Lee, and C. Z. Cheng, Phys. Fluids, 24, 1060, 1981.
Okuda, H., and M. Ashour-Abdalla, Geophys. Res. Lett., 8, 811, 1981.
Okuda, H., and M. Ashour-Abdalla, J. Geophys. Res., 88, 899, 1983.
Papadopoulos, K., J. D. Gaffey, Jr. and P. J. Palmadesso, Geophys. Res. Lett., 7, 1014, 1980.
Perkins, F. W., Phys. Fluids, 19, 1012 1976.
Potter, D., Computational Physics (Wiley, New York), 1973.
Pritchett, P. L., M. Ashour-Abdalla and J. M. Dawson, Geophys. Res. Lett., 8, 611, 1981.
Sagdeev, R. Z., V. D. Shapiro and V. J. Shevchenko, Sov. J. Plasma Phys., 4, 306 1978a.
Sagdeev, R. Z., V. D. Shapiro and V. J. Shevchenko, JETP Letters, 27, 340 1978b.
Sgro, A. G., and C. W. Nielson, Phys. Fluids, 19, 126, 1976.
Sharp, R. D., R. G. Johnson and E. G. Shalley, J. Geophys. Res., 82, 3324, 1977.
Singh, N., R. W. Schunk and J. J. Sojka, Geophys. Res. Lett., 8, 1249, 1981.
TFR Group, Phys. Rev. Lett, 41, 113, 1978.
Ungstrup, E., K. M. Klumpar and W. J. Heikkila, J. Geophys. Res., 84, 4289, 1979.
Yau, A. W., B. A. Whalen, A. G. MacNamara, P. J. Kellogg, and W., Bernstein, J. Geophys. Res., 88, 341, 1983.

PARTICLE SIMULATION OF ELECTROMAGNETIC WAVES
AND
ITS APPLICATION TO SPACE PLASMAS

Hiroshi Matsumoto and Yoshiharu Omura

Radio Atmospheric Science Center,
Kyoto University,
Uji, Kyoto 611, Japan

ABSTRACT

In this article, we present a basic description of particle simulation codes developed at RASC of Kyoto University for the study of wave phenomena in space plasmas. Out of the available codes at RASC, two most-frequently-used codes are presented here. One of them is the full electromagnetic code named KEMPO (Kyoto University ElectroMagnetic cOde) which is able to treat both electromagnetic and electrostatic wave modes simultaneously with a special time-filter technique using a multi-time-step (MTS) scheme. The other is the long time scale code (LTS) named KULTS (Kyoto University LTS) which is useful for studies of whistler mode wave-particle interactions over hundreds to thousands of wavelengths in both homogeneous and inhomogeneous plasmas. Principles and basic techniques of the codes as well as examples of their applications to space plasma problems are presented.

1. INTRODUCTION

Space exploration has been achieved by a rapid development of space vehicles and their use for scientific survey of our space environment. In 1960's and 1970's, lots of discoveries of new space phenomena were brought back from these spacecraft observations. Data analysis-phase followed after the discovery-phase yielding a variety of physical models of our space plasma environment and theories for the related plasma processes taking place therein. However, there still remains a large number of problems because of too many degrees of freedom of choice of physical models. Natural phenomena in space plasmas are often highly nonlinear and too complicated for theories which usually rely on linearity or weak nonlinearity, symmetry

H. Matsumoto and T. Sato, Computer Simulation of Space Plasmas,

characteristics, homogeneity or simple inhomogeneity and other simplistic assumptions. On the other hand, experiments by satellite observations are still too coarse: Measurements are limited in time and in space leaving a large amount of ambiguity and disputes among researchers as to the interpretation of the observed results and related theoretical models. To resolve these uncertainties among various models and theories and to find a hint or clue for understanding the unsolved physics underlying the observed phenomena, computer simulations have been high-lighted as a third promising research approach by bridging the traditional two approaches, theory and experiment, among space researchers in late 1970's and in 1980's.

Computer simulations for space physics may be divided into two categories: First is an MHD simulation which follows the nonlinear fluid motion of plasmas in their self-consistent fields (and applied fields if any). This class of simulations is useful in understanding macroscopic global-scale dynamics which cannot be understood by piecewise information from individual satellite observations. Second is a particle simulation which follows the nonlinear motion of many particles in their self-consistent fields (and applied fields if any). Particle simulations play a significant role in space physics in interpreting highly nonlinear kinetic effects like wave instabilities and associated plasma scattering, diffusion, heating and particle acceleration. Complicated and sometimes overlapping phenomena which nature generally exhibits can be decomposed into simpler elements of physics in simulations to obtain a clearer physical picture. One of the advantages of the computer simulation is that one can make as detailed diagnostics of plasma and field quantities as one desires. A precise visualization of the time evolution of the nonlinear micro-dynamics of interest is easily achieved in detail by graphic displays of the results of particle simulations. This could not be realized either by theory nor by satellite observations. Thus particle simulations provide useful data of wave-particle interactions which are inaccessible to satellite observations nor to theories. One such example is the rapid variation of the particle distribution function in velocity space. Such detailed information may often provide a hint and inspiration not only for further theoretical development but also for design of new satellite observations.

In this article, we will give an introductory review and guide of particle simulation of electromagnetic waves in plasmas and its application to space plasma phenomena. Basic concept of particle simulations is not given in detail because it is described elsewhere in this book [Okuda], and is found in the existing literature [e.g., Langdon and Lasinski, 1976; Hockney and Eastwood, 1981; Birdsall and Langdon, 1984]. Application to electrostatic wave phenomena is not presented either for the same reason. Two different codes are explained in this article which our research group have so far developed for the studies of electromagnetic wave phenomena in space. They are (1) Full Electromagnetic (FEM) code, and (2) Long Time Scale (LTS) code. These computer codes developed at Kyoto University are named "KEMPO" (Kyoto University ElectroMagnetic Particle cOde) and "KULTS" (Kyoto University LTS code), respectively.

In section 2, we briefly introduce a tutorial model and concepts of particle simulations for electromagnetic wave phenomena. Section 3 and 4 are devoted to the description of the basic equations and elements for coding of the KEMPO and KULTS, respectively. Development of the diagnostic tools is as important as coding of main programs in order to display various information in different ways so that one can extract meaningful physical results and can construct improved models and refine the related theory. Section 5 presents an outline of the diagnostic tools used in the KEMPO and KULTS particle codes. In section 6, applications of these codes to electromagnetic wave phenomena in space plasmas are presented.

2. ELEMENTARY CONCEPTS AND MODELS IN PARTICLE SIMULATIONS

In this section, we will give a short review of basic concept and models used in particle simulation in general. Readers are recommended to refer to existing textbook and literature for details [Morse and Nielson, 1969; Birdsall and Fuss, 1969; Dawson, 1970; Birdsall, Langdon and Okuda, 1970; Hockney, 1970; Langdon and Birdsall, 1970; Okuda and Birdsall, 1970; Okuda and Dawson, 1973; Langdon and Lasinski, 1976; Nielson and Lewis, 1976; Hockney and Eastwood, 1981; Birdsall and Langdon, 1984].

2.1 Superparticles

In particle simulation, plasma dynamics is studied by following a large number of particles in their self-consistent electric and magnetic fields. Naturally, one cannot emulate nature even with the today's super-computer. Only several orders of magnitude less number of particles are treatable in computer compared with the enormous number of particles in the real plasma of almost any size of interest. Therefore we are obliged to use an artificial model-particle called "superparticle" with a larger mass and charge than the real particle. The superparticle represents many particles in a real plasma, and has a finite size with its charge being distributed over a finite region of space.

The concept of the finite-sized superparticle was introduced historically by two research groups. First one was "Cloud-in-Cell(CIC)" [Birdsall and Fuss, 1969] and the other was "Particle-in-Cell(PIC) [Morse and Nielson, 1969]. Both were introduced as an improved model of zero-sized superparticle (plasma sheet mode) to suppress statistical fluctuations and short range collisions caused by a delta-functional nature of the zero-sized superparticles [Hockney,1966; Birdsall and Fuss, 1969; Morse and Nielson,1969; Okuda and Birdsall, 1970; Langdon and Birdsall, 1970]. At the same time the method of assignment of charge and current of the superparticle to its neighboring spatial grids was improved from NGP (Nearest Grid Point) scheme to AS (Area Sharing) scheme [Morse and Nielson,1969], MPE (Multipole Expansion) scheme and SUDS (Subtracted Dipole Scheme) [Kruer et al.,1973]. The theoretical works were developed on the modification of the plasma theory for the finite-size superparticle plasma [Langdon and Birdsall, 1970; Okuda and Birdsall, 1970]. Details such as modification of the dispersion characteristics and of the collisional effects are discussed elsewhere in this book [Okuda] and are not given here.

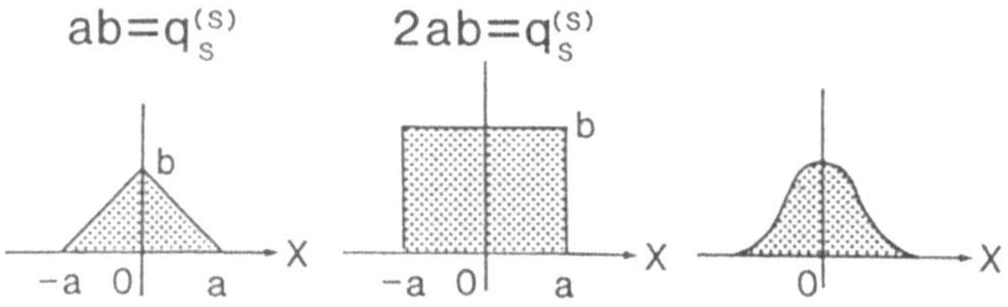

Figure 1. Three shapes of superparticle: (1) Square-shaped, (2) Triangular-shaped, (3) Gaussian-shaped.

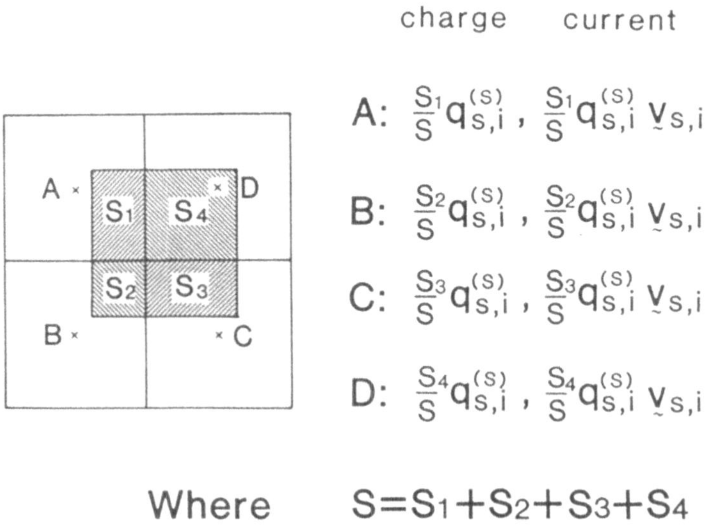

Figure 2. Schematic illustration of charge and current assignment for the square-shaped superparticle.

The shape of the superparticle can be of any form. Normally, however, one of the following three shapes is, for practical codings, hired in the particle simulation. These three shapes for one dimensional case are depicted in Figure 1. They are (1) Square-shaped, (2) Triangular-shaped, and (3) Gaussian-shaped. As for the scheme for the charge and current assignment to the neighboring grid points, the AS (area sharing) scheme is most frequently used . A schematic illustration of charge and current assignment for the square-shaped superparticle is given in Figure 2. In the figure the size of the square-shaped superparticle is assumed to be identical to that of the grid (or cell). Charge and current of a superparticle is shared by grid points where the superparticle is located. The charge and current assignment is made in such a way that the share to each grid is proportional to the overlapped area of that of the superparticle and the cell area of the grid. Other methods of the charge and current assignment have been proposed and are described in the literature [e.g., Hockney and Eastwood, 1981; Birdsall and Langdon, 1984; Okuda, 1984].

The charge and mass of the superparticle is much larger than those of the real particle. However the following three densities of the superparticles are kept the same as those of the real particles.

(1) Charge Density

$$N_s Q_s = N_r Q_r \tag{1}$$

(2) Mass Density

$$N_s M_s = N_r M_r \tag{2}$$

(3) Energy Density

$$N_s \kappa T_s = N_r \kappa T_r \tag{3}$$

where the suffixes, s and r, mean the quantities of the super- and real-particles, and N, Q, M, κ and T denote number density, charge, mass, Boltzmann's constant and temperature, respectively. Under these equalities, not only the charge to mass ratio but also the basic physical quantities such as the plasma and cyclotron frequencies, the Debye length, and the thermal velocity are kept identical to each other in both simulational and real plasmas. Thus the most of the physical properties of the plasma are reproduced in the particle simulation.

2.2 Discretization of Space and Time

In the particle simulation, both space and time are discretized. Spatial discretization is introduced by two reasons. One is to speed up the calculation of force acting on particles [Buneman, 1959; Dawson,1961]. Instead of calculating all contribution of Coulomb force from all particles, the force acting on a superparticle is calculated by field quantities defined on the grid points nearby the particle. Second reason is that the superparticle has a finite size over a certain region of space so that the spatial resolution smaller than the particle size is unnecessary and meaningless. Normally, the spatial grid spacing is taken to be from 1 to 3 times the Debye length. Numerical stability related to the choice of the grid size has been studied in detail and is found in the literature [Langdon and Birdsall, 1970; Okuda, 1970; Abe et al.,1975].

Time discretization is inevitable in any numerical approach to any partial differential equations. The problem is how large we can choose the time step keeping the numerical stability. To avoid the numerical instability, the Courant-Friedrichs-Lewy (CFL) condition should be satisfied. In case of the electromagnetic particle simulation, the CFL condition is

$$\Delta t < \Delta r/(c\sqrt{n}) \tag{4}$$

where n is the spatial dimension of the model, and c is the light speed.

Discretization of space and time modifies the dispersion characteristics. This modification comes from digital samplings of continuous quantities in space and time. As is well known from the sampling theorem, the frequency ω and wavenumber k should be replaced by $\Omega = \sin(\omega\Delta t/2)/(\omega\Delta t/2)$ and $K = \sin(k\Delta r/2]/(k\Delta r/2)$, respectively. Thus the dispersion equation for real plasma, $D(\omega,k) = 0$ is modified to $D(\Omega, K) = 0$. Because of this sampling function nature, high frequency and large k components are folded down in the low and small k domain in the ω-k space as "aliases". This effect should always be taken into account both in designing the model of simulation and in analyzing the simulation data.

2.3 Average Field and Self-Force

Since the charge of the superparticle with a finite size smears out in space, the force acting on it is the average one. In this subsection, we will discuss the procedure for the average and its related self-force problem. To this purpose, we will confine ourselves only to the simplest case of the square-shaped superparticle in one-dimensional case. Extension to the case of superparticles with other shapes and/or two-dimensional case is straightforward. We also limit our discussion only to the case of the electrostatic electric field as an example of the self-force. The magnetostatic self-force can be treated in a similar manner.

To guarantee the centered difference scheme for Maxwell's equation (Poisson's equation in this particular example), we normally adopt a staggered dual grid system composed of full-integer- and half-integer grid points as illustrated in Figure 3(a). The open circle points indicate full-integer-grid points on which the charge is defined. The open square symbols along the x-axis are half-integer grid points on which the electric field E_x is defined. Suppose that a superparticle with a size of the grid spacing Δx is located at point P in between F_n and F_{n+1} with a distance $a\Delta x$ from F_n as illustrated in Figure 3(b). Then the weighted average of the electric field $\bar{E}$ is given by

(a)

o Full Integer Grid Points for ρ

□ Half Integer Grid Points for E_x

$$\bar{E}x=(\tfrac{1}{2}-a)Ex(H_{n-1})+(\tfrac{1}{2}+a)Ex(H_n)$$

$$-\tfrac{1}{2} \leq a \leq \tfrac{1}{2}$$

$$Ex=Ex^{INT}=(\tfrac{1}{2}-a)Ex(H_{n-1})+(\tfrac{1}{2}+a)Ex(H_n)$$

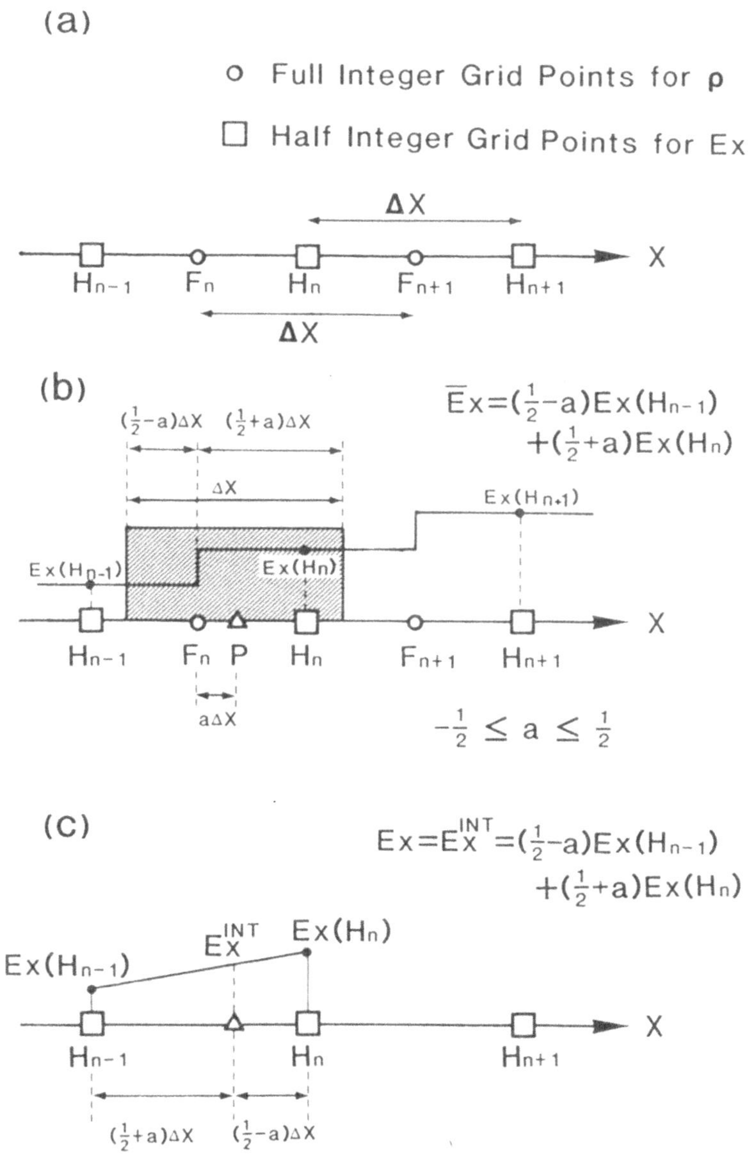

Figure 3.   Staggered dual grid system and effective electric field acting on a superparticle.

$$\bar{E} = \left(\frac{1}{2} - a \right) E_x(H_{n-1}) + \left(\frac{1}{2} + a \right) E_x(H_n) \ . \quad (5)$$

The effective electric field which the superparticle feels is this averaged value $\bar{E}$. It is formally computed by a simple linear interpolation from two values of E_x at two adjacent grid points H_{n-1} and H_n as shown in Figure 3(c). However, one should note that the simple interpolated value E^{INT} (P) from $E_x(H_{n-1})$ and $E_x(H_n)$ produces a nonphysical force acting on the superparticle. The nonphysical force

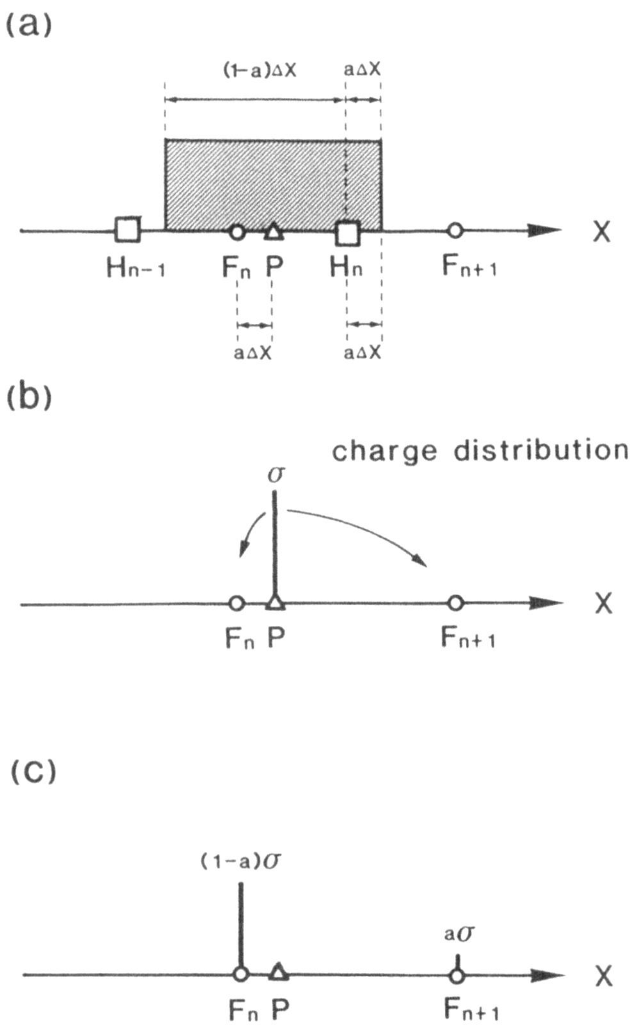

Figure 4. Charge distribution of superparticles to grid points.

is called a "self-force". Physically, the electric or magnetic field produced by a particle should not give any force back to the particle itself.

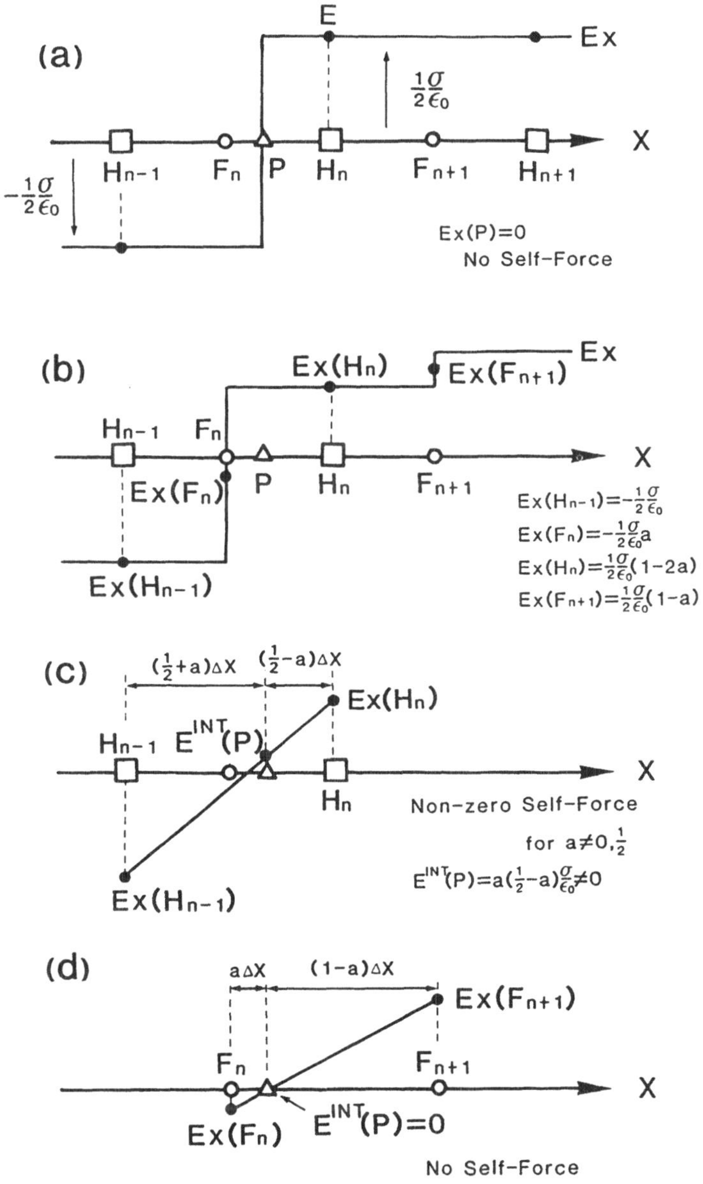

Figure 5. Explanation of self-force and its cancellation.

As far as we adopt the staggered dual grid system together with the area sharing scheme for charge and current distribution to grid points, the self-force comes in automatically and leads to violation of conservation of energy and momentum by the nonphysical acceleration of particles. To demonstrate how the self-force appears, let us consider a case where only one superparticle is placed in the system at point P as shown in Figure 4(a). The charge density σ of the superparticle is distributed to two adjacent grid points F_n and F_{n+1} by the area sharing scheme. This process is identical to divide the concentrated charge density σ at the point P onto F_n and F_{n+1} as illustrated in Figure 4(b). The resultant charges at F_n and F_{n+1} are $(1-a)\sigma$ and $a\sigma$, respectively as shown in Figure 4(c). The electrostatic field E_x produced by the particle is presented by a step function with a jump of amount of σ/ε_0 as shown in Figure 5(a). Due to the symmetry characteristics, $E_x = 0$ at the particle location P. However, in the computer simulation, the electric fields at half-integer grid points are calculated from the charge distribution at full-integer grid points. Thus as illustrated in Figure 5(b), the electric field at the half-integer points are

$$E_x(H_{n-1}) = -\frac{1}{2}\frac{\sigma}{\varepsilon_0}$$
$$E_x(H_n) = \frac{1}{2}\frac{\sigma}{\varepsilon_0}(1-2a) \quad . \tag{6}$$

These values are, in turn, used to compute the effective electric field acting on the superparticle at P using Eq.(5). Then, unfortunately, the result is given by

$$\bar{E}_x(P) = E_x^{INT}(P) = a\left(\frac{1}{2} - a\right)\frac{\sigma}{\varepsilon_0} \tag{7}$$

which is not zero except for special cases of $a = 0$ or $1/2$ as illustrated by Figure 5(c). This non-zero electric field acts on the particle and accelerates it nonphysically. Therefore, we cannot simply estimate the effective electric fields by Eq.(5) if we use the E_x values at half-integer points. However, as shown in Figure 5(d), the correct value can be computed by the same linear interpolation scheme if we use the electric field values at F_n and F_{n+1}, i.e., at two adjacent full-integer points. The values of E_x at these full-integer points are given by (see Figure 5(b))

$$E_x(F_n) = -\frac{1}{2}\frac{\sigma}{\varepsilon_0}a$$
$$E_x(F_{n+1}) = \frac{1}{2}\frac{\sigma}{\varepsilon_0}(1-a) \tag{8}$$

which yields

$$\bar{E}_x(P) = E_x^{INT}(P) = 0 \tag{9}$$

Therefore, to avoid the self-force, we need to relocate the values of

E_x onto the same grid points where the charge is defined. In the same way, the magnetic field should be relocated onto the grid points where the current is defined. Details of the grid assignment of various physical quantities are described in Section 3.3.

3. PARTICLE CODE VALID FOR BOTH ELECTROMAGNETIC AND ELECTROSTATIC WAVE MODELS

3.1 General Features of EM2 Code (KEMPO)

We have developed a two-and-half dimensional electromagnetic simulation code which solves Maxwell's equations and equations of motion of electrons and, if necessary, ions explicitly. We assume (x, y, z)-coordinates neglecting variation along the z axis, i.e., $\partial/\partial z = 0$. Three velocity components v_x, v_y, v_z and all three components of electromagnetic fields E and B are retained. Since no approximation is made to the basic equations except for $\partial/\partial z = 0$, the code is applicable to a wide rage of two-dimensional problems in plasma physics. The algorithm for integration of the field and particles is designed basically after Langdon and Lasinski [1976]. The fields are integrated using Maxwell's equations in a central difference form in space and time. Particles are integrated by the Buneman-Boris method [Buneman, 1967, Boris, 1970] which is accurate to the second order and time-reversible. The current density J and the charge density ρ are calculated using PIC (particle in cell) method which distributes the current and charge of a superparticle to the adjacent four grid points with area weightings. The electric and magnetic fields acting on a superparticle are interpolated to the particle position from the adjacent four grid points with the same area weightings.

In the following sections we describe several techniques we applied in order to make the simulation code more efficient and accurate. Especially, an Multiple-Time-Step (MTS) scheme we developed makes the code as efficient as the magnetostatic simulation code (with Darwin approximation), and makes it more applicable to low frequency problems of lower hybrid resonance region and of ion wave modes.

3.2 Multiple Time Step Scheme

The time step which satisfies the CFL condition is often too small for the wave modes of interest. Usually the wave frequencies of interest are of the order of the plasma frequency or the cyclotron frequency, while the maximum frequency in the system is $\pi/\Delta t$, and much higher than these wave frequencies of interest. Since most of the CPU time is used to solve the particle motion rather than the field integration, it is desired to take a larger time step for the particle calculations. Langdon and Lasinski [1976] proposed an algorithm in which the fields are integrated twice as often as the particles. However, they discovered that the algorithm can lead to a numerical instability. We developed a new algorithm where the particles are advanced less often than the field without leading to the numerical instability. A filtering of the field quantities in time is performed to avoid the numerical instability.

The high frequency parts of the fast extraordinary (FE) and ordinary (O) modes, i.e., the light modes, receive little contribution from the particle motion which is expressed as the conduction current density J in Maxwell's equations. This is justified by taking the ratio of the conduction current J and the displacement current $\varepsilon_0 \partial E/\partial t$ as

$$\frac{J}{\varepsilon_0 \partial E/\partial t} \leqslant \left(\frac{\omega_p}{\omega} \right)^2 \left(1 + \frac{v_{th} B_0}{V_p \; B_w} \right) \tag{10}$$

where v_{th} is the thermal velocity of the particles, and V_p is the phase velocity of the wave. B_w is a wave amplitude and B_0 is a static magnetic filed. If $\omega \gg \omega_p$ and $v_{th}/V_p \leqslant B_w/B_0$, the conduction current is negligibly small.

As the particle motion has little effects on the high frequency waves, we may eliminate the high frequency components of the current J, and treat the high frequency waves as the light mode in vacua. Then velocities and positions of particles may not necessarily be updated at as small time step as in the case of the wave integration. We adopt two different time steps for the field integration and particle integration, respectively. The time step for the field is called "field-time-step" and the time step for the particles is called "particle-time-step". We define the field-time-step as Δt and take the particle-time-step as an integer multiple of the filed-time-step, i.e., $m\Delta t$. We call the algorithm using the particle-time-step of $m\Delta t$ as "Multiple Time Step (MTS) scheme" hereafter.

In order to eliminate the high frequency component of the current density J it is necessary to perform time-filtering of the electromagnetic force acting on the particles. If the particles are not affected by the high frequency electromagnetic force, the conduction current density J, which is the summation of the particle motion, does not have high frequency components, yielding a consistent simulation. Since the current density J is calculated at each particle-time-step. the current density at each field-time-step is extrapolated and/or interpolated from the values available at the two nearest particle-time-steps.

In Figure 6 the algorithm of the MTS scheme is depicted as a time step chart. The time step charts are depicted for the case of m = [even number] in Figure 6(a) and for the case of m = [odd number] in Figure 6(b), respectively. E_p and B_p are electric and magnetic fields for particle pushing, which are the sums of the fields E and B at the m time steps indicated by the dashed rectangle in Figure 6. This operation automatically gives averages of the field E and B over the period of $m\Delta t$, and has an effect of the time filtering, which suppresses the high frequency components of the fields. These

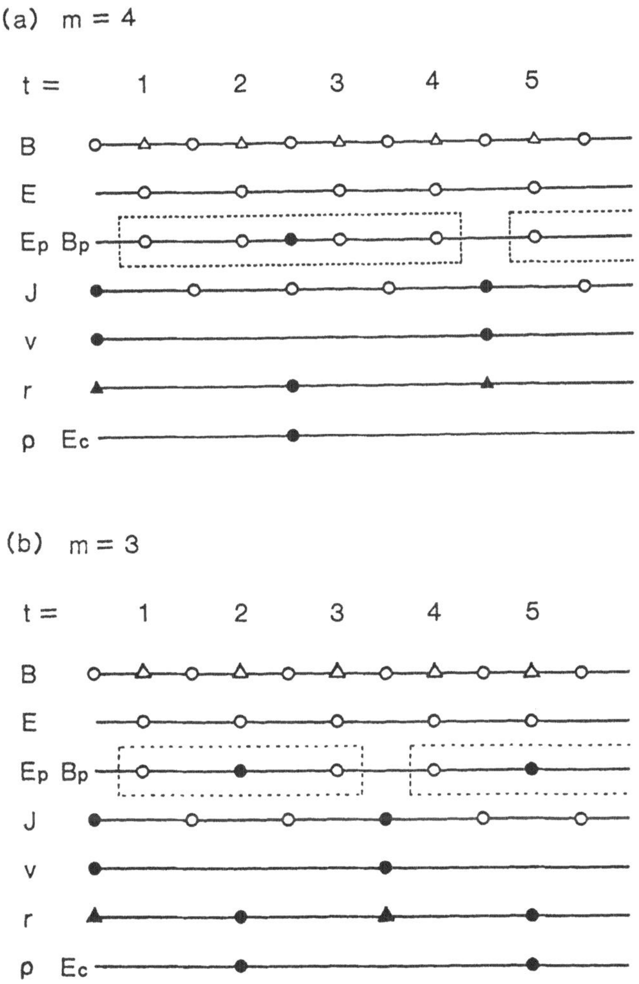

Figure 6. Time step chart of the MTS scheme : (a) m = 4 (b) m= 3

operations are depicted in the flow chart of the KEMPO as illustrated in Figure 7. Using these E_p and B_p, we advance the velocities of particles over the particle-time-step $m\Delta t$. The fields E and B are re-integrated from these E_p and B_p over the period $(m-1)\Delta t/2$ using the field-time-step Δt recursively. The current density J for the re-integration is interpolated in time from the values at the adjacent particle-time-steps. Then the fields E and B are integrated further over the period of $m\Delta t$ in order to obtain new E_p and B_p, where the current density J is extrapolated from the values at the previous two particle-time-steps.

The MTS scheme has an effect of digital filtering of the electromagnetic fields in time sequence. The digital filtering is m-point filtering with equal weightings, which is defined by the following equation.

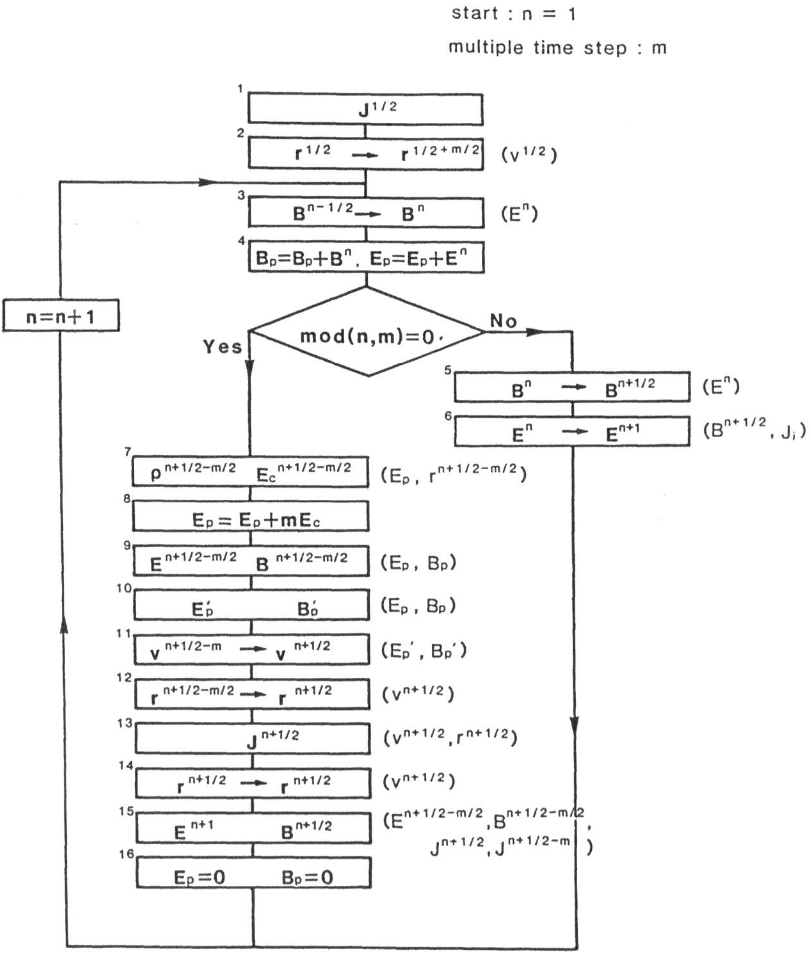

$$J_i = J^{m\cdot[n/m]+1/2} + \frac{\mathrm{mod}(n,m)}{m}\left(J^{m\cdot[n/m]+1/2} - J^{m\cdot[n/m]+1/2-m}\right)$$

Figure 7. Flow chart of the KEMPO : Quantities in boxes are calculated or advanced in time using quantities in parentheses.

$$\phi_f(T_j) = \frac{1}{m} \sum_{k=-(m-1)/2}^{(m-1)/2} \phi(T_{j+k}) \qquad (11)$$

High frequency waves are attenuated at every particle-time-step. The attenuation factor A of the wave of a frequency ω is obtained by assuming

$$\phi(\omega) = \frac{1}{N} \sum_{j=1}^{N} \phi(T_j)\, e^{i\omega T_j} \qquad (12)$$

where N is a number of time steps. From (11) and (12), we have

$$\phi_f(\omega) = \frac{1}{m} \sum_{k=-(m-1)/2}^{(m-1)/2} e^{-i\omega(k\Delta t)}\phi(\omega) \qquad (13)$$

which gives the attenuation factor for m = (even number) as

$$A(\omega\Delta t) = \frac{\phi_f(\omega)}{\phi(\omega)} = \frac{2}{m}\left[\sum_{k=1}^{m/2} \cos\{(k-\frac{1}{2})\omega\Delta t\}\right] \qquad (14)$$

and for m = (odd number) as

$$A(\omega\Delta t) = \frac{\phi_f(\omega)}{\phi(\omega)} = \frac{1}{m}\{1+2\sum_{k=1}^{(m-1)/2} \cos(\omega k\Delta t)\} \, . \qquad (15)$$

The attenuation factor A is plotted as a function of $\omega\Delta t$ for m = 2 to 8 in Figure 8. Since maximum frequency of the system is $\pi/\Delta t$, A is calculated in the range of $0 \leqslant \omega\Delta t \leqslant \pi$. As noted from the plot, high frequency modes undergo phase-reversal in the MTS scheme, but they also get damped and disappear from the system.

Electrons undergo oscillation with the plasma frequency ω_p in the direction of the static magnetic field. Since the plasma oscillation is a simple harmonic oscillation, the stability criterion for the leapfrog harmonic oscillator [Hockney and Eastwood, 1981] must be satisfied as

$$\omega_p m\Delta t \leqslant 2 \quad . \qquad (16)$$

In the transverse plane electrons follow the cyclotron motion with cyclotron frequency ω_c. The Buneman-Boris method for particle integration is unconditionally stable for the cyclotron motion. However, if the frequency of the waves of interest are less than or nearly equal to ω_p or ω_c, the following condition is generally satisfied in order to obtain a reasonable resolution and accuracy.

$$\mathrm{Max}(\omega_p,\omega_c)\cdot m\Delta t \leqslant 0.2 \qquad (17)$$

We define an attenuation rate Γ_A by $\log A(\omega) / (m\Delta t)$. In Figure 9 the ratio of the attenuation rate and a wave frequency Γ_A/ω is plotted as a function of $\omega\Delta t$ for m=1~8. The MTS factor m and field-time-step Δt must be chosen so that Γ_A is much less than physical growth/damping rate of electromagnetic waves in a simulation model.

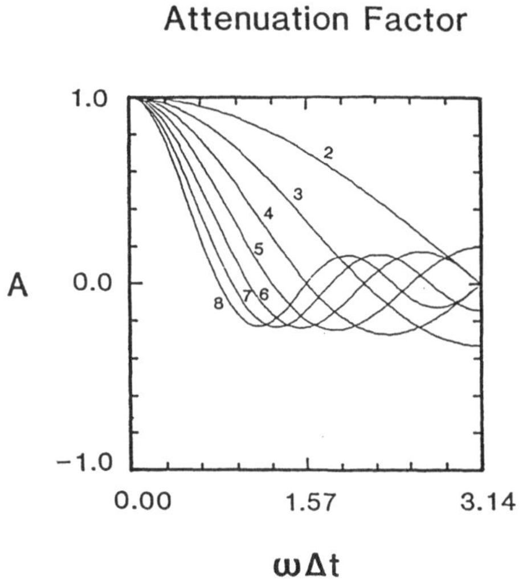

Figure 8. Attenuation factor A(ωΔt) of the MTS scheme for different multiple time steps.

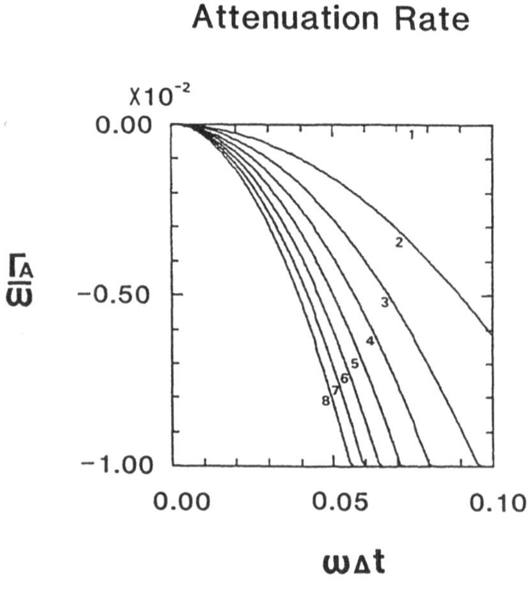

Figure 9. Attenuation rate Γ_A/ω of the MTS scheme for different multiple time steps.

3.3 Grid Location for Integration of the Field

For the two dimensional system Maxwell's equations are reduced to the following two independent sets of equations.

(A)

$$\frac{\partial B_z}{\partial t} = -\frac{\partial E_y}{\partial x} + \frac{\partial E_x}{\partial y} \tag{18}$$

$$\frac{\partial E_x}{\partial t} = c^2 \left[\frac{\partial B_z}{\partial y} - \mu_0 J_x \right] \tag{19}$$

$$\frac{\partial E_y}{\partial t} = c^2 \left[-\frac{\partial B_z}{\partial x} - \mu_0 J_y \right] \tag{20}$$

(B)

$$\frac{\partial B_x}{\partial t} = -\frac{\partial E_z}{\partial y} \tag{21}$$

$$\frac{\partial B_y}{\partial t} = \frac{\partial E_z}{\partial x} \tag{22}$$

$$\frac{\partial E_z}{\partial t} = c^2 \left[\frac{\partial B_y}{\partial x} - \frac{\partial B_x}{\partial y} - \mu_0 J_z \right] \tag{23}$$

These equations (A) and (B) are solved as central difference equations in space and time by defining the E, B and J at the location shown in Figure 10-(a) and -(b) [Langdon and Lasinski, 1976]. For further reference we call the grid with label "i" a full grid (F) in x and the grid with label "i+1/2" a half grid (H) in x. Likewise, we call the grid with label "j" a full grid (F) in y and the grid with label "j+1/2" a half grid (H) in y. Combining x and y locations, we express the locations of the grid points with (i,j), (i+1/2,j), (i,j+1/2) and (i+1/2,j+1/2) by FF, HF, FH and HH.

The grid locations shown in Figure 10 are summarized as follows.

E_x	B_x	J_x	:	HF
E_y	B_y	J_y	:	FH
E_z	B_z	J_z	:	HH

The current densities J_x, J_y and J_z are defined at different locations. To assign contributions of particles to the grid points, we have to calculate three different area-weightings. In order to reduce the CPU time in calculating area-weightings, we first calculate all J_x, J_y and J_z at HH grid points, and then relocate J_x and J_y to HF

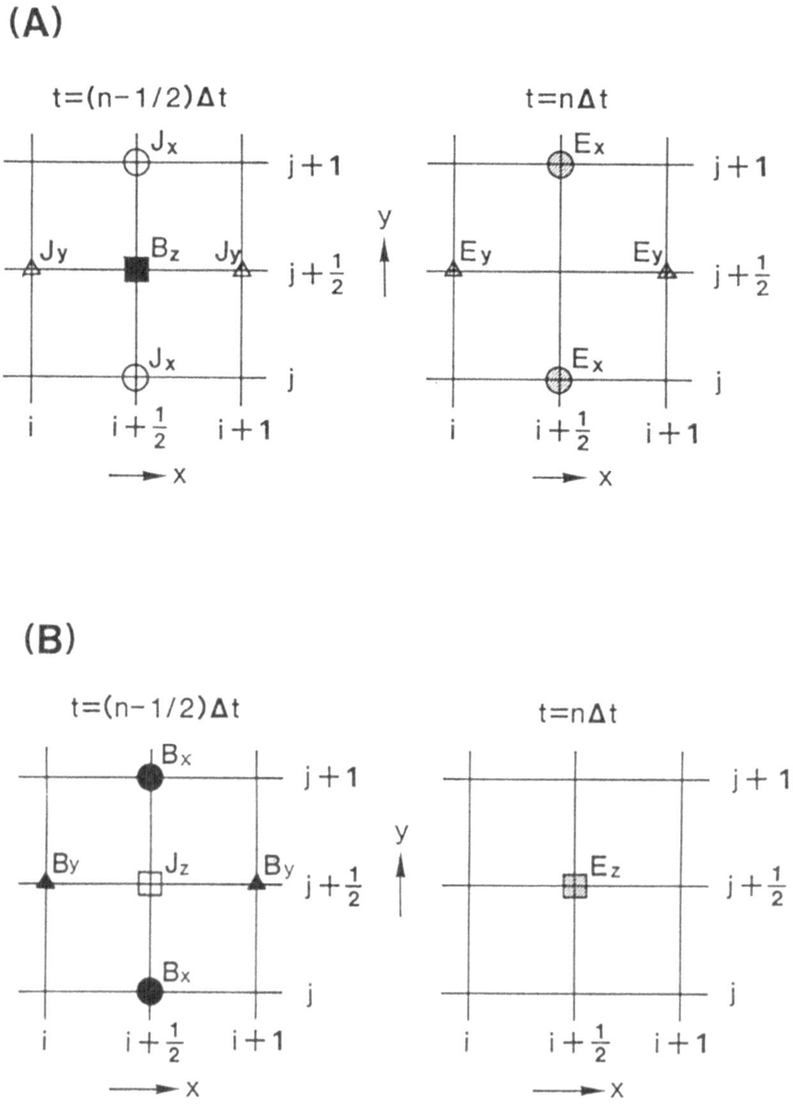

Figure 10. Allocation of field quantities to grid points: (a) E_x, E_y, B_z J_x, J_y (b) E_z, B_x, B_y, J_z.

and FH grid points by the following operations, respectively.

$$J'_{x\ i+1/2,\ j} = \frac{1}{2} (J_{x\ i+1/2,\ j-1/2} + J_{x\ i+1/2,j+1/2}) \qquad (24)$$

$$J'_{y\ i,\ j+1/2} = \frac{1}{2} (J_{y\ i-1/2,\ j+1/2} + J_{y\ i+1/2,j+1/2}) \qquad (25)$$

The relocation procedure has the effect of spatial filtering (see Section 3.4). This filtering reduces electromagnetic radiation loss at small wavelengths. As J_z is not relocated, it is necessary to apply the three point filtering to J_z both in x and in y directions. It is noted that filtering of J_x in x-direction and J_y in y-direction is not necessary at all. Because, fluctuations of J_x in x direction and J_y in y direction are not responsible to the electromagnetic radiations. In addition, The current density J and E must satisfy

$$\text{div } (J + \varepsilon_0 \frac{\partial \rho}{\partial t}) = 0 \qquad (26)$$

Since the electrostatic components E_x and E_y are calculated from the current distribution via Poisson's equation, the relocation of J_x in x and J_y in y only increases the inconsistency of J_x and J_y with E_x and E_y, respectively.

The electric fields E_x and E_y obtained by integrating Eqs. (19) and (20) are corrected by solving Poisson's equation using the charge density defined at FF grid points. If we calculate electric force acting on particles from the electric field E_x and E_y defined at HF and FH grid points, it results in a electrostatic self-force acting on particle, violating the momentum conservation as discussed in Section 2.1. The area weighting in calculation of the charge density and that in calculation of the force acting on particles must be identical and must be done using the quantities defined at the same grid locations. Since the charge density is defined at FF grid points, the electric field E_x and E_y obtained at FH and HF grid points must be relocated to the FF grid points before calculating the particle force. The above discussion of the electrostatic self-force is also applied to the magnetostatic force induce by Ampere's low $\nabla \times B = \mu_0$, which is rewritten as

$$\frac{\partial B_z}{\partial y} = \mu_0 J_x \qquad (27)$$

$$-\frac{\partial B_z}{\partial x} = \mu_0 J_y \qquad (28)$$

$$\frac{\partial B_y}{\partial x} - \frac{\partial B_x}{\partial y} = \mu_0 J_z \qquad (29)$$

As the B_z is defined at HH grid points where Jx and Jy are also calculated by the area weighting scheme, The relations of Eqs. (28)

and (29) do not yield the self-force. However, B_x and B_y are defined at HF and FH grid points, while J_z is defined at HH grid points. In calculation of magnetic force acting on particles, B_x and B_y must be relocated to HH grid points.

Let us summarize the relocation procedure for elimination of the self-force.

E_x	:	HF	→	FF
E_y	:	FH	→	FF
B_x	:	HF	→	HH
B_y	:	FH	→	HH

E_z and B_z are not relocated.

Owing to the relocation procedure, computing time for calculating the particle forces is reduced because the area weighting for FH and HF grid points are not computed any more. In calculation of electromagnetic forces acting on particles, E_x and E_y are interpolated from FF grid points and E_z, B_x, while B_y and B_z are interpolated from HH grid points.

3.4 Digital Filtering

Spatial filterings of the quantities defined at grid points are generally used in order to eliminate nonphysical noises at short wavelengths, where the finite difference of the fields becomes most inaccurate. Two kinds of digital filtering schemes are used in the KEMPO. One is a filtering of two-point averaging as defined by

$$\phi'_{i+1/2} = \frac{1}{2} (\phi_i + \phi_{i+1}) \tag{30}$$

which is implicitly involved in the relocation procedure of the field quantities. The other is three-point digital filter or a binomial digital filter [Birdsall and Langdon, 1984] given as

$$\phi'_i = \frac{1}{4} (\phi_{i-1} + 2\phi_i + \phi_{i+1}) \tag{31}$$

which is explicitly used in the code. The attenuation factor of these two filters are given by the following equations.

Two-point digital filter :

$$\phi'(k) = \cos \frac{k\Delta x}{2} \phi(k) \tag{32}$$

Three-point digital filter :

$$\phi'(k) = \cos (\frac{k\Delta x}{2})^2 \phi(k) \tag{33}$$

These filters are used in the following two manners. Firstly they are applied to the particle-pushing fields E_p and B_p in order to eliminate nonphysical random forces which causes stochastic heatings of particles. Since the relocation procedures implicitly involving the two-point filter are different for the components (E_x, E_y, etc.) and for the directions (x and y), it is necessary to apply the three-point filter to the components and directions which are not affected by the relocation procedure. Secondly these filters are applied to the current density calculated from the particles. The current density is the source of electromagnetic radiations. Fluctuations at short wavelengths correspond to high frequency electromagnetic waves which undergo a damping by the time filtering of MTS scheme as time goes on. Due to the electromagnetic radiations which are to be damped by both spatial and time filtering, particles lose their thermal energy. In order to prevent the energy loss, it is necessary to eliminate fluctuations at short wavelength by the digital filters.

3.5 Unit System used in the KEMPO

It is noted that there is no necessity to stick to a real unit system like CGS or MKS unit system in simulations. What is important in simulations are ratios of quantities in the system, i.e., a ratio of a wave magnetic field to the static magnetic field, or a ratio of a kinetic energy to the total energy, etc.. In most of simulations physical quantities are normalized to the basic parameters in the system where the basic parameters are assumed to be unity. However, selections of basic parameters are different depending on physical models. In order to make the simulation code applicable to various problems of a wide range of parameters, we do not normalize any quantities, or rather we adopt a relative unit system where all parameters are calculated from a set of independent basic parameters via basic equations. These basic parameters are as follows.

1. Angular frequency (plasma, cyclotron, wave frequency etc.)
   ω_{pi}, ω_{ci}, ω
2. System length L_x, L_y
3. Charge-to-mass ratio $(q/m)_i$
4. Number of superparticles in the system N_i

where the subscript "i" denotes the i-th species of particles. Values of these four quantities are given arbitrary, except that ratios of quantities in the same units like ω_{pi}/ω_{ci}, L_y/L_x or $(q/m)_2/(q/m)_1$, are kept the same as those of the real physical quantities. Especially, it is noted that numbers of superparticles have no relation to the real number densities in the plasma, and N_1 and N_2 are independent of each other.

The basic equations are written in the following form which is identical to those in MKS unit system.

(1) Equations of motion

$$\frac{dr}{dt} = v \tag{34}$$

$$\frac{dv}{dt} = \frac{q_i}{m_i} (E + v \times B) \tag{35}$$

(2) Maxwell's equations

$$\text{rot } B = \mu_0 J + \frac{1}{c^2} \frac{\partial E}{\partial t} \tag{36}$$

$$\text{rot } E = - \frac{\partial B}{\partial t} \tag{37}$$

$$\text{div } B = 0 \tag{38}$$

$$\text{div } E = \frac{\rho}{\varepsilon_0} \tag{39}$$

where c, ε_0 and μ_0 are the light speed, electric permittivity and magnetic permeability, respectively.

It is noted that ε_0 and μ_0 may be chosen arbitrary except that $\varepsilon_0 \mu_0 = 1/c^2$. For simplicity, we define $\varepsilon_0 = 1$ and $\mu_0 = 1/c^2$, and the units of charge, mass, electric and magnetic fields are given based on their definition.

Other physical quantities are calculated via the following relations obtained from the basic equations. The cyclotron frequency of species 1 and the plasma frequencies of species i are given as

$$\omega_{c1} = \frac{q_1}{m_1} B_0 \quad , \quad \omega_{pi} = \left(\frac{n_i q_i^2}{m_i \varepsilon_0} \right)^{\frac{1}{2}} \tag{40}$$

where n_i is a particle density of species i defined as a density in the two dimensional system of $L_x \times L_y$.

-Particle density

$$n_i = \frac{N_i}{L_x L_y} \tag{41}$$

From (40) and (41) we can calculated the following physical quantities.

-Charge of a particle

$$q_i = \frac{\varepsilon_0 L_x L_y \omega_{pi}^2}{N_i (q/m)_i} \tag{42}$$

-Mass of a particle

$$m_i = \frac{\varepsilon_0 L_x L_y \omega_{pi}^2}{N_i (q/m)_i^2} \tag{43}$$

-Static magnetic field

$$B_0 = \frac{\omega_{c1}}{(q/m)_1} \tag{44}$$

It is noted that the mass m_i and charge q_i have little physical meaning in the relative unit system. What are physically meaningful are the mass density $m_i n_i$ and charge density $q_i n_i$ given as

$$q_i n_i = \frac{\varepsilon_0 \omega_{pi}^2}{(q/m)_i}, \quad m_i n_i = \frac{\varepsilon_0 \omega_{pi}^2}{(q/m)_i^2} \tag{45}$$

The values of ω_{pi} and $(q/m)_i$ of different species must be given based on the ratios in the physical model.

3.6 Rescaling of Physical Quantities

In order to attain computational efficiency, it is necessary to reduce the number of operations involved in difference equations of fields and particles. Since operations of multiplying and dividing by Δr, $\Delta t/2$ and (q/m) are frequently used in the difference equations, we rescale the distance, time and charge-to-mass-ratio expressed in the relative unit by Δr, $\Delta t/2$ and $(q/m)_1$, respectively. Other physical quantities are rescaled as follows.

distance	$r^* = (1/\Delta r)\ r$
time	$t^* = (2/\Delta t)\ t$
velocity	$v^* = (\Delta t/2)(1/\Delta r)\ v$
number density	$n^* = (\Delta r)^2\ n$
charge	$q^* = (q/m)_1 (\Delta t/2)^2 (1/\Delta r)^2\ q$
mass	$m^* = (q/m)_1^2 (\Delta t/2)^2 (1/\Delta r)^2\ m$
Electric field	$E^* = (q/m)_1 (\Delta t/2)^2 (1/\Delta r)\ E$
Magnetic field	$B^* = (q/m)_1 (\Delta t/2)\ B$
Current density	$J^* = (q/m)_1 (\Delta t/2)^3 (1/\Delta r)\ J$
Electric potential	$\phi^* = (q/m)_1 (\Delta t/2)^2 (1/\Delta r)^2\ \phi$
Charge density	$\rho^* = (q/m)_1 (\Delta t/2)^2\ \rho$
Energy density	$\sigma^* = (q/m)_1^2 (\Delta t/2)^4 (1/\Delta r)^2\ \sigma$

quantities with * are rescaled ones.

The difference equations of particle motion of species 1 are written as in the following simple forms. The *'s on r, t, v, n, q, m, E, B, J, ϕ, ρ and σ are omitted hereafter.

$$v^- = v^{n-1/2} + E^n \tag{46}$$

$$v' = v^- + v^- \times B^n \tag{47}$$

$$v^+ = v^- + \frac{2}{1 + (B^n)^2} v' \times B^n \tag{48}$$

$$v^{n+1/2} = v^+ + E^n \tag{49}$$

$$r^{n+1/2} = r^n + v^{n+1/2} \tag{50}$$

$$r^{n+1} = r^{n+1/2} + v^{n+1/2} \tag{51}$$

For the i-th species of particles ($i \neq 1$), E and B must be multiplied by $(q/m)_i/(q/m)_1$.

The difference equations for fields are written as

$$B^n = B^{n-1/2} - \nabla^* \times E^n \tag{52}$$

$$B^{n+1/2} = B^n - \nabla^* \times E^n \tag{53}$$

$$E^{n+1} = E^n + 2c^2 \nabla^* \times B^{n+1/2} + 2J^{n+1/2} \tag{54}$$

where the operation $\nabla^* \times A_{i,j}$ (A = E or B) are realized by the following simple operations.

$$(\nabla^* \times A_{i,j})_x = A_{z\ i,j+1} - A_{z\ i,j} \tag{55}$$

$$(\nabla^* \times A_{i,j})_y = A_{z\ i,j} - A_{z\ i+1,j} \tag{56}$$

$$(\nabla^* \times A_{i,j})_z = A_{y\ i+1,j} - A_{y\ i,j} - A_{x\ i,j+1} + A_{x\ i,j} \tag{57}$$

The rescaling is performed after all the input parameters are given to the simulation code, and all the output data are reconverted again to the values in the relative unit system. Since all input and output data are expressed in the relative unit system, users of the code do not have to be familiar with the rescaling inside the code.

4. LONG TIME SCALE CODE FOR MONOCHROMATIC WHISTLER INTERACTION

Simplification and reduction of CPU time is attainable in particle simulations by limiting the allowed wave modes in the simulation to a monochromatic wave with the assumption of slow variations in amplitude and frequency. This code is called the Long-Time-Scale (LTS) code [Rathmann et al., 1978, Omura and Matsumoto, 1981]. Instead of solving the Maxwell equations and equations of motion of all particles as is usually done in the conventional electromagnetic codes, only the wave amplitude, frequency and wavenumber of a monochromatic wave is incremented forwards in time with a self-consistent updating of charge and current density calculated from numerical solutions to the equation of motion obtained solely for resonant particles in the monochromatic wave.

Basic equations for the monochromatic whistler mode wave and resonant electrons are described in Sections 4.1 and 4.2, respectively [Omura and Matsumoto, 1981]. These equations are coded into difference equations in Sections 4.3 and 4.4. The flow chart of LTS whistler simulation is given in Sections 4.5.

4.1 Equations for the Wave Fields

We assume a purely transverse whistler mode wave which propagates along the geomagnetic field line and interacts with counter-streaming high energy resonant electrons existing in the magnetosphere. We use a right-handed coordinate system; z being a distance along the field line from the equatorial plane, and x and y axes being perpendicular to the external magnetic field. For convenience the wave fields are expressed in complex variables with dot marks on their top. The x and y components are represented by real and imaginary parts of the complex quantities, respectively.

$$\dot{B}_w = B_{wx} + iB_{wy} \quad , \qquad \dot{E}_w = E_{wx} + iE_{wy} \tag{58}$$

Neglecting the term for the displacement current in Maxwell's equation, and dividing a conduction current into a cold plasma current $\dot{J}_c$ and an energetic resonant electron current $\dot{J}_R$, we have Maxwell's equations in the form

$$i\frac{\partial \dot{E}_w}{\partial z} = -\frac{\partial \dot{B}_w}{\partial t} \quad , \qquad i\frac{\partial \dot{B}_w}{\partial z} = \mu_o(\dot{J}_c + \dot{J}_R) \quad . \tag{59}$$

The linearized momentum equation of the cold electrons is

$$\frac{\partial \dot{J}_c}{\partial t} - i\Omega_e \dot{J}_c = \varepsilon_o \Pi_e^2 \dot{E}_w \tag{60}$$

where Ω_e and Π_e are the electron cyclotron and plasma frequencies, respectively. Eliminating $\dot{J}_c$ and $\dot{E}_w$ from (59) and (60), we have

$$\left[\frac{\partial}{\partial z} \left(\frac{\partial}{\partial t} - i\Omega_e \right) \frac{\partial}{\partial z} - \frac{\Pi_e^2}{c^2} \frac{\partial}{\partial t} \right] \dot{B}_w = -i\mu_o \left(\frac{\partial}{\partial t} - i\Omega_e \right) \frac{\partial}{\partial z} \dot{J}_R \tag{61}$$

where we have assumed that the inhomogeneity of the medium is sufficiently small within a distance of the wave length, i.e.,

$$\left| \frac{\partial \Omega_e}{\partial z} \right| \ll |k(\omega - \Omega_e)| \quad , \quad \left| \frac{\partial \Pi_e^2}{\partial z} \right| \ll |k\Pi_e^2| \quad . \tag{62}$$

The quantities $\dot{B}_w$ and $\dot{J}_R$ are expressed in terms of the amplitude B_w, J_R and phase ψ and $\psi + \psi_R$ as

$$\dot{B}_w = B_w \exp\{i\psi\} \quad , \quad \dot{J}_R = J_R \exp\{i(\psi + \Psi_R)\} \tag{63}$$

where Ψ_R is a relative phase angle between J_R and B_w, and is assumed to be a slowly varying function with respect to time and space.

The frequency ω and wavenumber k for the right-handed polarized wave are defined by

$$\omega = \frac{\partial \psi}{\partial t} \quad , \quad k = -\frac{\partial \psi}{\partial z} \quad . \tag{64}$$

We assume here that the terms involving the first-order time and space derivatives of k, J_R and Ψ_R as well as the second-order derivatives of B_w are negligible. Substitution of (63) into (61) yields

$$\left(k^2 + \frac{\Pi_e^2}{c^2} \right) \frac{\partial B_w}{\partial t} + 2k\,(\,\Omega_e - \omega\,)\,\frac{\partial B_w}{\partial z} - i\,\left\{ k^2(\,\Omega_e - \omega\,) - \frac{\Pi_e^2}{c^2}\,\omega \right\} B_w$$
$$= \mu_o k\,(\,\Omega_e - \omega\,)\,(\,J_R \sin \Psi_R - i J_R \cos \Psi_R\,) \tag{65}$$

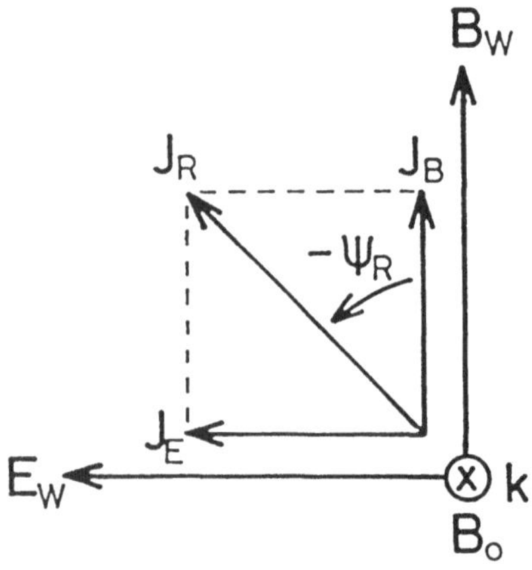

Figure 11. Configuration of the wave magnetic (B_w) and electric (E_w) fields, wavenumber vector (k), external magnetic field (B_w), and resonant current (J_R) and its components J_B and J_E.

Separating imaginary and real parts in (65), we have

$$\left(k^2 + \frac{\Pi_e^2}{c^2}\right)\frac{\partial B_w}{\partial t} + 2k(\Omega_e - \omega)\frac{\partial B_w}{\partial z} = -\mu_0 k(\Omega_e - \omega)J_E \tag{66}$$

$$\left\{k^2(\Omega_e - \omega) - \frac{\Pi_e^2}{c^2}\omega\right\}B_w = \mu_0 k(\Omega_e - \omega)J_B \tag{67}$$

where

$$J_E = J_R \sin(-\psi_R) \quad , \qquad J_B = J_R \cos(-\psi_R) \quad , \tag{68}$$

and J_E and J_B are the components of the transverse resonant current J_R parallel to the wave electric field E_w, and to the wave magnetic field B_w, respectively (see Figure 11). These J_E and J_B are calculated by the following equations

$$J_E = -e\sum_i v_{\perp i}\sin(-\zeta_i) \qquad J_B = -e\sum_i v_{\perp i}\cos(-\zeta_i) \tag{69}$$

where $v_{\perp i}$ is a velocity component perpendicular to the external magnetic field and ζ_i is a phase angle between $v_\perp i$ and B_w, and Σ is taken over resonant electrons in a unit cell. Since the group velocity is expressed by

$$V_g = \frac{2k(\Omega_e - \omega)}{k^2 + (\Pi_e^2/c^2)} \tag{70}$$

we have from (66)

$$\frac{\partial B_w}{\partial t} = -V_g\left(\frac{\partial B_w}{\partial z} + \frac{1}{2}\mu_0 J_E\right) \tag{71}$$

The second order differentiation of the wave phase ψ gives a relation of k and ω as

$$\frac{\partial k}{\partial t} = -\frac{\partial \omega}{\partial z} \tag{72}$$

Using (71) and (72), we can follow the time evolution of Bw and k, while the frequency ω is calculable from (67) which is rewritten as

$$\omega = \frac{k(k - \mu_0 J_B/B_w)}{k(k - \mu_0 J_B/B_w) + \Pi_e^2/c^2}\,\Omega_e \tag{73}$$

As is seen in (71) and (73), J_E causes a change of the wave amplitude B_w, and J_B modifies the frequency ω.

4.2 Equations of Motion of Resonant Electrons

We have derived the wave equations showing how the wave is modified by the presence the resonant current JR, which is formed as a result of cyclotron resonance between hot electrons and the wave. The equations of motion of resonant electrons are expressed in the following form. We introduce polar coordinates $(v_\parallel, v_\perp, \phi)$ for the velocity, where $v_\parallel$ and $v_\perp$ are the velocity components parallel and perpendicular to the external geomagnetic field respectively, and ϕ is the Larmor phase angle of $v_\perp$ in the rest frame. Equations of motion are then expressed as [e.g. Dysthe, 1971]

$$\frac{dv_\parallel}{dt} = \Omega_w v_\perp \sin(\phi - \psi) - \frac{v_\perp^2}{2\Omega_e}\frac{\partial \Omega_e}{\partial z} \tag{74}$$

$$\frac{dv_\perp}{dt} = \Omega_w\left(\frac{\omega}{k} - v_\parallel\right)\sin(\phi - \psi) + \frac{v_\parallel v_\perp}{2\Omega_e}\frac{\partial \Omega_e}{\partial z} \tag{75}$$

$$\frac{d\phi}{dt} = \frac{\Omega_w}{v_\perp}\left(\frac{\omega}{k} - v_\parallel\right)\cos(\phi - \psi) + \Omega_e \tag{76}$$

$$\frac{dz}{dt} = v_\parallel \tag{77}$$

where the wave amplitude B_w is replaced by $\Omega_w = eB_w/m$. The first terms in the right-hand sides of (74) ~ (76) correspond to the effect of the wave, while the second terms express the effect of the external geomagnetic field. The effect of the inhomogeneity of the medium is expressed by the second terms in (74) and (75), which disappear in a homogeneous case, as well as by k and Ω_e in (74) ~ (76), which vary slowly with the distance in the present model.

4.3 Particle Pushing Algorithm

The equations for the wave field (71) and (73) shows that the resonant currents J_E and J_B control the evolution of the monochromatic whistler wave. The resonant currents are calculated by solving the equations (74) ~ (77) of motion for a large number of hot electrons. An algorithm used in a "long-time-scale (LTS)" code for a whistler simulation first proposed by Rathmann et al. [1978] is adopted for particle pushing in the present simulation. The increments of the quantities $v_\parallel$, $v_\perp$, ϕ and z over a time step δt are separated into increments due to the wave and increments due to the cyclotron motion in the external magnetic field. The particle pushing algorithm takes the form of

$$f(t_{n+1}) = f(t_n) + \delta f_w + \delta f_c \quad , \tag{78}$$

where $t_{n+1} = t_n + \delta t$, and f is a quantity such as $v_\parallel$, $v_\perp$, ϕ and z. δf_w and δf_c are increments due to the wave and due to the cyclotron motion, respectively, and they are obtained by integrating (74) ~ (77) over the time interval from t_n to t_{n+1} under the conditions that $|\delta v_\parallel| \ll |v_\parallel|$, $|\delta v_\perp| \ll v_\perp$, $|\delta k| \ll k$, $|\delta\Omega_e| \ll |\Omega_e|$, $|\delta\omega| \ll \omega$, and $|\delta\Omega_w| \ll \Omega_w$. The increments due to the wave are expressed in terms of the quantities at $t = t_n$ as

$$\delta v_{\parallel w} = \frac{\Omega_w v_\perp}{\zeta'} \delta\cos\zeta \tag{79}$$

$$\delta v_{\perp w} = -\frac{\Omega_w}{\zeta'} \left(\frac{\omega}{k} - v_\parallel \right) \delta\cos\zeta \tag{80}$$

$$\delta \phi_w = \frac{\Omega_w}{v_\perp \zeta'} \left(\frac{\omega}{k} - v_\parallel \right) \delta\sin\zeta \tag{81}$$

$$\delta z_w = \frac{\Omega_w v_\perp}{\zeta'^2} \left(\zeta'\delta t \cos\zeta - \delta\sin\zeta \right) \tag{82}$$

where

$$\delta\cos\zeta = \cos(\zeta + \zeta'\delta t) - \cos\zeta , \quad \delta\sin\zeta = \sin(\zeta + \zeta'\delta t) - \sin\zeta \tag{83}$$

and ζ is a relative phase angel between $v_\perp$ and B_w, i.e., $\phi - \psi$. ζ' denotes a time derivative of ζ obtained from (76), i.e.,

$$\zeta' = \frac{\Omega_w}{v_\perp} \left(\frac{\omega}{k} - v_\parallel \right) \cos\zeta + \Omega_e - \omega + kv_\parallel \tag{84}$$

When $\zeta' \cong 0$, (79) ~ (82) are not appropriate because the right hand sides of these equations shows an overflow in the calculation process. In this case the first terms of (74), (75) and (76) are directly integrated assuming ζ is almost constant, and we have

$$\delta v_{\parallel w} = \delta t \, \Omega_w \sin\zeta \tag{85}$$

$$\delta v_{\perp w} = \delta t \, \Omega_w \left(\frac{\omega}{k} - v_\parallel \right) \sin\zeta \tag{86}$$

$$\delta \phi_w = \frac{\delta t \, \Omega_w}{v_\perp} \left(\frac{\omega}{k} - v_\parallel \right) \cos\zeta \tag{87}$$

$$\delta z_w = -\frac{1}{2} \delta v_{\parallel w} \delta t \tag{88}$$

The increments due to the cyclotron motion are expressed as

$$\delta v_{\parallel c} = -\frac{v_\perp^2}{2v_\parallel} \frac{\Omega_e(z + v_\parallel \delta t) - \Omega_e(z)}{\Omega_e(z)} \tag{89}$$

$$\delta v_{\perp c} = \frac{v_\perp}{2} \frac{\Omega_e(z + v_\parallel \delta t) - \Omega_e(z)}{\Omega_e(z)} \tag{90}$$

$$\delta\Phi_C = \frac{\delta t}{2} \{ \Omega_e(z) + \Omega_e(z + v_\parallel \delta t) \} \tag{91}$$

$$\delta z_C = \delta t \{ v_\parallel + \frac{\delta v_{\parallel C}}{2} \} \tag{92}$$

In the simulation, w, ω, k and Φ are calculated at the spatial grid points, and to obtain these quantities at particle position z, we interpolated from values of the neighboring grid points Z_m and Z_{m+1} in the manner as

$$F(z) = F(Z_m) + \frac{F(Z_{m+1}) - F(Z_m)}{\Delta Z} (z - Z_m) \tag{93}$$

where F corresponds to quantities such ase, w, ω or k, and ΔZ is the grid spacing defined by $\Delta Z = Z_{m+1} - Z_m$. Integration of k(z) expressed by (93) gives the phase of the wave,

$$\Psi(z) = \Psi(Z_m) - k(Z_m)(z - Z_m) - \frac{k(Z_{m+1}) - k(Z_m)}{2\Delta Z} (z - Z_m)^2 \tag{94}$$

In the simulations presented in Section 6, we take into account the inhomogeneity of the geomagnetic field which is approximated by a dipole magnetic field given by the following equations.

$$B_O = B_{OS} (\frac{R_E}{R})^3 \sqrt{1 + 3\sin^2\Phi} \tag{95}$$

$$R = R_E L \cos^2\Phi \tag{96}$$

$$\frac{z}{R} = \frac{1}{2\sqrt{3}} (x + \sinh x \cosh x) \tag{97}$$

$$x = \sqrt{3} \sin\Phi \tag{98}$$

where Φ and R are a geomagnetic latitude and a distance to the concerned point from earth's center. The quantities Bo and Bos are magnetic fields at (R, Φ) and at the earth's surface, respectively. R_E and L are the radius of the earth (6370km) and geocentric distance at the equatorial plane normalized to R_E. The quantity z is a distance measured along the geomagnetic field line from the equatorial plane to the concerned point. Let B_{OEQ} denote the magnetic field at R = LR_E in the equatorial plane, then we have from (95)

$$B_{OEQ} = L^{-3} B_{OS} \tag{99}$$

Substituting (96), (97) and (98) into (95), we express Bo in term of B_{OEQ} and z as,

$$B_O = B_{OEQ} \frac{\sqrt{1 + x^2}}{(1 - x^2/3)^3} \tag{100}$$

where x is given as a solution of the following equation.

$$F(x, z) = x + \sinh x \cosh x - \frac{2\sqrt{3}\, z}{LR_E} = 0 \tag{101}$$

From (100) and (81) we can calculate an external magnetic field at each grid point.

4.4 Field Updating Algorithm

In the previous section the algorithm for updating physical quantities associated with each particle was discussed. This section presents the method to update the quantities which are assigned to the spatial grid points. These quantities are the wave amplitude B_w and its time derivative $\partial Bw/\partial t$, the wavenumber k and its time derivative $\partial k/\partial t$, the wave frequency ω and the wave phase ψ in the rest frame, and the resonant current J_E and J_B.

The wave amplitude and the wavenumber are advanced over the time step δt using their time derivatives at $t = t_n$, i.e.,

$$B_{wm}(t_{n+1}) = B_{wm}(t_n) + \left(\frac{\partial B_w}{\partial t}\right)_m (t_n)\ \delta t \tag{102}$$

$$k_m(t_{n+1}) = k(t_n) + \left(\frac{\partial k}{\partial t}\right)_m (t_n)\ \delta t \tag{103}$$

where the subscript m denotes a value at a grid point Z_m. The wave phase is first advanced at the boundary grid Z_1 and extended successively to the next grid by integrating k over the grid spacing as

$$\Psi_1(t_{n+1}) = \Psi_1(t_n) + \omega(t_n)\ \delta t \tag{104}$$

$$\Psi_{m+1}(t_{n+1}) = \Psi_m(t_{n+1}) - \frac{1}{2}\,(\,k_m(t_{n+1}) + k_{m+1}(t_{n+1})\,\Delta Z \tag{105}$$

After the quantities z_i, $v_{\perp i}$, $v_{\parallel i}$ and ϕ_i of each resonant particle and the wave phase ψ are updated, the resonant current J_E and J_B can be calculated. An electron in the particle simulation is not a "real electron", but it is a superparticle which represents a number of electrons in the real plasma. Therefore, we assign a density n_s to each superparticle (we may regard it a super-electron with a charge $-n_s e$), and each super-particle forms resonant currents j_{Ei} and j_{Bi}

$$j_{Ei} = -n_s e \sin(-\zeta_i) \quad , \quad j_{Bi} = -n_s e \cos(-\zeta_i) \tag{106}$$

where i denotes a quantity of each particle; and $\zeta_i = \phi_i - \psi(z_i)$ and $\psi(z_i)$ is calculated by (94). These j_{Ei} and j_{Bi} are assigned to the neighboring grid points Z_m and Z_{m+1} ($Z_m < z_i < Z_{m+1}$) with the first order particle weighting, or particle-in-cell (PIC) model [e.g. Birdsall and Langdon, 1984]. Resonant currents assigned to grid Z_m and Z_{m+1} are thus obtained by

$$J_{Em,i} = j_{Ei} \frac{Z_{m+1} - z_i}{\Delta Z} , \qquad J_{E(m+1),i} = j_{Ei} \frac{z_i - Z_m}{\Delta Z} \tag{107}$$

and $J_{Bm,i}$ and $J_{B(m+1),i}$ are given in the same manner as above. Summing the currents J_{Emi} and J_{Bmi} over all particles between Z_{m-1} and Z_{m+1} gives the resonant currents J_{Em} and J_{Bm} at a grid point Z_m.

When the resonant currents are obtained, the wave frequency ω and the time derivatives $\partial Bw/\partial t$ and $\partial k/\partial t$ are calculated through (71), (72) and (73) as follows.

$$\omega_m = \frac{k_m(k_m - \mu_0 J_{Bm}/B_{wm})}{k_m(k_m - \mu_0 J_{Bm}/B_{wm}) + \Pi_{em}^2/c^2} \tag{108}$$

$$\left(\frac{\partial B_w}{\partial t} \right)_m = - \frac{2k_m(\Omega_{em} - \omega_m)}{k^2 + \Pi_{em}^2/c^2} \left(\frac{B_{wm} - B_{w(m-1)}}{\Delta Z} + \frac{1}{2} \mu_0 J_{Em} \right) \tag{109}$$

$$\left(\frac{\partial k}{\partial t} \right)_m = - \frac{\omega_{m+1} - \omega_{m-1}}{2\Delta Z} \tag{110}$$

As (109) and (110) contain the terms of spatial finite differentiation, the boundary conditions must be chosen carefully. In the present simulation we assume the wave propagating from the left boundary Z_1 to the right boundary Z_M. The wave field at Z_M is reasonably extrapolated from the field at Z_{M-1} and Z_{M-2} as

$$B_{wM} = 2B_{w(M-1)} - B_{w(M-2)} \tag{111}$$

and ω_M and k_M are also extrapolated in the same manner. At the left boundary Z_1, however, extrapolation from Z_2 and Z_3 is not applicable, because the wave propagates from Z_1 to Z_2 and Z_3. In the simulation where the wave is externally injected from the boundary, the wave field at Z_1 is given arbitrarily. In the simulation where a self-evolution of a uniform and periodic wave-particle interaction is followed, the field at Z_1 is set equal to the field at Z_2.

J_{Bm}/B_{wm} in (108) causes a non-physical fluctuation of ω, when the wave amplitude B_w becomes very small, because the resonant current J_{Bm} calculated from a limited number of particles necessarily contains some numerical fluctuations. This difficulty is relieved by assuming a thermal magnetic noise level B_{th} in the magnetospheric plasma, and setting the term J_{Bm}/B_{wm} equal to J_{Bm}/B_{th} when $B_{wm} < B_{th}$. In the simulation presented in Section 6.3 we follow evolution of a wave whose amplitude is $B_w = 10^{-5} \sim 10^{-6} B_0$, and B_{th} is chosen as $B_{th} = 10^{-6} B_0$.

4.5 Flow Chart of Long Time Scale Whistler Simulation

Recurrent use of the particle pushing and field updating algorithms enables us to follow a self-consistent evolution of the whistler wave-particle interaction. Motion of cold plasma particles which support the wave propagation is not solved, because its effect is implicitly included in the wave equation (71) and (73). Computing time is greatly saved by not solving equations of motion for cold electrons. It is noted that there is no necessity to choose the time step as $\delta t \cdot \mathrm{Max}(\Omega_e, \Pi_e) << 1$ which is usually a necessary condition in the usual particle simulation of plasmas. One requirement of the time step is $\delta t \cdot \omega_t << 1$, where ω_t is the trapping frequency given by $\omega_t = (kv_\perp \Omega_w)^{1/2}$. The relative phase angles ζ's of the resonant particles trapped in the wave potential change their phase angle ζ's with a time scale of ω_t^{-1} and thus the resonant currents controlling the wave evolution vary with ω_t. Another requirement is that $V_{max}\delta t \leqslant \Delta Z$, where V_{max} is the maximum velocity both in the wave group velocity and in the velocities of particles in the system. This is the loosest condition that must be satisfied in the difference scheme used in the algorithm.

The present simulation method is basically the long-time-scale (LTS) algorithm first formulated by Rathmann et al. [1978]. In their algorithm, however, they follow the time evolution of the wave frequency ω using the equation

$$\frac{\partial \omega}{\partial t} + V_g \frac{\partial \omega}{\partial z} = - \frac{\mu_o}{2k} V_g \frac{\partial}{\partial t} \left(\frac{kJ_B}{B_w} \right) \tag{112}$$

which is obtained by differentiating (67) with time t under the assumption that the cold plasma dispersion relation without the contribution of hot resonant electrons is valid. We found this equation vulnerable to a numerical instability because of the time differentiation of kJ_B/B_w which picks up numerical fluctuations of J_B, where J_B is calculated from a limited number of particles and correspondingly involves statistical errors. Actually, (112) is not necessary in the algorithm if the phase relation (72) is made use of to follow the time evolution of the wavenumber k. In our method the frequency ω is determined by (73) as a function of k and J_B/B_w after advancing k in time by (72). In the method of Rathmann et al., however, the wavenumber k is determined by (67) or (73) as a function of ω and J_B/B_w after advancing in time by (112). The flow chart of the LTS simulation is shown in Figure 12.

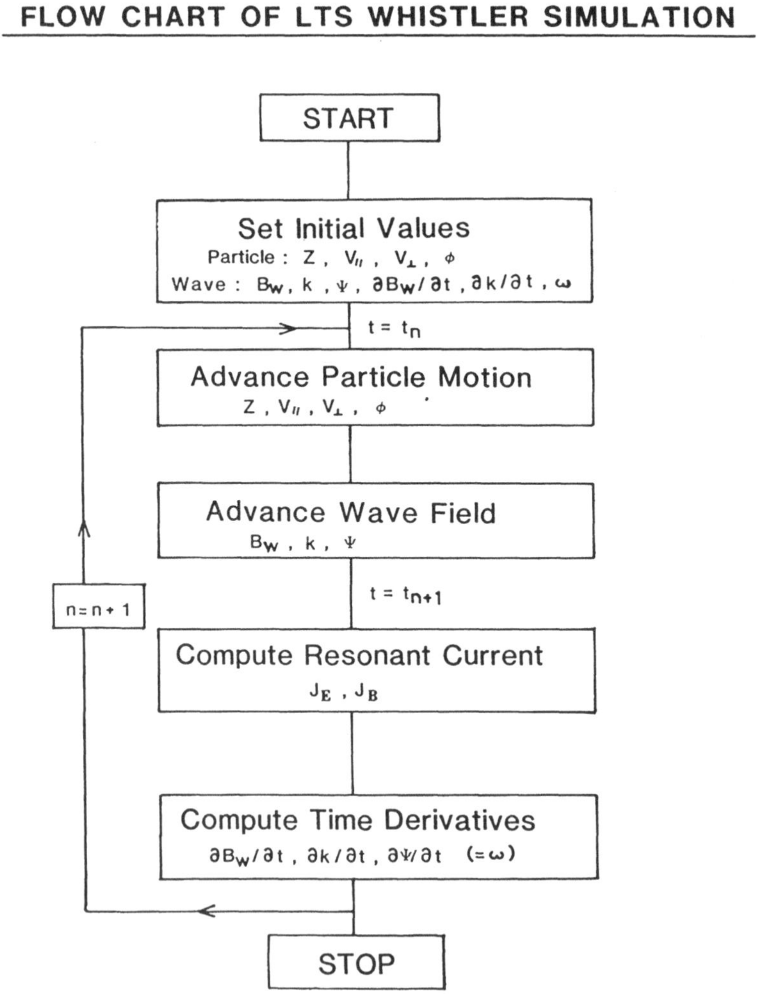

Figure 12. Flow chart of LTS whistler simulation.

5. DIAGNOSTIC TOOLS

5.1 Diagnostic Routines and Post-processors

We understand physical processes in a simulation run only through various diagnoses of the simulation data. Good and systematic diagnoses are needed for quick and good understanding of the simulated process. Diagnostic processes of a simulation consist of two steps. One is to use a diagnostic routine in a simulation code (main program), and the other is to use a group of diagnostic programs called post-processors.

The diagnostic routine stores simulation data into data files at a specified time interval. Data to be stored are specified depending on physical problems to be simulated, such as one of three components of the electric fields at grid points or part of particle velocities and positions. It is not realistic to store all the physical quantities at every time step due to a limitation of available data storage systems. The diagnostic routine also calculate wave and particle energies and distribution functions of particles simultaneously in a simulation run, and print them out or store them into data files. These in-run diagnoses become very important in a large scale of simulations, because they can extract necessary information from a large amount of physical quantities all of which are unable to be stored to files at each time step.

Post-processors give analyses of simulation data stored by the diagnostic routine in a simulation code and produce various graphic outputs such as a time history of wave and particle energies or phase space plots of particles at a certain time. Some typical post-processors of KEMPO and KULTS codes are introduced with their graphic outputs in the following sections. Well organized graphic libraries and sophisticated graphic systems are necessary for development and efficient use of post-processors.

5.2 Energy History

If a simulation code assumes a periodic boundary condition, total energy which is a summation of wave and particle energies must be conserved. Therefore, the energy diagnosis is a very basic and important step in checking a validity of the simulation code. We can also extract information from the data about physical process of energy transfer between wave and particles or among different particle species.

In the KEMPO we calculate average energy densities of electric, magnetic and kinetic energies in a simulation system.

(1) Electric energy density

$$W_E = W_{Ex} + W_{Ey} + W_{Ez}$$
$$= \frac{\varepsilon_0}{2} \frac{1}{N_x N_y} \sum_{i=1}^{N_x} \sum_{j=1}^{N_y} (E^2_{xi,j} + E^2_{yi,j} + E^2_{zi,j}) \qquad (113)$$

(2) Magnetic energy density

$$W_B = \frac{1}{2\mu_0} \frac{1}{N_x N_y} \sum_{i=1}^{N_x} \sum_{j=1}^{N_y}$$
$$\cdot \{(B_{xi,j} - B_{ox})^2 + (B_{yi,j} - B_{oy})^2 + (B_{zi,j} - B_{oz})^2\} \qquad (114)$$

It is noted that we do not include the static magnetic energy density in the magnetic energy density.

(3) Kinetic energy density

$$K_s = K_{sx} + K_{sy} + K_{sz}$$
$$= \frac{m_s}{2L_x L_y} \sum_{i=1}^{N_s} (v^2_{xi} + v^2_{yi} + v^2_{zi}) , \qquad (115)$$

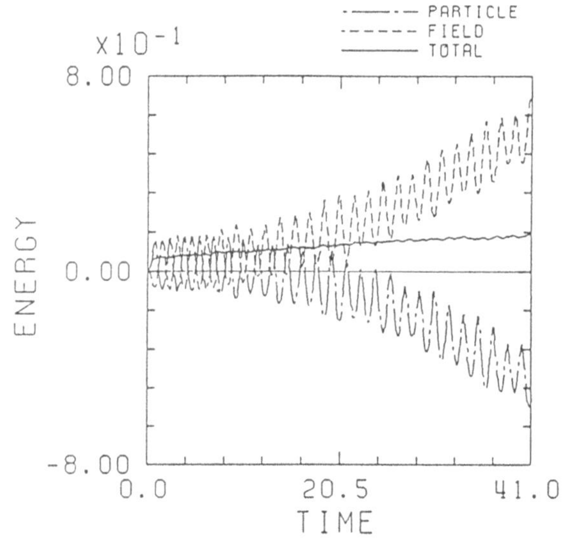

Figure 13. Time history of variations of total, field and particle energy densities.

where subscript "s" denotes quantities of particle species s. The total energy density ($W_E + W_B + \Sigma K_s$) must be conserved in a periodic model.

The kinetic energy density K_s is divided into two components, i.e., a drift energy density K_{sd} and a thermal energy density K_{st} calculated by

$$K_{sd} = K_{sdx} + K_{sdy} + K_{sdz}$$
$$= \frac{m_s}{2L_xL_y} \{ (\sum_{i=1}^{N} v_{xi})^2 + (\sum_{i=1}^{N} v_{yi})^2 + (\sum_{i=1}^{N} v_{zi})^2 \} \quad (116)$$

$$K_{st} = K_{stx} + K_{sty} + K_{stz}$$
$$= (K_{sx} - K_{sdx}) + (K_{sy} - K_{sdy}) + (K_{sz} - K_{sdz})$$

In the diagnostic routine in the KEMPO, all three components of W_E, W_B, K_s and K_d are calculated and stored into files at specified time intervals. The post processor combines these data and plots a time history of various energy densities. An example of history plots of energy variations is given in Figure 13.

5.3 ω -k Diagram

Fourier analysis of wave field data in space and time yields wave spectra in frequency ω and wavenumber k. For simplicity, we limit spatial variations to one dimension, i.e., the x-direction in the following explanation. The wave field data are one of three components of electric or magnetic fields, which are stored into files by the diagnostic routine with a format as

$$[\ a(x_i,\ t_j)\ ] \quad (i = 1, 2, \ldots, N_x; \quad j = 1, 2, \ldots, M) \quad (118)$$

where N_x is a number of grid points in a system and M is a number of data in time sequence.

We assume a presence of a wave mode with frequency ω and wavenumber k in the field data as

$$A_1\sin (\omega t - kx + \theta_1) + A \sin (\omega t + kx + \theta_2) \quad (119)$$

which is a summation of a forward traveling wave and backward traveling wave. Fourier analysis of [$a(x_i, t_j)$] in position x gives the cosine coefficients and sine coefficients which contain the following a_x and b_x, respectively.

$$a_x = A_1\sin (\omega t + \theta_1) + A_2\sin (\omega t + \theta_2) \quad (120)$$

$$b_x = - A_1\cos (\omega t + \theta_1) + A_2\cos (\omega t + \theta_2) \quad (121)$$

Fourier analyzing further the cosine and sine coefficients, we obtain the following four elements for the frequency ω and wavenumber k.

$$c_{11} = A_1 \sin\theta_1 + A_2 \sin\theta_2 \tag{122}$$

$$c_{12} = A_1 \cos\theta_1 + A_2 \cos\theta_2 \tag{123}$$

$$c_{21} = -A_1 \cos\theta_1 + A_2 \cos\theta_2 \tag{124}$$

$$c_{22} = A_1 \sin\theta_1 - A_2 \sin\theta_2 \tag{125}$$

Solving (122-125) for A_1 and A_2, we have

$$A_1 = \frac{1}{2} \sqrt{(c_{12} - c_{21})^2 + (c_{11} + c_{22})^2} \tag{126}$$

$$A_2 = \frac{1}{2} \sqrt{(c_{12} + c_{21})^2 + (c_{11} - c_{22})^2} \tag{127}$$

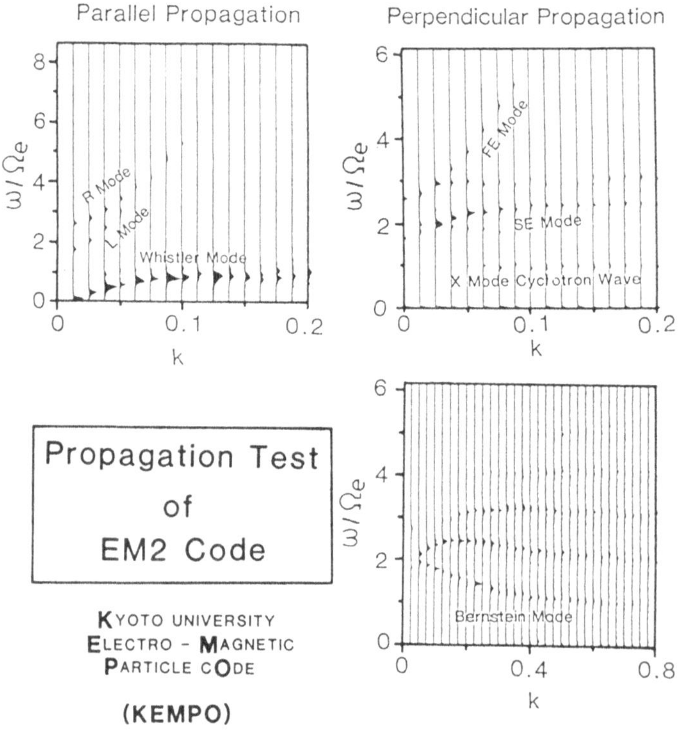

Figure 14. ω-k diagram of plasma dispersion relations observed in particle simulations by the KEMPO.

In the post-processor of the ω -k diagram, we calculate A_1 and A_2 for backward and forward traveling modes. In Figure 14 we show some of the outputs of the post processor. In this particular example, we applied the KEMPO to one dimensional problem with a model in which only thermal electrons are present initially. Out of thermal fluctuations of electrons, only electromagnetic as well as electrostatic waves satisfying the plasma dispersion are left over. Spectra shown in Figure 14 are summation of forward and backward propagation. The upper left panel shows spectra of the transverse magnetic field B_y in a case where the wavenumber vector k is parallel to the static magnetic field. We can recognize the R mode, L mode and whistler mode. The upper right panel shows spectra of the magnetic field B_Z in the case of perpendicular propagation where the wavenumber vector is in the x-direction and the static magnetic field is in the y-direction. We see fast and slow extraordinary modes and x mode electromagnetic cyclotron waves. The lower right panel shows the electrostatic component E_X in the same case of perpendicular propagation. We see very clear Bernstein modes.

We have described the one-dimensional ω - k analysis where the wavenumber k is in the x-direction. Extension to the two-dimensional ω -(k_X, k_y) analysis is simple in algorithm, but it requires a large size of memory because the data to be analyzed become three-dimensional array [a(x, y, t)].

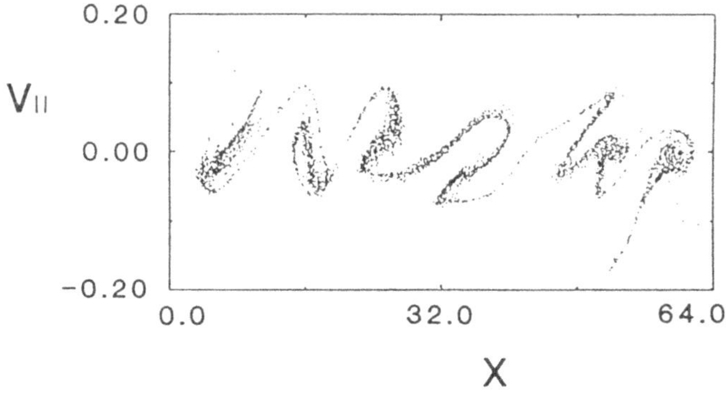

Figure 15. Phase space plot in the $v_\parallel$ - x plane.

5.4 Miscellaneous Diagnoses

Other than diagnoses of energy and ω -k spectra, various analyses are needed depending on physical problems under consideration. However, most of them does not require any special techniques in data processing. They only depends graphic routines for a (x, y) axis plot, a contour map, a three-dimensional perspective plot, etc., which are not directly related to nor inherent solely to particle simulations. Therefore, we briefly introduce examples of miscellaneous diagnoses.

(1) Phase Space Plot

A typical diagnosis of particle simulations is a phase space plot of particles. Locations of particles at a certain time is plotted by dots in a two-dimensional phase space where two of five components (x, y, v_x, v_y and v_z) are selected. An example of the phase space plots in the $v_{\parallel}$ -x plane is shown in Figure 15.

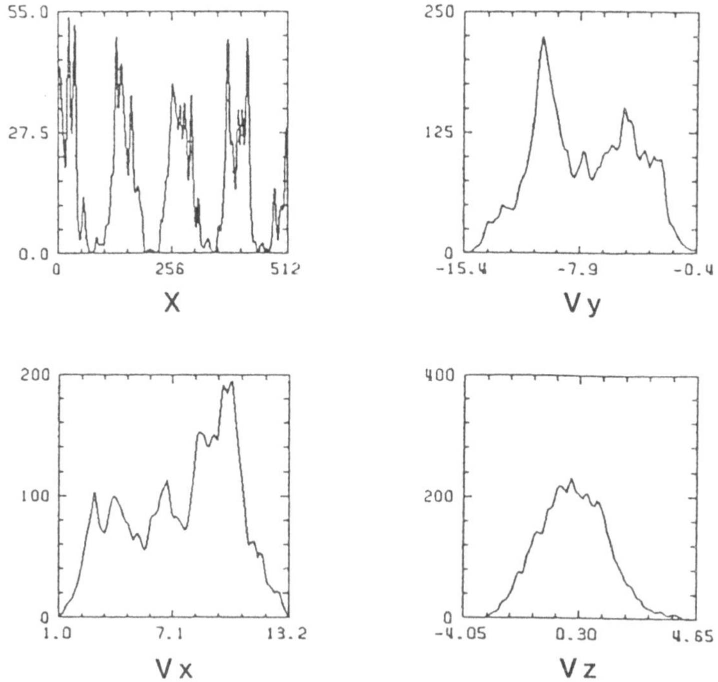

Figure 16. Particle distribution function in x, v_x, v_y and v_z.

(2) Distribution Function

In the diagnostic routine in the KEMPO, distribution functions of particles in a simulation run are calculated and stored into a data file. Reading the data from the file, one of the post processors plots the distribution functions in x, v_x, v_y and v_z as shown in Figure 16. Another post processor plots a two-dimensional particle distribution in the x-y plane using a three-dimensional perspective graphic routine as shown in Figure 17, which shows a hole formed in the plasma due to an electron beam injection.

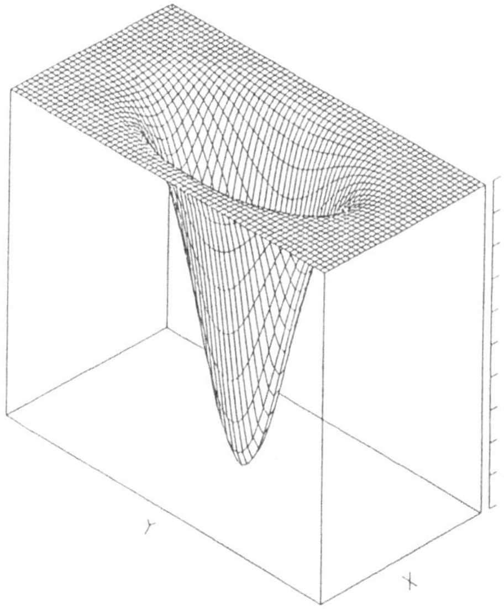

Figure 17. Particle distribution function in the x - y plane : a three-dimensional perspective plot.

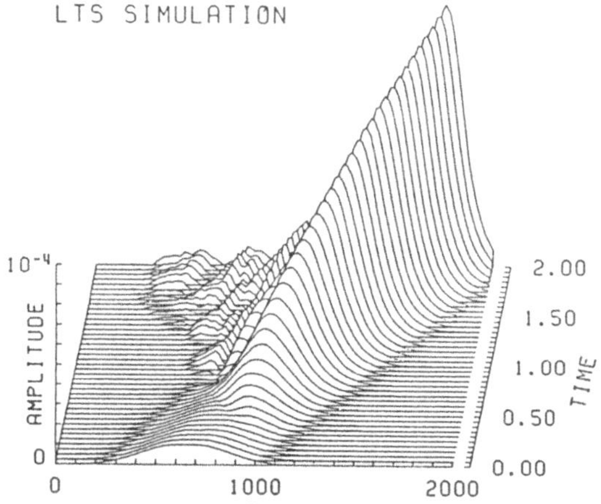

Figure 18. Spatial variation of wave amplitude and its time evolution.

(3) Spatial Variation and Time Evolution

By plotting one-dimensional spatial variation of field quantities with a slight shift at each time, we can see time evolution of the spatial variations. In Figure 18 we plot an output of KULTS code where we can see propagation of a whistler pulse with its growth and damping.

(4) Two-dimensional Vector Plot

Two-dimensional vector quantities defined in the x-y plane such as a current density (J_x, J_y), or a electrostatic force (E_x, E_y) are effectively indicated by arrows in the x-y plane. Intensities are indicated by the size of arrows. Figure 19 shows a flow pattern of current densities observed in a beam injection simulation by the KEMPO.

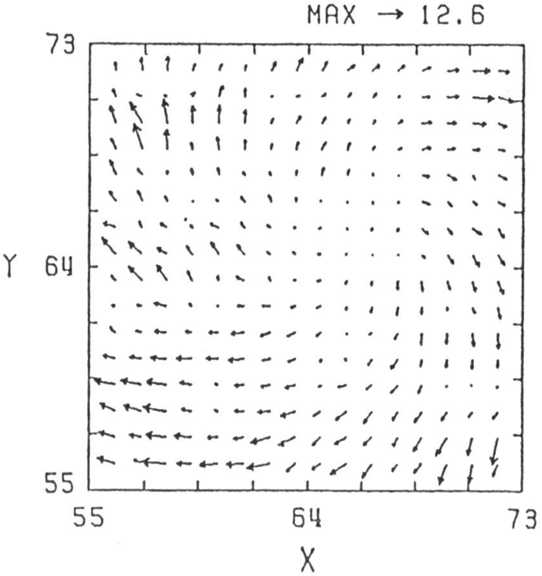

Figure 19. Two-dimensional vector plot : Flow pattern of current densities in x - y plane.

6. APPLICATION TO SPACE PLASMA WAVE PHENOMENA

The particle simulation codes developed at RASC such as the KEMPO and KULTS have so far been used to interpret nonlinear plasma dynamics associated with space wave phenomena and related particle events observed by space vehicles. In this section, we briefly describe three examples out of them.

Space observations are limited by two main reasons. First is due to limited power, volume and weight available to scientific payload in almost any space vehicles. Complicated and fancy measurements are therefore given up in most cases. This causes lack of information on detailed structures of field and particles blocking theoreticians to construct a satisfactory theory or scenario for the observed phenomenon. For example, very few simultaneous data of density of cold plasmas, velocity distribution functions of energetic particles and fields have been available from spacecraft. The other drawback of space vehicle observations is due to a fact that the observation is made only on a limited region in the vast area of space plasmas. Sometimes an interesting electromagnetic phenomenon is detected by wave receivers but not by particle detectors. This often means that the space vehicle missed a chance to pass the most interesting area where the corresponding wave-particle interaction takes place and picked up only the propagated wave effect.

Section 6.1 and 6.2 are examples of applications of the KEMPO to problems in which both electromagnetic and electrostatic waves play key roles in the wave-particle interactions. These examples are appropriate ones to demonstrate how computer simulations can provide detailed information which could never be obtained by spacecraft observations nor by analytic theories, compensating the first limitation of space observations. Section 6.3 is an example of simulation by the KULTS applied to whistler mode nonlinear wave-particle interactions in the inhomogeneous magnetic field. This is a good example to demonstrate the usefulness of the LTS code to handle a nonlinear interaction which takes place in a wide are of space over thousands of wavelengths along the inhomogeneous plasma medium such as that in the magnetospheric plasma, which is not observable as a whole by spacecraft either.

All examples shown in this section have already been published in journals so that only the background philosophy and motivation of each simulations are given together with a brief summary.

6.1 Hook-induced Electrostatic Bursts

This particular example is a particle simulation by the KEMPO on nonlinear wave-particle interactions involving obliquely propagating whistler mode wave and electrostatic waves. Reinleitner et al. [1982, 1983] discovered by ISEE satellite that strong electrostatic (ES) burst emissions are often associated with a coherent whistler mode chorus hook element, as shown in Figure 20, and that a high energy electron beam with an energy of the order of several hundred electron volts is always found at the time of occurrence of these ES bursts.

Motivated by the discovery, we started our particle simulation by the use of the KEMPO to shed some light into its physical mechanism [Matsumoto et al., 1984]. The KEMPO is useful to this kind of problems because both electromagnetic and electrostatic waves are

ISEE Observation of Chorus-induced Electrostatic Burst

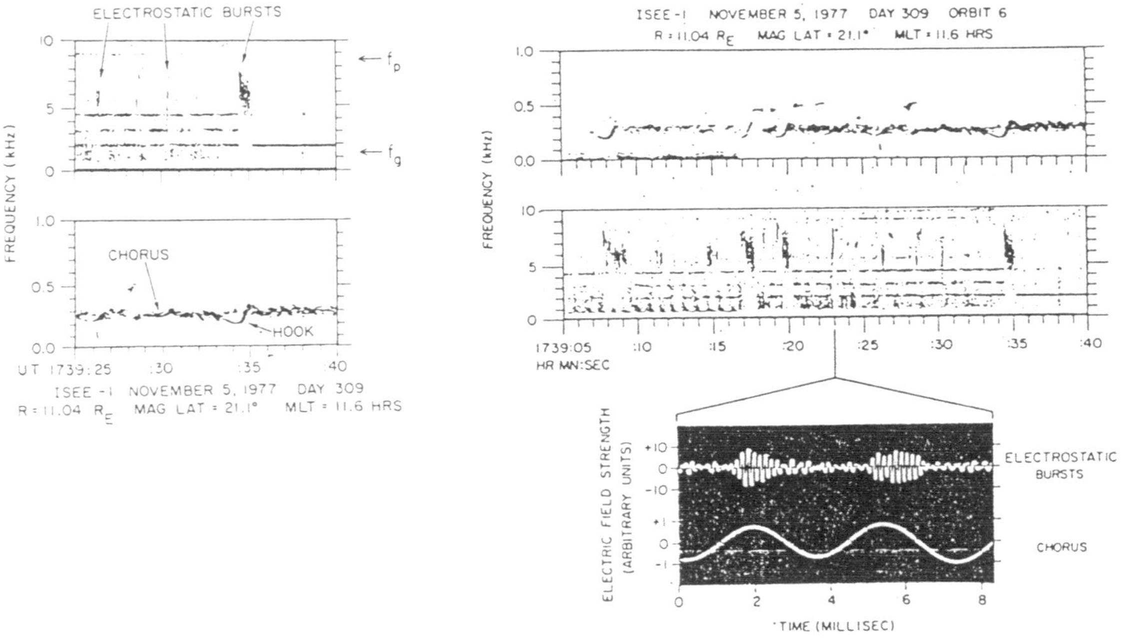

Figure 20. Observation data by ISEE satellite : Strong electrostatic (ES) burst emissions are associated with a coherent whistler mode chorus hook element.

automatically treated simultaneously. Two-and-half dimensional model was hired instead of easier one-and-two half dimensional model because we need to know which kind of ES mode is preferentially excited by the chorus element as a result of interaction with enhanced population of high energy electrons, and to see which direction of the k vector of the excited waves is to be as a result of the nonlinear interaction. Thus a model shown as Figure 21 was adopted.

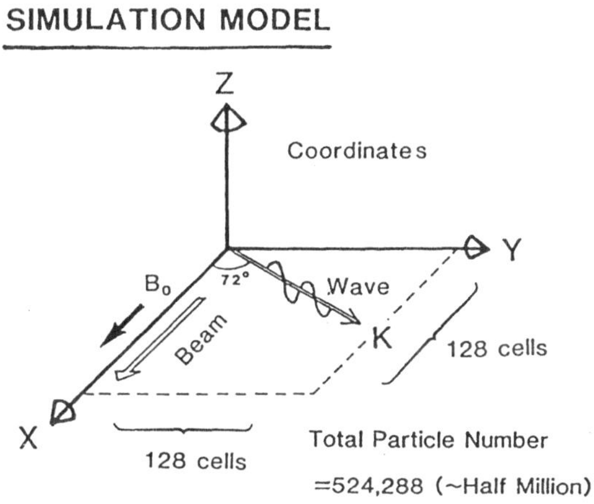

SIMULATION PARAMETERS

$\Pi_e=2$	$\omega=0.27$	$\Delta t=0.01$
$\Omega_e=1$	$\theta=72°$	$\Delta r=1.0$
$N_b/N_a=10^{-2}$	$B_w/B_o=10^{-2}$	$\lambda_D=1.0$
$V_{th}=2.0$	$M_x=1$	
$c=50.0$	$M_Y=3$	

Figure 21. Simulation model and parameters.

Prior to our simulation, we assumed that the ES bursts are produced by a coherent nonlinear interaction of a monochromatic whistler wave with resonant electrons in the enhanced bump in the tail in the velocity distribution function seen by ISEE at the time of the event. Therefore we carried out two different runs; one for a simple beam-wave interaction without a monochromatic whistler wave and the other with a monochromatic whistler wave. Like this example, to make

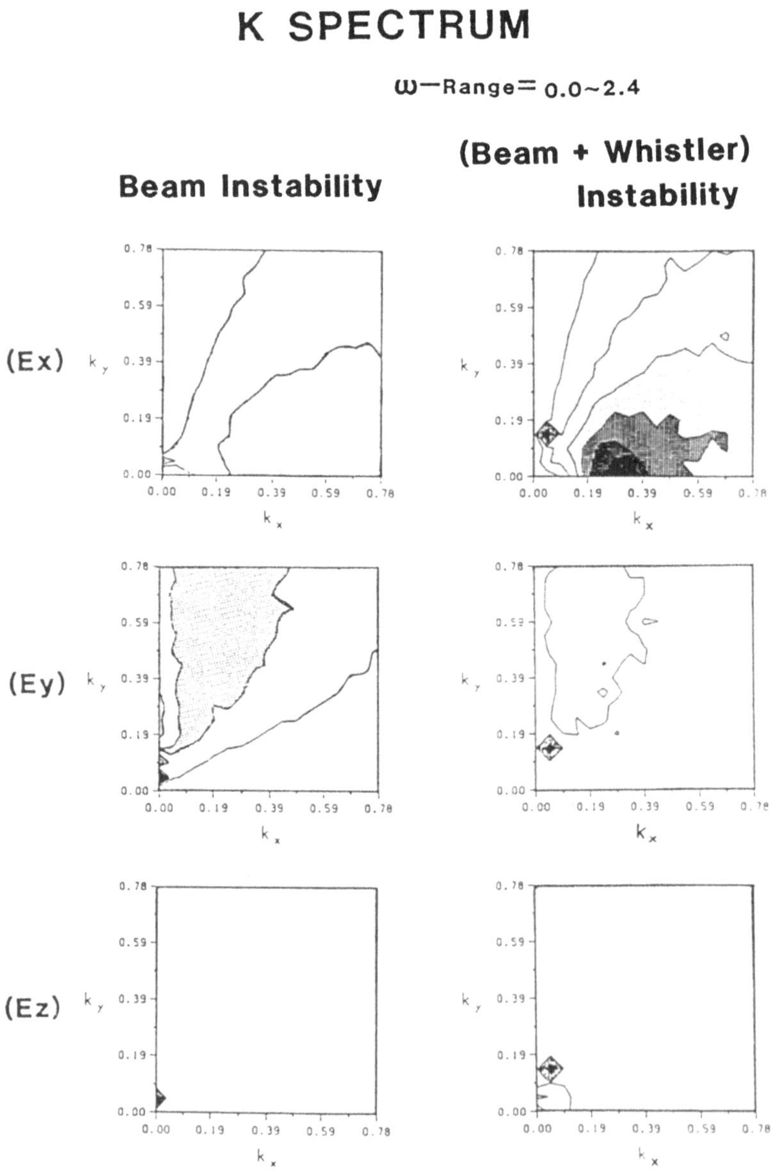

Figure 22. Spectrum of wavenumber k in k_x-k_y plane : (a) Beam only, (b) Beam + Whistler

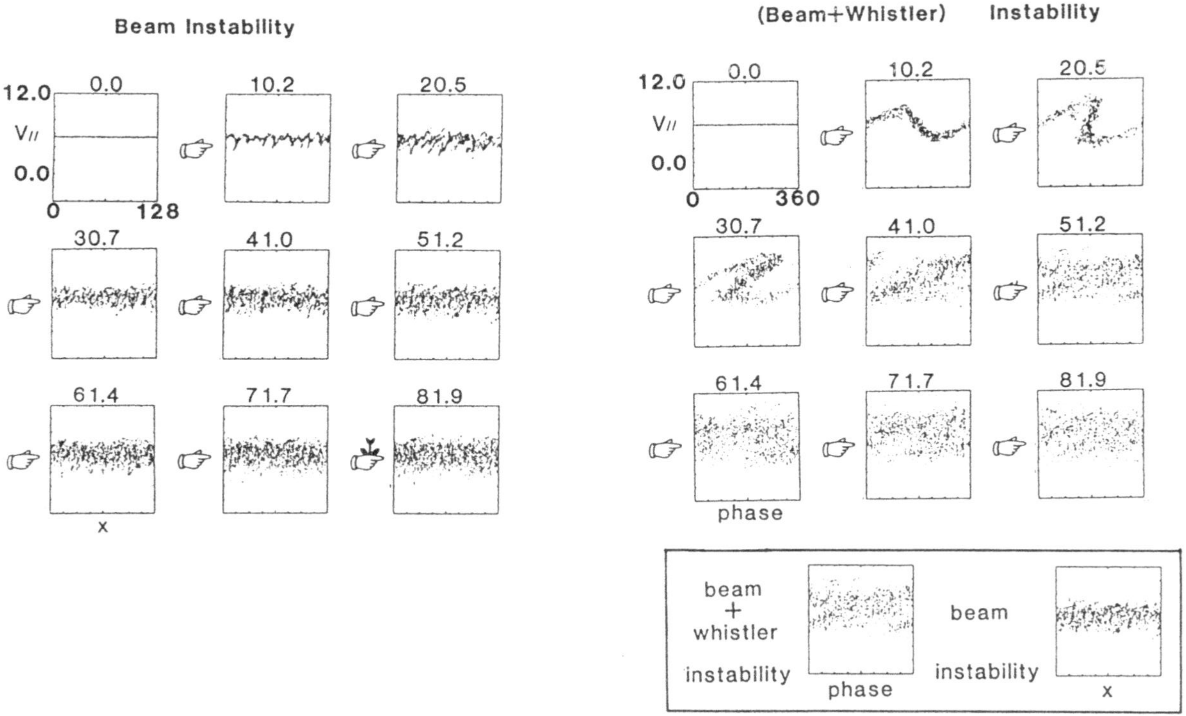

Figure 23. Phase space plot for beam electrons in $V_{\parallel}$ - phase plane. (a) Beam only, (a) Beam + Whistler

different numerical experiments by changing some key parameters or by eliminating or adding some elements participating in the interaction of interest is one of the advantages of computer simulations.

Such contrasting runs revealed the role of the coherent whistler wave. Figures 22 and 23 show comparisons of results of the two contrasting case A (beam only case) and case B (beam + whistler case) for a contour map of electric fields E_x, E_y and E_z in $k_{\parallel}$-$k_{\perp}$ plane, and for time evolution of $v_{\parallel}$-x phase diagrams. These diagnostic figures together with other diagnostic tools as described in Section 5 lead to the following conclusion.

"The generation mechanism of the electrostatic burst emissions are essentially a beam instability boosted under the action of the electrostatic component of the obliquely propagating whistler wave. The role of the whistler wave is to produce two beams via action of nonlinear trapping and subsequent detrapping process. One beam has higher velocity than the whistler phase velocity along the external magnetic field B_0 (i.,e., resonance velocity) and the other has lower one. The two beam jetting enhances the growth rate of the electrostatic waves propagating along B_0."

6.2 Wave Excitation and Particle Acceleration by Artificial Electron Beam Injection

Active experiments in space plasmas have recently been carried out extensively to investigate dynamical nature of space plasmas. Principle of these active experiments is to give artificial perturbations to the plasma and then measure its response and reaction. As the perturbation is given locally (and strongly in most cases), the response is transient, strongly inhomogeneous and highly nonlinear. It is generally difficult to treat by analytic theories such transient phenomena with highly inhomogeneous and nonlinear nature. Computer simulations are very effective to such problems. We will show an example of application of the KEMPO to one of the active experiments.

Electron beam injection is one of the most frequently used methods in active experiments. We performed two computer simulations with two different models in order to give physical interpretation of the phenomena obtained in the SEPAC (Space Experiment with Particle ACcelerators) onboard the Space Shuttle/ Spacelab-1; one is with a one-and-two-half dimensional model taking the model plane normal to the electron beam, and the other is with a one-and half dimensional model taking the 1D axis perpendicular to the electron beam. In both models, the beam is assumed to be parallel to the external magnetic field B_0 as the SEPAC electron gun launched the electron beam along B_0.

The first simulation with two-and-half dimensional model was performed mainly to investigate radiation of electromagnetic and electrostatic waves with k vector perpendicular to the beam and B_0

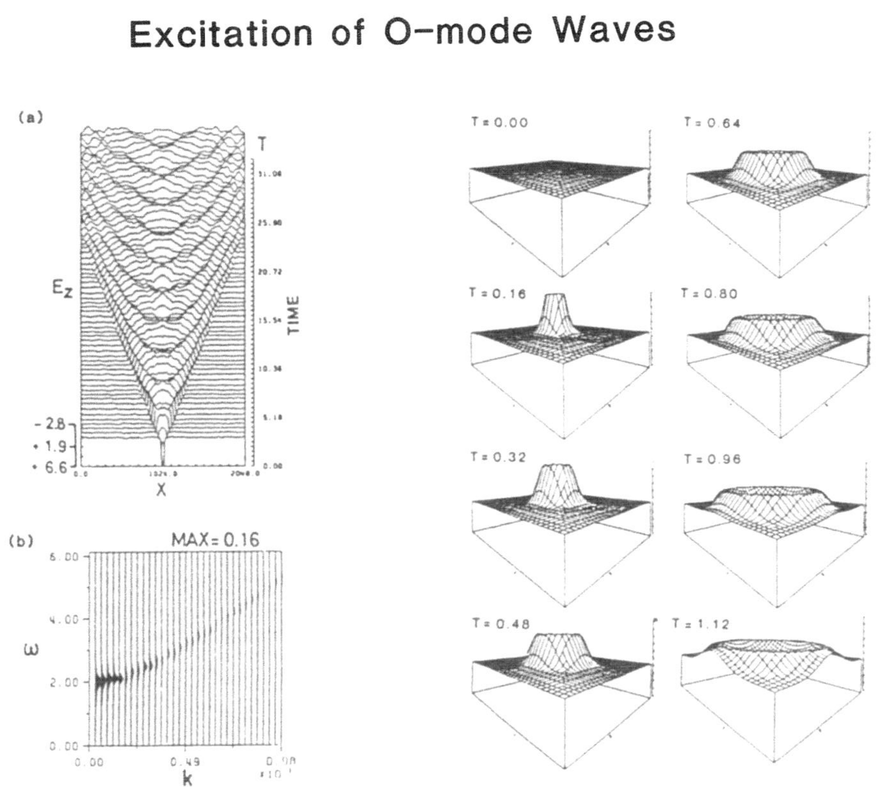

Figure 24. Excitation of O mode waves

direction [Omura and Matsumoto, 1984a]. Figures 24 and 25 are a part of the results showing the O-mode and X-mode wave radiation, respectively. In each figure, the time evolution of the wave amplitude (showing propagation effect), and ω-k diagram (compared with linear theory shown by a solid line in Figure 25) are given. In Figure 24 a three-dimensional display of the E_Z component is shown as well, while time evolution of contour maps of densities of background and beam electrons are shown in Figure 25.

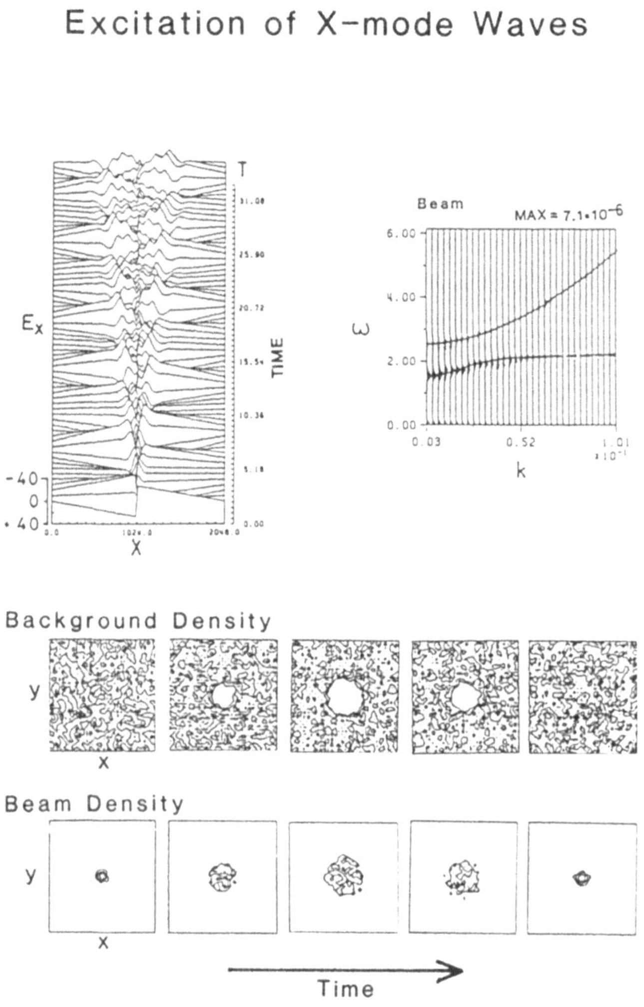

Figure 25. Excitation of X mode wave

It is expected before simulation that the O mode wave is radiated from the beam column and propagates in the radial direction as a result of a sudden injection of the beam current. However, in addition to the first impulsive radiation of the O mode electromagnetic wave, many pulses are launched from the beam successively as seen in the left upper panel and in the right 3D display of Figure 24. This comes from a fact that the beam current intensity is modulated by the change of the size of the beam column due to electrostatic oscillations.

Figure 25 shows that the hybrid mode oscillation of beam and plasma density is caused by excess charge injection in a localized area at the position of the beam and is a dominant source of the X mode radiation. The ω-k diagram indicates that the predominant mode is the slow extraordinary mode waves in the frequency range below the upper hybrid frequency ($(\Pi_e^2 + \Omega_e^2)^{1/2} \cong 2.25$). The lower panel shows that the background ionospheric plasma responds to the beam injection in such a manner that a variable-sized hole is created at the location of the beam.

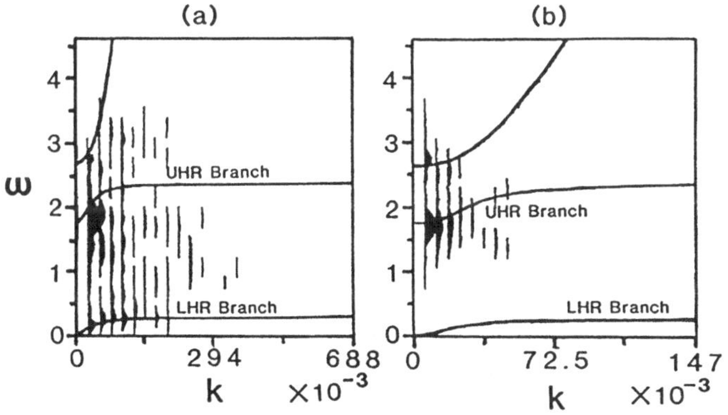

Figure 26. ω-k diagram : (a) wave spectra in the localized area in the vicinity of the beam, (b) wave spectra in the whole simulation region.

In the second simulation [Matsumoto and Fukuchi, 1984], ions are taken into account as a mobile constituent on top of background and beam electrons. In addition to the high frequency UHR waves on the UHR branch in the ω-k diagram as shown in Figure 26(b) which was made by FFT analysis from field data in the whole region in the simulation model, lower hybrid waves (LHR waves) are also excited in the vicinity of the beam location as seen in Figure 26(a) which is the ω-k diagram

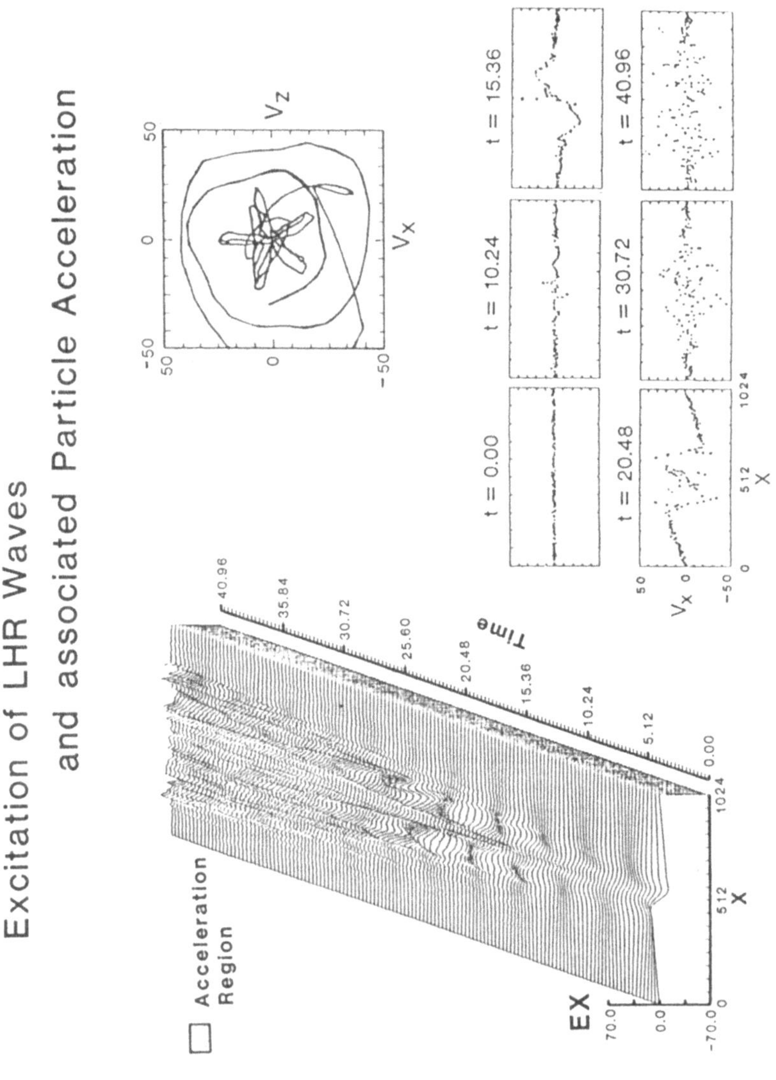

Figure 27. Excitation of LHR Waves and associated particle acceleration

made from field data only from a localized area in the vicinity of the beam. Figure 27 shows the corresponding electric field structure of the LHR waves as a function of time (left panel) together with a time evolution of the V_X-X phase diagram of background electrons (lower right panels) and a typical orbit in the $V_\perp$ plane of an accelerated electron. Ions start to move responding to the slowly-varying component of the electric field after t ~ 15 which is about half of the LHR period under the simulation parameters. Therefore we can see the low frequency LHR waves with pulsive nature and with short wavelengths of the order of 100 is excited first in the vicinity of the beam location at the central area of the system, and then propagate outward at later time. Some of the background electrons are accelerated by the local Landau resonance with the LHR waves when its velocity component parallel to the direction of propagation of the LHR waves becomes close to their phase velocity. This result of simulation gave an interpretation to the SEPAC result that very high energy electrons with pitch angles more than 88 degrees are found in the vicinity of the Shuttle when the electron beam was injected.

It is possible to predict qualitatively some of expected responses to such artificial perturbations by analytic theories. However, most of the complicated responses can neither be predicted by analytic theories nor even be interpreted properly without assistance of computer simulations. It is easy to understand and interpret the simulation results and thereby corresponding experimental results in terms of fundamental elements of physics by making as detailed diagnoses as one desires. The reverse process, however, is hardly possible. Reader may recall the famous tale about Columbus's Egg in this regard. In other words, it is extremely difficult for theories without simulations to predict the response of the plasma to the active perturbations and even to give proper interpretations of the experimental data because the data set itself is far from complete and only a fractional reflection of the whole complicated physical processes. Japanese and Chinese (and probably some other countries') proverb says "GUNMOH-ZOH-NADERU" (which means many blinds surrounding an elephant guess wrong by touching only its fractional part without any knowledge of this animal). We might fall into pit-holes in interpreting plasma processes if we make our interpretation scenario without qualitative <u>and</u> quantitative investigation. Simulations are, of course, not almighty but at least can be a good guide for researchers.

6.3 Nonlinear Whistler Interaction

One of the unsolved open problems in space plasma physics is the generation mechanism of VLF triggered emissions. This section is devoted to demonstrate an application of the KULTS to the VLF emission problem. It is now widely recognized that the VLF triggered emissions are the manifestation of nonlinear whistler interaction in the inhomogeneous magnetic field structure. This means it is uneconomical and practically impossible to simulate the whole emission process by a full electromagnetic (EM) particle code such as the KEMPO because the interaction region is of length of thousands of wavelengths and is much longer than the grid size which is the order of the Debye length in such EM particle code. The LTS code is useful to such a problem

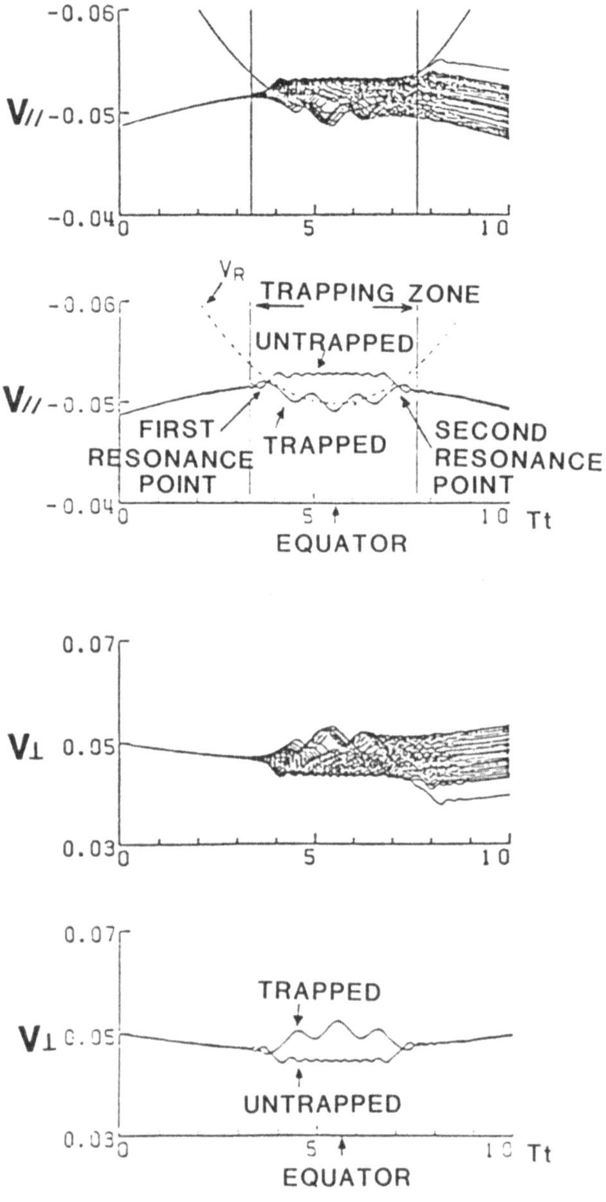

Figure 28. Trajectories of resonant electrons in a dipole magnetic field.

because the grid size can taken much larger than the Debye length as it does not solve the phase variation of the wave but solve only the slowly varying amplitude, frequency and wave number. The drawback of the LTS as a compensation of this gain is as follows.

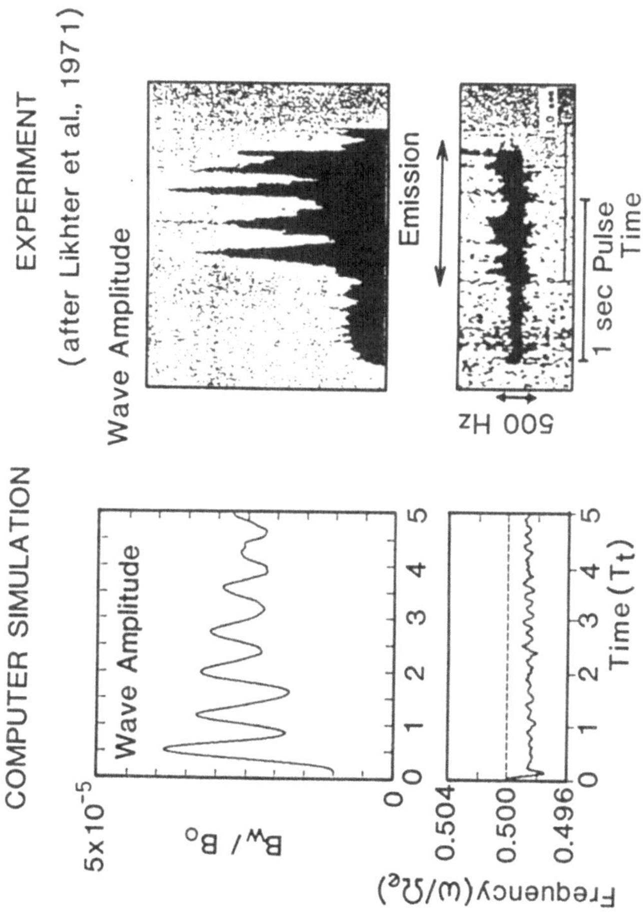

Figure 29. Comparison between simulation and experiment.

1. It can handle only one single wave mode and hence must neglect wave-wave couplings or competing process related to other wave modes.

2. Linear response of the background (non resonant) particles should be assumed. This means a highly nonlinear problem cannot be treated where the wave amplitude is such large as to drive the background plasma particles to nonlinear orbit.

3. The code is invalidated when the wave amplitude becomes zero or extremely small value.

The LTS code has often been used in coherent nonlinear whistler problems in disregard of these drawbacks [Rathmann et al., 1978; Vomvoridis et al., 1982; Omura and Matsumoto, 1982; Matsumoto and Omura, 1983; Omura and Matsumoto, 1984b].

Examples of the results of the KULTS simulations are shown in Figures 29, 30 and 31. Figure 28 shows orbits, of resonant electrons in the velocity-vs-time plot. By this diagnosis we can clarify dynamical behavior of trapped and untrapped particles under the nonlinear force due to the wave and mirror force due to the inhomogeneous magnetic field. Figure 29 is an example of comparison between simulation and experiment. The characteristics of amplitude oscillation and associated small frequency variation are well reproduced in the simulation. The third example obtained by the KULTS simulation shown in Figure 30 is the time evolution of the wave amplitude of the triggering wave (TW) pulse and its frequency variation with time. Followed by the amplified TW, triggered emissions are clearly seen associated with a frequency variation as shown in the upper right panel. The last example in Figure 31 is the result of a run which examined the contribution of untrapped but resonant electrons to the wave growth in the inhomogeneous medium. In this particular example, we artificially input a special group of electrons which are not trapped by the wave in order to investigate the role of untrapped resonant electrons. The result shows that these electrons contribute to give energy to the wave leading to a large amplification at the interaction point around $z \cong 1000$. The bottom panel shows the locations of these particles in the phase diagram. Obviously, as intended, the particles move outside the eye-shaped trapping area in the phase-space. This kind of artificial setting of initial conditions is one of the advantages of simulation. One can extract one or few important factors among various competing processes, which in the real plasma is not separable, yielding clearer understanding of the roles of the selected element(s) in the interaction.

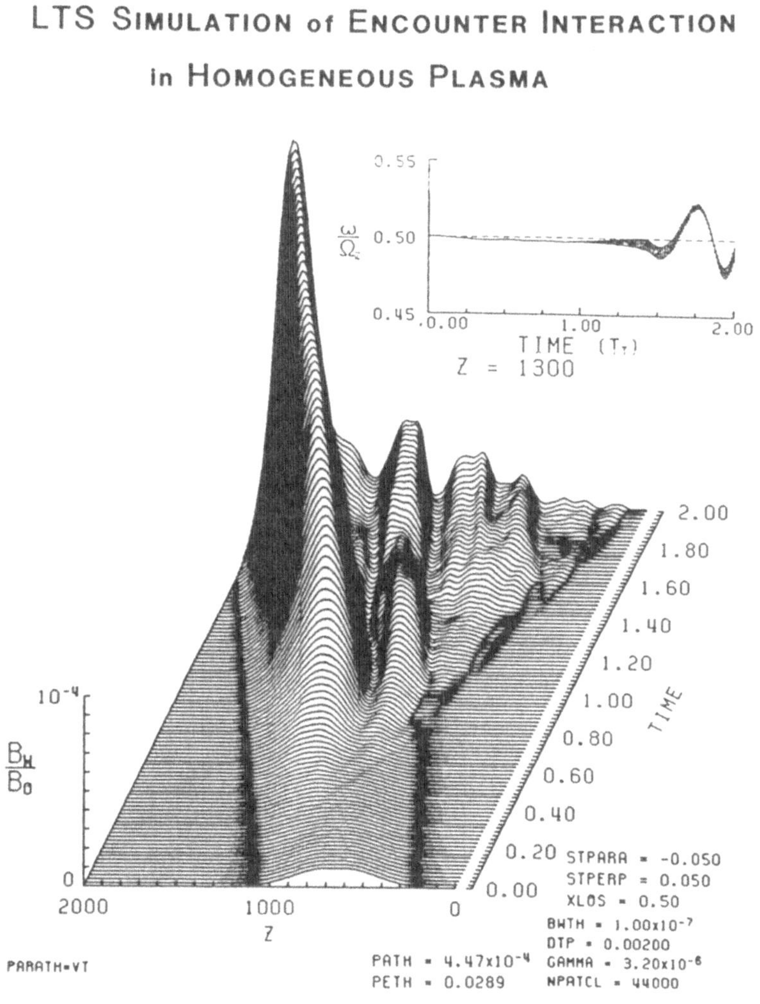

Figure 30. Time evolution of the wave amplitude of the triggering wave and its frequency variation with time.

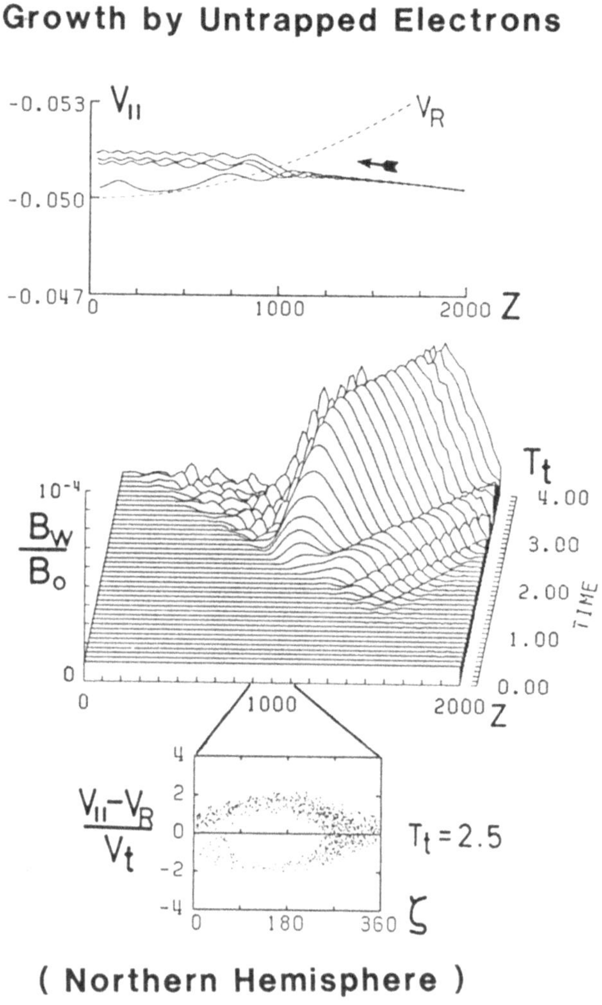

Figure 31. Wave growth due to the interaction with untrapped resonant electrons.

7. CONCLUDING REMARKS

In this article, we have given the outlines of the two kinds of particle simulation codes KEMPO and KULTS used at RASC for research of electromagnetic and electrostatic waves and of related wave-particle interactions together with examples of their application to space plasma phenomena. Some of the basic concept and techniques of electromagnetic simulations have also been described.

Readers are recommended to write their own code, not hiring or copying the others, because simulationists should be aware of the inherent limit of the code one uses, otherwise one may fall in a pit-hole of erroneous conclusion by simply making a superficial comparison of the simulation results with observations. The needed attitude is to perform a simulation based on both sound physical insight into the physical process of interest and numerical techniques which one uses. We hope this article may be a help to the scientists and students who are going to make simulation studies in space plasma physics.

ACKNOWLEDGEMENTS

We thank K. Miwa for typing the manuscript and students T. Yamada, T. Kimura, N. Komori, Y. Matsumoto, H. Tanaka, T. Tanaka, K. Inagaki and T. Ohashi at RASC for their help in preparing figures. Part of the simulations are supported by the Grant-in-Aid #56420016 of the Ministry of Education, Culture and Science, Japan.

REFERENCES

Abe, H., J.Miyamoto, and R. Itatani, J. Comput. Phys., 19, 134, 1975.

Birdsall, C. K., and D. Fuss, J. Comput. Phys., 3, 494, 1969.

Birdsall, C. K., and A. B. Langdon, Plasma Physics via Computer Simulation, McGraw-Hill, New York, in press, 1984.

Boris, J. P., Proc. 4th Conf. on Numerical Simulation of Plasma, Stock 085100059, 3, U.S. Govt. Printing Office, Washington, D.C., 1970.

Buneman, O., Phys. Rev., 115, 503, 1959.

Buneman, O., J. Comput. Phys., 1, 517, 1967.

Dawson, J. M., Phys. Fluids, 4, 869, 1961.

Dawson, J. M., Methods Comput. Phys., 9, 1, 1970.

Dysthe, K. B., J. Geophys. Res., 76, 6915, 1971.

Hockney, R. W., Phys. Fluids, 9, 1826, 1966.

Hockney, R. W., Methods Comput. Phys., 9, 135, 1970.

Hockney, R. W., and J. W. Eastwood, Computer Simulation Using Particles, McGraw-Hill, New York, 1981.

Kruer, W. L., M. Dawson, and B. Rosen, J. Comput. Phys., 13, 114, 1973.

Langdon, A. B., and C. K. Birdsall, Phys. Fluids, 13, 2115, 1970.

Langdon, A. B., and B. F. Lasinski, Methods Comput. Phys., 16, 327, 1976.

Matsumoto, H., and Y. Omura, Geophys. Res. Lett., 10, 607, 1983.
Matsumoto, H., M. Ohashi, and Y. Omura, J. Geophys. Res., 89, 3873, 1984.
Matsumoto, H., and K. Fukuchi, Geophys. Res. Lett., in press, 1984.
Morse, R. L., and C. W. Nielson, Phys. Fluids, 12, 2418, 1969.
Nielson, W.C., and H. R. Lewis, Methods Comput. Phys., 16, 367, 1976.
Okuda, H., and C. K. Birdsall, Phys. Fluids, 13, 2123, 1970.
Okuda, H., Proc. 4th Annu. Conf. on Numerical Simulation of Plasma, Office of Naval Research, Arlington, Va., 515, 1970.
Okuda, H., and J. M. Dawson, Phys. Fluids, 16, 408, 1973.
Okuda, H., Computer Simulation of Space Plasma, 3, Terra Scientific Publishing Company, 1984.
Omura, Y., and H. Matsumoto, J. Geophys. Res., 87, 4435, 1982.
Omura, Y., and H. Matsumoto, Radio Sci., 19, 496, 1984a.
Omura, Y., and H. Matsumoto, submitted to J. Geomag. Geoelectr., 1984b.
Rathmann, C. E., J. L. Vomvoridis, and J. Denavit, J. Comput. Phys., 26, 408, 1978.
Reinleitner, L. A., D. A. Gurnett, and D. A. Gallagher, Nature, 295, 46, 1982.
Reinleitner, L. A., D. A. Gurnett, and T. E. Eastman, J. Geophys. Res., 88, 3079, 1983.
Vomvoridis, J. L., T. L. Crystal, and J. Denavit, J. Geophys. Res., 87, 1473, 1982.

RELATIVISTIC CODE APPLIED TO RADIATION GENERATION

A. T. Lin

Department of Physics
University of California, Los Angeles
Los Angeles, California, U. S. A.

ABSTRACT

In this chapter, a general algorithm for simulating a relativistic electromagnetic plasma is given. Emphasis is placed on how to handle the electron mass dependent on its energy and to treat boundary conditions for outgoing electromagnetic waves. The results of mode conversion from an extraordinary wave into an electrostatic Bernstein wave and radiation generation which exploits the relativistic effects of electrons such as gyrotron and auroral kilometric radiation are presented.

1. INTRODUCTION

The subject of this chapter is the interaction between charged particles and propagating electromagnetic waves. Of particular concern are the situations in which the relativistic effect of electron mass dependent on its energy plays influential roles in generating the electromagnetic wave and the wave amplitude is large such that nonlinear effects are important. Under these circumstances, the most effective tool in dealing with these phenomena is particle simulations.

In the early fifties electonic engineers[1] had already used particle simulations to investigate nonlinear state of various microwave devices. In their approaches, a steady state is assumed to exist. An electromagnetic wave with a specified frequency is injected into the interaction region and its spatial behaviour is followed. On the other hand, the nonlinear state in plasma physics is far more complicated. For instance, an electromagnetic wave can decay into other waves with different dispersive characteristics. A more general approach to particle simulation[2,3] was introduced in the late fifties to study two-stream instability. Since that time a great deal of progress in computing power and simulation techniques has been made. Undoubtedly, particle simulations have made significant contribution to deeper understanding of plasma behavior.

In particle simulation the computer is employed to advance the motion of large number of charged partices, moving under the influence of the self-consistent and externally applied fields. The most general equations to represent these interactions are the Maxwell's equations

$$\frac{\partial \vec{E}}{\partial t} = c\nabla \times \vec{B} - 4\pi\vec{J}$$

$$\frac{\partial \vec{B}}{\partial t} = - c\nabla \times \vec{E} \qquad (1)$$

$$\nabla \cdot \vec{B} = 0$$

$$\nabla \cdot \vec{E} = 4\pi\rho$$

and the relativistic equation of motion

$$\frac{d\vec{P}_j}{dt} = q_j \left\{ \vec{E} + \frac{\vec{P}_j \times \vec{B}}{m_j c \left(1 + \frac{P_j^2}{m_j^2 c^2} \right)^{1/2}} \right\} \qquad (2)$$

where most of the symbols appearing in these equations have been defined in the previous chapters. Since computr time and storage are limited, discrete time step and spatial grids have to be used to change Eqs.(1) and (2) into finit difference equations. The basic schemes for solving these equations[4] have been described extensively in the previous chapters and will not be repeated. Here only the difference from the previous approach will be emphasized.

Instead of advancing the electromagnetic fields on each spatial grid[5], one can make use of Fast Fourier Transform (FFT) and advances their Fourier components[6]. To facilitate this we split $\vec{E}$ and $\vec{j}$ into transverse and longitudinal components; of course $\vec{B}$ only has transverse components. The longitudinal component of $\vec{E}$ is obtained as in the electrostatic case from Poisson's equaiton and the transverse $\vec{E}$ and $\vec{B}$ fields are obained by solving the Maxwell equations which now become

$$\frac{\partial \vec{E}_T(\vec{k},t)}{\partial t} = i c\vec{k} \times \vec{B}_T(\vec{k},t) - 4\pi\vec{j}_T(\vec{k},t)$$

$$\frac{\partial \vec{B}_T(\vec{k},t)}{\partial t} = -i c\vec{k} \times \vec{E}_T(\vec{k},t) \ , \qquad (3)$$

where the transverse components of the current are defined by

$$\vec{j}_T(\vec{k},t) = \vec{j}(\vec{k},t) - \frac{\vec{k}\vec{k} \cdot \vec{j}(\vec{k},t)}{k^2} \tag{4}$$

Here again Eqs. (3) should be converted to finite difference equations in time and a leap frog scheme should be employed to solve for $\vec{E}_T$ and $\vec{B}_T$.

Since the Fourier transformation of a Gaussian distribution is again a Gaussian, the advantage of using a finite sized particle of Gaussian shape is easily included. Furthermore, as has been shown[6] that in contrast to the spatial finite differnce scheme, the phase velocity of electromagnetic wave is greater than c for all Fourier modes. This is also an important attribute of this algorithm. Especialy, if one is simulating a relativistic plasma as will be done here, and this is not the case, relativistic particles can exceed the light velocity in the spatial finit difference scheme. As consequence they emit spurious Cherenkov radiations which can easily swamp the pheneomenon we are interested in.

In treating the fully relativistic equations of motion for the charged particles, one now uses the particle momentum $\vec{P}j$ rather than the velocity $\vec{V}j$. However, in advancing $\vec{P}j$ as can be seen from Eq.(2), we need to know the average value of $\vec{P}j$ during a time step. Since $\vec{P}j$ also appears in the denominator of Eq.(2), it is very difficult to use the straightforward implicit scheme. Instead, one takes advantage of the fact that the magnetic force causes only a rotation of the particle about magnetic field but does not alter the particle energy. The scheme commonly adopted is the following, using only the electric field to advance $\vec{P}j$ by half time step, then use the updated $\vec{P}j$ to determine$|\vec{P}j\ ^2|$. Equation (2) now resembles the nonrelativistic equations of motion and the usual implicit method can be used to advance $\vec{P}j$.

Various schemes of treating boundary conditions for outgoing electomagnetic waves will be discussed in Sec.2. The simulation results of radiation generation which exploit the relativistic effects of electrons will be presented in Sec 3 .

2. BOUNDARY CONDITIONS FOR OUTGOING ELECTROMAGNETIC WAVES

An electromagnetic wave originated in a plasma region can propagate into vacuum regions. If there are no reflicting walls to confine the radiation, the electomagnetic wave can reach infinity. In order to use a finite space to simulate an infinite region, some approximations have to be imposed on the Maxwell equaitons. There are three algorithms which have been successfully implemented in simulating various electomagnetic phenomena in plasma physics. They will be briefly described in this section. The first two algorithms are appropriate for the scheme which advances electromagnetic fields on each spatial grid while the last algorithm is convenient for the scheme using FFT.

2.A. Decomposing Radiation Into Left-and Right-Going Waves

Assmue that space variations are in one direction (x) only and the electromagnetic wave is linearily polarized in y-direction. Let $F^{\pm}=E_y \pm B_Z$. The electromagnetic components of Eq. (1) can be written as

$$\left(\frac{\partial}{\partial t} \pm c\frac{\partial}{\partial x}\right) F^{\pm} = \mp 4\pi J_y \tag{5}$$

In vacuum F^+ represents the wave traveling with velocity c in the negative x-direction wheareas F^- denotes the wave traveling with velocity c in the positive x-direction. Equation (5) can be solved by integrating separately the functions $F^{\pm}$ along the characteristic line of the light wave ($x \mp ct$=const.) which eliminates the spurious Chrenkov radiaitons. The grid spacing and time step are linked by the vacuum characteristics ($\Delta x/\Delta t=c$)[7]. The electromagnetic fields required for pushing particles can be determined from $E_y=(F^+ + F^-)/2$ and $B_Z=(F^+ - F^-)/2$. The outgoing boundary condition for electromagnetic waves is simulated by assuming that the fictitious walls of the simulation system are radiation transparent.

The algorithm for more general cases involving variations in more than one direction is not quite so simple. Some attempts have been made without much success. A different algorithm which is commonly adopted in the field of laster fusion[8] will be discussed in the next sub-section.

2.B. Projection Operator

Consider a slab geometry which is periodic in the direction y but is finite in the direction x. In vacuum the wave equation for the electromagnetic field can be written as

$$\left(\frac{\partial}{\partial t} \pm c\cdot\hat{G}\frac{\partial}{\partial x} \right) A = 0 \tag{6}$$

where A is the wave vector potential and $\hat{G}$ is a projection operator which will be defined later. With properly chosen $\hat{G}$, Eq. (6) with the plus sign describes left-going waves (A_L) and with the minus sign represents right-going waves (A_R).

In order to demonstrate how the out-going waves can be absorbed by a properly chosen $\hat{G}$, consider a solution for Eq. (6) of the form

$$A = \left(A_R e^{ik_x x} + A_L e^{-ik_x x} \right) \exp[i(k_y y - \omega t)] \tag{7}$$

Substituting Eq. (7) into Eq. (6) gives

$$A_R + A_L = \frac{\hat{G}}{G_o} (A_R - A_L) \tag{8}$$

where

$$G_o = \frac{\omega}{ck_x} = (1 - \frac{c^2 k_y^2}{\omega^2})^{-1/2} \tag{9}$$

From Eq. (8) the reflection coefficient A_L/A_R is zero if $\hat{G}$ is chosen to be equal to Go. Empirically, Lindman[8] has found the best approximated expansion for $\hat{G}$ as follows

$$\hat{G} = 1 + \sum_{n=1}^{N} \hat{g}_n \tag{10}$$

where

$$\hat{g}_n = \frac{\alpha_n c^2 \hat{D}^2}{1 - \beta_n c^2 \hat{D}^2} \tag{11}$$

where the operator $\hat{D}=\partial_y/\partial_t$. Substitution of Eq.(11) into Eq.(6) and retaining only the plus sign yields

$$(\frac{\partial}{\partial t} + c\frac{\partial}{\partial x}) A = - c\sum_{n=1}^{N} \hat{h}_n \tag{12}$$

where $\hat{h}_n=\hat{g}_n\partial A/\partial x$. For each Fourier mode (ky), the usual finite difference scheme can be used to solve Eq. (12). Extra storage space which is small compared with the storage space for plasma region is required to implement this algorithm because it is only operative in the vacuum region. Lindman found that by choosing N=3, α=0.3269, 0.1272, 0.0309, and β=0.7375, 0.9838, 0.9966, the reflection coefficient is at its minimal value for a range of propagation angles from 0° to 89°.

2.C. Resistive boundary Layers

An electromagnetic wave propagates through a resisitive layer can be completely absorbed if the thickness of the layer is large enough. The property of a resistive layer can be modeled in particle code by adding an absorbing region on both sides of the plasma system to absorb the electromagnetic radiation leaving the plasma region. The schematic diagram of the simulation model is shown in Figure 1. The radiation field $\vec{E}_T(x,ky)$ is left unchanged between $x=0.5\,L_a$ and $x=0.5L_a+L_s$ but is multiplied by a function $f_a(x)$ which falls off from unity quadratically to zero in the absorbing region. Since the derivative of $f_a(x)$ is zero at x=0.5 La and (0.5 La+Ls), the reflection of an electromagnetic wave passing through these transition points is minimized. This algorithm can be easily implemented. However, it suffers the disadvantage of requiring more storage space to model the absorption regions than the previous two algorithms.

As an example, a mode conversion process[9] will be given to

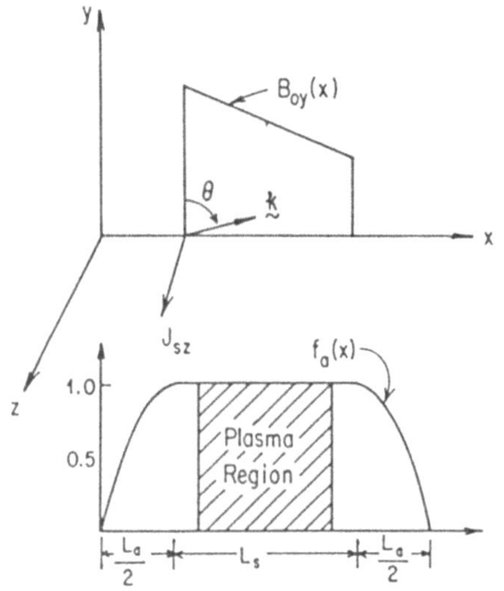

Figure 1. Schematic diagram of bounded simulation model. Radiation is absorbed by multiplying transverse electric field $E_T(x,ky)$ by the absorbing function $f_a(x)$ at each time step.

demonstrate the effectiveness of the outgoing wave boundary condition. For convenience, the resistive layer algorithm is chosen for our particle model. Consider an extraordinary wave of given ω_o propagates to the upper hybrid resonant layer, both its phase and group velocities approach zero in a collisionless cold plasma and the wave energy is converted into upper hybrid oscillations. For simplicity, consider a normal incident (Figure 1, $\theta=90^0$) extraordinary wave which is launched from the high field side into a plasma with uniform density. The magnetic field is assumed to vary linearly

$$\frac{B_o(x)}{B_o(0)} = 1 - \frac{x}{L_m} \tag{13}$$

where L_m is magnetic field scale length. The propagation of the extraordinary wave is governed by Maxwell's equations and the equation of motion which can be simplified to

$$\frac{d^2E_z}{dx^2} + \frac{\omega_o^2}{c^2}\left\{1 - \frac{\omega_{pe}^2}{c^2}\,\frac{(\omega_o^2-\omega_{pe}^2+i\nu\omega_o)}{\omega_o^2-\omega_{pe}-\omega_{ce}^2+i\frac{\nu}{\omega_o}(2\omega_o^2-\omega_{pe}^2)}\right\}E_z = 0 \tag{14}$$

$$E_x = -i\,E_z\,\frac{\omega_{ce}}{\omega_o}\,\frac{\omega_{pe}^2}{\omega_o^2-\omega_{pe}^2-\omega_{ce}^2+i\frac{\nu}{\omega_o}(2\omega_o^2-\omega_{pe}^2)} \tag{15}$$

where ν is a phenomenological damping rate. In the computer

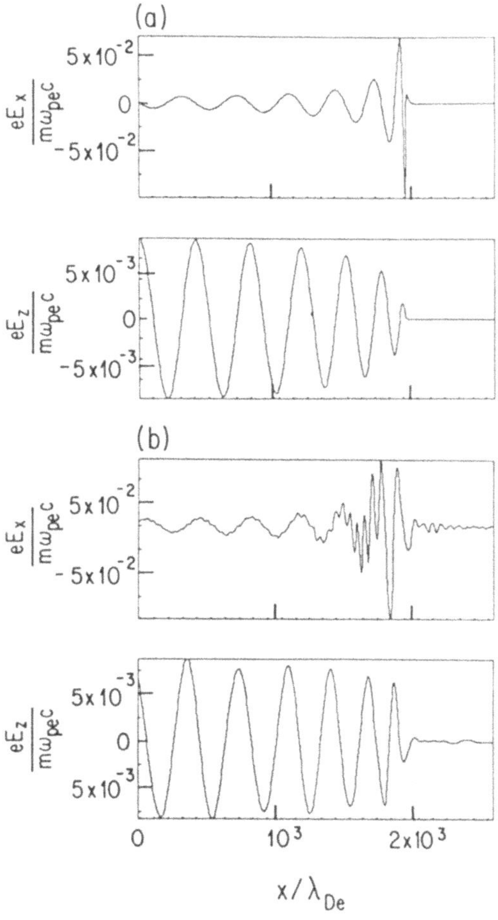

Figure 2. The electrostatic (E_x) and electromagnetic (E_z) field spatial distribution at ω_{pe}=500, (a) from cold plasma theory, (b) from computer simulation with immobile ions.

simulation, a current sheet is located at the high field side of the vacuum region. The current flows along the $\hat{z}$ direction and oscillates at a frequency ω_o to generate an extraordinary wave which when it reaches the plasma region will induce an electron to oscillate with a velocity $V_o=eE_o/m\omega_o$. The parameters for the simulation are chosen to be $Te/mc^2=1.7\times10^{-4}$, $Vo/c=0.01$, $\omega_o/\omega_{pe}=1.4$, $L_m/\lambda_{De}=6500$, $\omega_{ce}(0)/\omega_{pe}=1.53$

and the ions are immobile. Substituting this set of parameters into Eq.(14) and (15) and choosing $\nu/\omega_{pe}=4.5\times10^{-3}$ yield the spatial distribution of both the electromagnetic component (E_z) and the electrostatic component (E_x). The theoretical result and the simulation result are displayed in fig.(2a) and Figure (2b) (at $\omega_{pe}t=500$) respectively. The agreements are quite good except that the simulation result shows that there is a short wavelength electrostatic Bernstein's wave which propagates out of the resonant region into the high field side with decreasing wavelength. This is a kinetic effect which the cold plasma theory is unable to reveal.

In the next section, the simulation results of two physical phenomena which rely on the relativistic effect of the electrons as their bunching mechanisms will be described.

3. ELECTRON CYCLOTRON MASER INSTABILITIES

The cyclotron radiation at the fundamental or at the first few harmonics of the cyclotron frequency by weakly relativistic electrons with non-equilibrium energy distribution and rotating in a magnetic field had been investigated by Twiss[10] in 1958. Based on his theory, in the past few years, a new class of high power and high efficiency microwave tubes[11] (Gyrotron) has emerged and the auroral kilometric radiation has been satisfactorily explained.

The basic energy transfer process in electron cyclotron maser instabilities can be visualized as follows: those electrons being accelerated by the wave gain energy and rotate slower whereas those electrons being decelerated by the wave lose energy and rotate faster; as a result of the electron inertia electrons bunch together azimuthally; if the initial mismatch $\Delta\omega=\omega_0-\omega_{ce}$ is greater than zero, the electron bunch is in the decelerating phase of the electromagnetic wave and a net transfer of enrgy from the electrons to the wave takes place.

3.A. Gyrotron Amplifiers

In order to simulate the performance of gyrotron amplifiers, what is called a stretched one and two halves dimensional code is used. The code is stretched in the sense that the transverse dependence of the electromagnetic field is assumed to be that of the fundamental empty waveguide mode and the transverse position of the electrons which have very small Larmor radius is fixed at the location where the maximum coupling between the electrons and fields takes place. Electrons are injected at the entrance to the interaction region and removed at the exit. An oscillating external current in front of the beam entrance models a matched RF input of arbitrary waveform. The resulting equaitons to be solvd in the circular waveguide configuration become

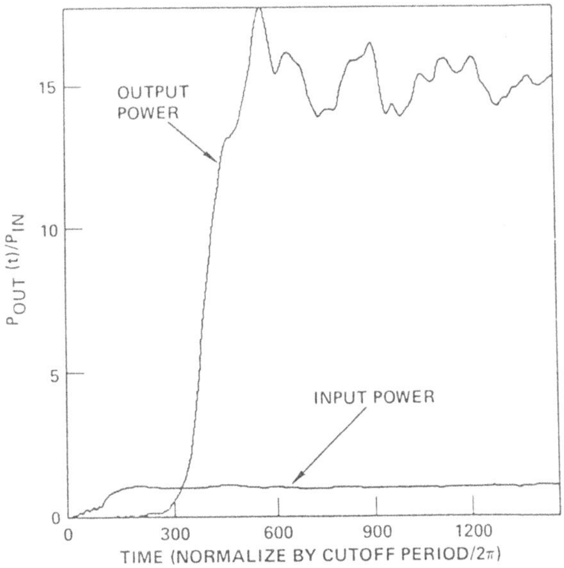

Figure 3. Output and input power monitored at the electron beam exit and entrance by calculating Poynting flux in time.

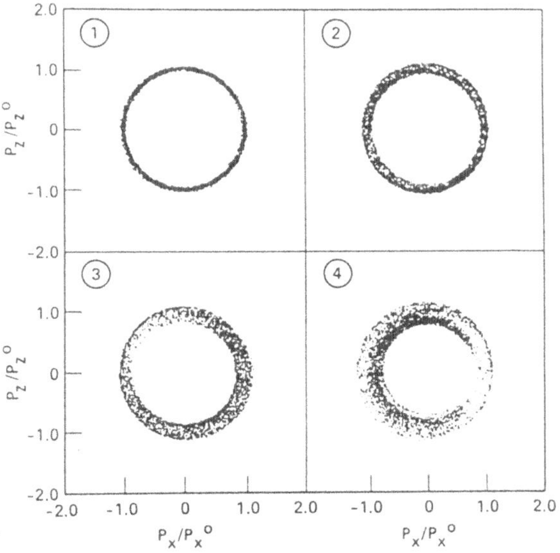

Figure 4. Electron positions in normalized transverse momentum space showing bunching at different axial beam locations going from beam entrance to beam exit in the order as labelled.

$$\frac{\partial B_r}{\partial t} = ick_x E_\phi$$
$$\frac{\partial B_x}{\partial t} = -\omega_{co} E_\phi \qquad (16)$$
$$\frac{\partial E_\phi}{\partial t} = ick_x B_r + \omega_{co} B_x - 4\pi J_\phi$$

where ω_{co} is the cut-off frequency of the waveguide. Outside the beam region the radiation is absorbed by using the resistive layer algorithm described in the previous section.

Simulation results for a 70 kev, $V_\perp/V_\parallel=1.5$, and 1 amp electron beam coupled to TE_{01} mode of a circular waveguide with radius Ro=0.535 cm and length Lo=10 cm under grazing condition will be discussed. A series of simulations was carried out. Figure 3 shows the time evolution of the input (500 watts) and output Poynting flux (power) which clearly indicates the desired amplification process. Figure 4 gives the spatial evolution (4 quarters along the beam) of the transverse momentum which illustrates that the electrons do give up their transverse energy along the interaction region and that phase trapping is the saturation mechanism. By changing the input power and frequency, a comparison of the nonlinear and liner bandwidth can be obtained (Figure 5). If bandwidth is defined as 3dB down, then the linear badwidth (80 watts input) is 4.5% versus a nonlinear bandwidth (500 watts input) of 7.5%.

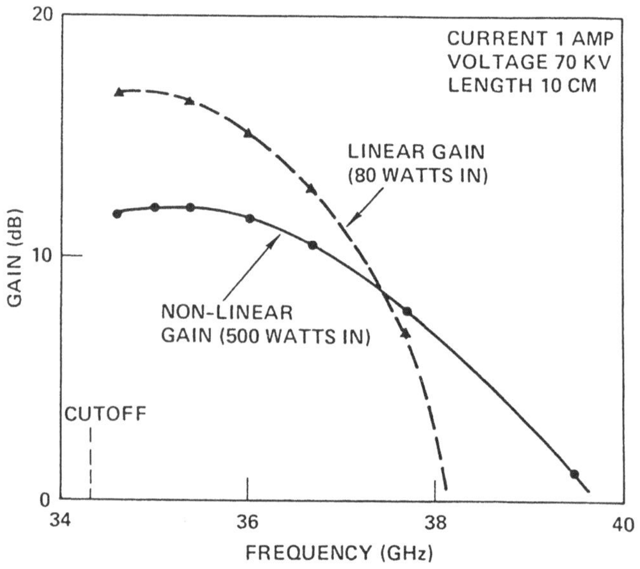

Figure 5. Linear and nonlinear gain versus frequency which determines the band width of an amplifier.

3.B. Auroral Kilometric Radiation

The electron cyclotron maser mechanism has recently been used to interpret the auroral kilometric radiation[12]. The theory predicts direct amplification of fast extraordinary mode radiation around the electron cyclotron frequency which agrees with the observed polarization and frequency of the radiation in the source region. The energy source for the instability comes from the anisotropic loss cone distribuiton of the electron energy which results from the reflection of enrgetic electrons (~1 kev) originated from the plasma-sheet region by the earth mirror magnetic fields.

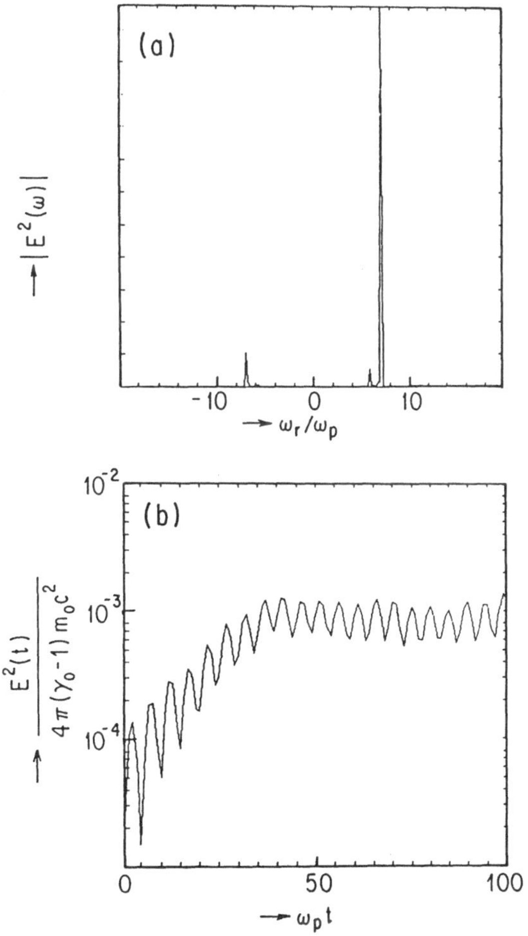

Figure 6. Simulations of electron cyclotron maser instability due to a loss cone distribution, (a) power spectrum, (b) time evolution, for the electronmagnetic field of the most unstable mode.

Most of the theoretical calculations are in the linear region. The only saturation mechanism which has been discussed is the quasi-linear flattening of the distribuiton function at the resonant region. A one and two-halves dimensional relativistic electromagnetic particle code with periodic boundary conditions is used to address the saturation mechanism of the electron cyclotron maser instability with parameters appropriate for the auroral region. The electron distribuiton function is taken to be

$$f_o(P_\perp^2,P_y) = AP_\perp^2 \exp(- \frac{P^2}{\Delta P^2}) \tag{17}$$

where A is a normalizaiton constant such that $\int d^3\vec{P} f_o=1$. $P_\perp$ and P_y are respectively the momentum of the electron perpendicular and parallel to the external magnetic field $B_o\hat{e}_y$. ΔP is a measure of the electron temperature

$$T_e = mc^2 \{ 1 + \frac{(\Delta P)^2}{m^2c^2} \}^{1/2} - mc^2 \tag{18}$$

The electron distribution peaks at $P_\perp^2=(\Delta p)^2$ which gives a population inversion in $P_\perp^2$. The following parameters are used: $\omega_{ce}/\omega_{pe}=7.5$, $m^2c^2/(\Delta p)^2=50$, $\vec{K}\perp\vec{B}_o$, and ions are immobile.

Figure 6a gives the power spectrum of the most unstable extraordinary mode which shows two peaks in the region of $\omega_r>0$. From the cold plasma dispersion relation, there are two frequencies for each kx. The higher frequency mode corresponds to the fast wave ($v_p>c$) and the lower frequency mode corresponds to the slow wave ($v_p<c$). They are separated by the right-hand cutoff frequency. Initially, both modes are excited by the thermal noise. However, only the fast mode can be amplified by the electrons. Figure 6b displays the the time evolution of the most unstable mode which shows the desired exponential growth as well as the beat disturbance between the fast and slow waves. The growth rate determined from the simulation is $\omega_i/\omega_{ce}=6.6\times10^{-3}$ which is very close to the theoretical prediction of 6.7×10^{-3}. The unstable mode is saturated at $\omega_{pe}t=40$ and the conversion efficiency is about 1.5×10^{-3}.

At the time of saturation for the most unstable mode, the electron distribution function (Figure 7) still maintains a small positive slope. It is until $\omega_{pe}=70$ that the loss cone is completely filled by the electron energy diffusion due to the unstable waves. This agrees with the simulation result which shows that the total electromagnetic energy is saturated at $\omega_{pe}t=70$ and reveals that it might not be correct to retain only the most unstable mode in evaluating the energy of the saturated radiation. The change in the electron distribution in time could render some initially stable modes unstable which would also contribute to the total saturated radiation energy.

Before comparing the simulation results with the observational results, a more realistic boundary condition which allows electrons to

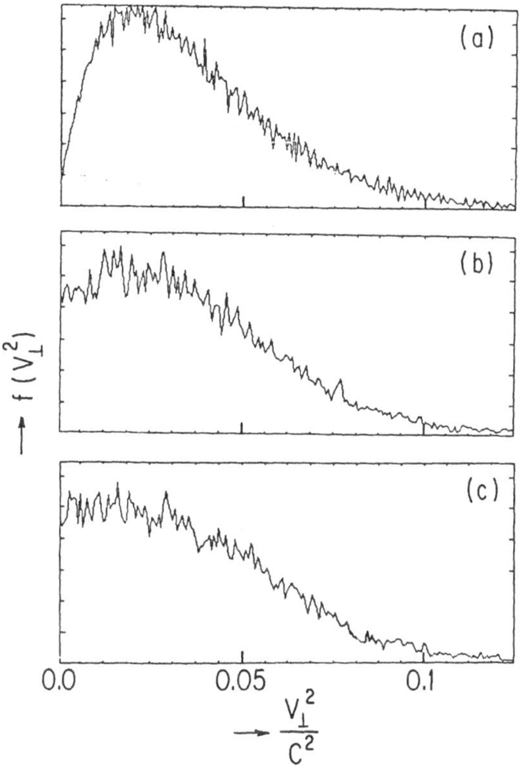

Figure 7. Time evolution of the electron energy distribution at (a) $\omega_{pe}t=0$, (b) $\omega_{pe}t=40$, (c) $\omega_{pe}t=70$.

go out if they are scattered into the loss cone should be implemented. A cold component of electrons originated from the ionosphere should also be included in the distribution function.

Ackowledgments

I wish to thank Mrs. Chih-Chien Lin for her valuable numerical support. This work has been supported by the National Science Foundation under Contract No. ECS81-15653 and PHY 80-26048.

REFERENCES

1. J. E. Rowe, Nonlinear Electron-Wave Interaction Phenomena (Academic Press), 1965.
2. O. Buneman, Phys. Rev., 115, 503, 1959.
3. J. M. Dawson, Phys. Fluids 5, 445, 1962.
4. J. M. Dawson and A. T. Lin, "Particle Simulation," Handbook of Plasma Physics, edited by A. Galeev and R. N. Sudan, North Holland Publishing Co., Amsterdam ,1983.
5. A. B. Langdon and B. F. Lasinski, Methods of Computational Physics, edited by J. Killen (Academic Press), 1976.
6. A. T. Lin, J. M. Dawson, and H. Okuda, Phys. Fluids, 17, 1995, 1974.
7. A. B. Langdon, Ph.D. Thesis, Princeton University, 1969.
8. E. L. Lindman, J. Comput. Phys., 18, 66, 1975.
9. A. T. Lin and Chih-Chien Lin, Phys. Rev. Lett. 47, 981, 1981.
10. R. Q. Twiss, Australian J. Phys., 11, 564, 1958.
11. M. Caplan, A. T. Lin, and K. R. Chu, International J. of Electronics, 53, 659, 1982.
12. C. S. Wu and L. C. Lee, Astrophys. J., 230, 621, 1979.

MODERN DEVELOPMENT IN PARTICLE SIMULATION

J. C. Adam

Centre de Physique Theorique de L'Ecole Polytechnique
Plateau de Palaiseau - 91128 Palaiseau - Cedex - France

ABSTRACT

We review some of the extension that has been made recently to particle codes to allow them to handle low frequency phenomena economically.

1. INTRODUCTION

Over the past few years several attempts have been made to extend the applicability of particle simulation to low frequency phenomena in large systems. Plasma experiments provide many examples of low frequency phenomena in which kinetic effects are important. These examples can be found in inertial confinement experiments ("laser fusion") as well as in magnetic confinement experiments (Tokamac, Mirrors, etc...). The data shown at this school also show that space plasma simulations will also required the use of such models.

The purpose of this paper is to review the diverse approaches that have been used by different authors to allow the use of large time steps in particle simulation. Let us recall that the stability of standard particle models is subject to the constraint $\omega_p\Delta t<1$ due to the use of explicit difference scheme in the equation of motion of particles as well as in the field equation (for the electromagnetic case, which in fact yield also $\omega_o\Delta t<1$, where ω_o is the frequency of the electromagnetic wave). To eliminate this constraint an implicit formulation must be used. This means that the solution is advanced from time $n\Delta t$ to $(n+1)\Delta t$ by backward time differencing from $(n+1)\Delta t$ to $n\Delta t$ instead of forward marching in an explicit scheme.

The organization of this chapter is the following. In the next section we shall review different implicit algorithms because the choice of an efficient one has to be made at the very beginning of the

H. Matsumoto and T. Sato, Computer Simulation of Space Plasmas,

design of an implicit particle code. In section 3 the moment implicit method will be presented and in section 4 we shall present the so called direct approach.

2. IMPLICIT TIME INTEGRATION

This topic has been discussed in details in the literature [Denavit,1981; Cohen et al.,1982; Barnes er al., 1982]. Our presentation will be based on Cohen et al.[1982] which is the most exhaustive.

An implicit time integration scheme may be thought as a low pass filter : one wishes to have an accurate solution for frequencies lower than some value ω_0 and suppress as efficiently as possible modes with frequencies higher than ω_0 because they cannot be accurately described with the time step chosen. In this optic the ideal implicit scheme should be like an ideal low pass filter i.e. a step function. Much in the same way a low pass filter approaches the step function when the number of coefficient is increased, the more past information is used in an implicit algorithm the sharper is the transition between frequencies correctly described and the ones that are damped.

2.1 First order damping implicit scheme

Such scheme have used in Mason[1981] and in Brackwill and Forslund [1982]. Because the scheme of Brackwill and Forslund[1982] is presented in another chapter of this book and at the end of section 3 we shall restrict our presentation to the scheme of Mason [1081]. For particle motion it writes :

$$X^{n+1} - X^n = V^{n+1/2}\Delta t,$$

$$V^{n+1/2} - V^{n-1/2} = \theta\Delta t\alpha^{n+1} + (1-\theta)\Delta t\, \alpha^{*n} \tag{1}$$

$$a^{*n} = \frac{1}{4}(a^{n+1} + 2a^n + a^{n-1})$$

$$0 < \theta < 1$$

Very general technics can be applied to study the stability of this scheme but a "feeling" of its behaviour can be obtained by applying it to the problem of the harmonic oscillator where acceleration and position are related by $\lambda=-\omega_0^2$. Writing all the quantities in (1) as A=exp(-iωt) one can obtain the dispersion relation of this scheme. This dispersion relation can be solved by expansion for $\omega_0\Delta t \ll 1$. It gives :

$$\mathrm{Re}\,\frac{\omega}{\omega_0} = +\left[1 - \frac{1}{24}\,\omega_0\Delta t^2\,[2 + 3\theta(\theta + 1)]\right],$$

$$\mathrm{Im}\,\frac{\omega}{\omega_o} = -\theta\frac{\omega_o\Delta t}{2}$$

For $\theta=0$ this scheme gives phase errors that are twice as large as the standard leapfrog scheme and no damping. The other extreme is $\theta=1$ which correspond to the fully implicit scheme which yields :

$$\mathrm{Re}\,\frac{\omega}{\omega_o} = \pm[1 - \frac{1}{3}(\omega_o\Delta t)^2\,] \quad \mathrm{Im}\,\frac{\omega}{\omega_o} = -\,\frac{\omega_o\Delta t}{2} \tag{3}$$

A change in θ allows a control on the damping rate. It should however be emphasized that $\theta\neq 1$ implies the storage of field quantities at two times level for a dependence of the damping rate linear in $\omega_o\Delta t$. We shall see in the next subsection that with the same information schemes with a stronger dependence on $\omega_o\Delta t$ can be built. This is desirable since it provides stronger damping at large value of $\omega_o\Delta t$ while keeping it low for small value of $\omega_o\Delta t$.

2.2 Third order damping implicit scheme.

In this section we shall see that by keeping information on enough time levels it is possible to design implicit schemes, with transfer functions that approximate more closely the ideal low pass filter than the preceding scheme.

The first class of algorithms we shall consider is called class C scheme in Cohen et al.[1982]. For the equations of motion of particles it can be written:

$$V^{n+1/2} - V^{n-1/2} = \alpha^n$$

$$X^{n+1} - X^n = V^{n+1/2} + C_o\Delta t(a^{n+1} - a^n) + C_1\Delta t^2(a^n - a^{n-1})\ldots \tag{4}$$

Insight in the behaviour of this scheme can be gained as in the preceding section by appling it to the problem of the harmonic oscillator. The roots of the dispersion relation for small values of $\omega_o\Delta t$ are given by :

$$\mathrm{Re}\,\frac{\omega}{\omega_o} = \pm\,(1 + \frac{1}{2}\,(\omega_o\Delta t)^2(\frac{1}{12} - C_o - C_1\,\ldots) + O(\Delta t^3)$$
$$\mathrm{Im}\,\frac{\omega}{\omega_o} = -\frac{1}{2}(\omega_o\Delta t)^3(C_1 + 2C_2 + \ldots) + O(\Delta t^4) \tag{5}$$

At this stage the constants C_o, C_1, C_2 ... have to be chosen to give optimal properties to the scheme. It is obvious from (4) that for $C_o=C_1=C_2=\ldots=0$ one recover the standard explicit leapfrog scheme. The algorithm used by Denavit [1981] for advancing particles is

$$v^{n+1} - v^n = \left(\frac{3}{4}a^{n+1} + \frac{1}{4}a^{n-1} \right)\Delta t$$

$$x^{n+1} - x^n = \left(\frac{3}{4}v^{n+1} + \frac{1}{4}v^{n-1} \right)\Delta t \qquad (6)$$

which can be cast in the form of equations (4) with $C_0=9/16$, $C_1=1/8$, $C_2=1/16$. Then from (5) one has :

$$\mathrm{Re}\,\frac{\omega}{\omega_0} = \pm \left(1 - \frac{(\omega_0\Delta t)^2}{3} + \ldots \right) \qquad \mathrm{Im}\,\frac{\omega}{\omega_0} = - \frac{(\omega_0\Delta t)^3}{8} \qquad (7)$$

Another scheme proposed by Denavit writes :

$$v^{n+1} - v^n = \frac{1}{16} \left(9a^{n+1} + 6a^n + a^{n-1} \right)$$

$$x^{n+1} - x^n = \frac{1}{16} \left(9v^{n+1} + 6v^n + v^{n-1} \right)\Delta t \qquad (8)$$

The corresponding coefficient are $C_0=81/256$, $C_1=14/256$. $C_2=1/256$ and (5) gives

$$\mathrm{Re}\,\frac{\omega}{\omega_0} = \pm \left(1 - \frac{7}{48} (\omega_0\Delta t)^2 \right); \quad \mathrm{Im}\,\frac{\omega}{\omega_0} = - \left(\frac{\omega_0\Delta t}{32} \right)^3 \qquad (9)$$

It is instructive to compare the phase properties and damping properties of (6) and (8). In the second scheme the error on the phase (i.e. the real part of ω) has been reduced by almost a factor 2, but simultaneously the damping of unwanted frequencies has been reduced by four.

In Denavit[1981] another scheme is proposed called "C_1 optimized" scheme which is obtained by maximizing the damping of the least damped mode of the dispersion relation of (1) subject to the constraint C_s ($s \geqslant 2$)=0. One then find that for $C_0=0.302$, $C_1=0.04$:

$$\mathrm{Re}\,\frac{\omega}{\omega_0} = \pm \left(1 - 0.13 (\omega_0\Delta t)^2 \right) \qquad \mathrm{Im}\,\frac{\omega}{\omega_0} = -0.02(\omega\Delta t)^3) \qquad (10)$$

which is not very different of (9) for less storage of information. A last scheme proposed in Denavit[1981] is the "D_1 scheme" which is based on a second order stiffly stable scheme given by Gear[19**] for first order differential equation. It writes :

$$x^{n+1} - x^n = v^{n+1/2}\Delta t$$

$$v^{n+1/2} - v^{n-1/2} = 1/2a^{n+1}\Delta t + 1/2(v^{n-1/2} - v^{n-3/2}). \qquad (11)$$

Solving the dispersion the dispersion relation for a harmonic oscillation as precedingly yields :

$$\mathrm{Re}\left(\frac{\omega}{\omega_0}\right) = \pm\left(1 - \frac{11}{24}(\omega_0\Delta t)^2\right); \quad \mathrm{Im}\left(\frac{\omega}{\omega_0}\right) = 1/2\,(\omega_0\Delta t^3) \tag{12}$$

Comparison of (12) and (7) shows that the damping rate of high frequencies ($\omega_0\Delta t << 1$) in this scheme is 50 times larger than for the optimized C_1 scheme. However the phase error for ($\omega_0\Delta t << 1$) is also multiply by 4 and becomes much larger than in the case of the standard explicit leapfrog scheme. This means that if one limits the phase error to some maximum value for a given frequency of interest a time step smaller by a factor 2 has to be used. This in turn reduces the damping by 8. Even under this constraint the D_1 scheme remains superior to C_1 scheme but then it becomes twice as expensive.

A recursive formulation of (11) given in Barnes[1982] makes it very attractive in term of storage requirement. It writes:

$$X^{n+1} = X^n + V^{n+1/2}\Delta t$$

$$V^{n+1/2} = V^{n-1/2} + 1/2\bar{a}_n$$

$$\bar{a}^n = 1/2\,(a^{n+1} + a^{n-1}) \tag{13}$$

The equivalence of (11) and (13) can be readily shown by factoring equation (15a) of Denavit[1981] into two equations (A.B. Langdon private communication);

- the first one corresponds to a low pass filtering, it is the introduction of $\bar{a}$
- a leapfrog step define by the two first of (13).

This scheme is probably one of the most attractive in term of global behaviour (i.e. : phase error, damping, storage requirement)

The schemes that have been described in this section are not directly applicable to a particle simulation. A fully self-consistent implicit particle code would require solving iteratively the system of equations describing the N particles plus the field equations. Such an algorithm is prohibited by storage requirement as well as cost of going iteratively through particle data. Methods have to be devised that uncoupled implicit equations for particle and implicit equations for field.

3. MOMENT METHOD

3.1 Electrostatic case

The essence of the method outlined in Mason[1981] and Cohen et al.[1982] is that the microscopic (particle) and macroscopic (moment equations) representation of plasmas should not differ significantly over a time because field equations depend only on the two first moment i.e., density and current. The method devised by R. Mason uses the first order damping described in the preceding section : Fluid equations are first integrated using :

$$\tilde{J}_\alpha^{n+1/2} = \tilde{J}_\alpha^{n-1/2} - \frac{1}{m_\alpha}\frac{\partial P_\alpha}{\partial x} - q_\alpha n_\alpha^n E^* \Delta t \qquad (14)$$

$$\tilde{n}_\alpha^{n+1} = n_\alpha^n - \frac{\partial \tilde{J}_\alpha^{n+1/2}}{\partial x}\Delta t \qquad (15)$$

α denotes the particles specifies. E^* is the same linear combination as (1). J_α and P_α are obtained from the particle data by :

$$J_\alpha = \Sigma_p V_{p,\alpha} \qquad P_\alpha = m\Sigma_{\alpha p} V_{p,\alpha}^2$$

Poisson equation is then integrated using the predicted value of $\tilde{n}^{n+1}$. Note however that the integration of (14) necessitates the knowledge of E^* which is a function of E^{n+1} which in turn is a function of $\tilde{n}^{n+1}$

$$\tilde{E}^{n+1} = 4\pi\int_0^x \Sigma_\alpha q_\alpha \tilde{n}_\alpha^{n+1} dx + E^{n+1}(0) \qquad (16)$$

Substitution of (14, 15) into (16) yields an equation for E^{n+1} which can be explicitly in 1D (see Mason[1981]). In a 2D system, (16) must remain in differential form yielding an equation of the following type :

$$\nabla\tilde{E}^{n+1} = f(\, n^n, J^n, E^n, E^{n-1}\,) + g(\tilde{E}^{n+1}) \qquad (17)$$

In finite difference representation (17) is a matrix equation as the usual Poisson equation but its coefficients are no longer constant. A method for solving that kind of equation is described in Brackwill and Forslund [1982] for the same algorithm.

It should also be pointed that a difficulty arise due to the pressure gradient term. The only quantity that one can build explicitly at the end of a time step is obtained from particle data with positions define at level n and velocities at level n-1/2. This explicit pressure term introduces a stability condition $\Delta t < \Delta x/V_T$ where V_T is thermal velocity of the plasma. In Mason [1981] this difficulty is alleviated by the use of a predicted value of the pressure term obtained by a local adiabatic approximation in the following way :

- $n_e^{n-1/2}$ is accumulated at half time step
- n^* is built as $n_i(n) - (1/4\pi e)(\partial E^*/\partial x)$ then :
- $P_\alpha = P^{n-1/2}(n^*/n_e^{n-1/2})^3$

this pressure is then used to integrate equation (16) and an iteration over E^* is necessary. In Brackwill and Forslund [1982] a fully implicit pressure term define as $p=n^{n+1}T^n$ is used for stability.

After the E^{n+1} field has been obtained particles are advanced using Eq. (1) and a new cycle initiated.

Denavit [1981] approach to the same problem is slightly different. One start again with fluid equations

$$\frac{\partial n_\alpha}{\partial t} = -\nabla(\, n_\alpha U_\alpha \,) \tag{17a}$$

$$\frac{\partial}{\partial t}(\, n_\alpha U_\alpha \,) = \frac{q_\alpha}{m_\alpha} n_\alpha E - \nabla\cdot P_\alpha \tag{17b}$$

$$\nabla\cdot E = 4\pi\Sigma_\alpha\, n_\alpha\, e_\alpha \tag{17c}$$

In 1D (17a) and (17c) can be combine to give

$$\frac{\partial E}{\partial t} = 4\pi e \sum_\alpha n_\alpha\, u_\alpha\, e_\alpha \tag{18}$$

the algorithm define by equation (6) is then applied to the resolution of the system of equations (17) and (18). It becomes :

$$(\, n_\alpha\, u_\alpha \,)^{n+1} = n_\alpha\, u_\alpha^n + \frac{3\Delta t}{4}(\, \delta_\alpha n_\alpha\, E + F_\alpha \,)^{n+1} + \frac{\Delta t}{H}(\, \delta_\alpha\, n_\alpha\, E + F_\alpha \,)^{n-1} \tag{19a}$$

$$E^{n+1} = E^n + \frac{3}{4}\Delta t\; \Sigma\; n_\alpha u_\alpha e^{n+1} - \frac{\Delta t}{H}\; \Sigma\; n_\alpha u_\alpha e_\alpha^{n-1} \tag{19b}$$

where $F_\alpha = -\nabla\cdot P_\alpha$ and δ_α depends on the charge and mass of each particle specie.

From (19a) and (19b) an explicit expression for the predictor field at E^{n+1} is obtained that depend only on fluid quantities at level n and n-1 except for the pressure term which appear at time level (n+1). This imply that an iterative procedure has to be apply to obtain the field. Fluid quantities are obtained as in Mason method from the particle including the pressure so that the iteration must involve the particle. At each time step the value needed to start the iteration are the ones obtained at the preceding time level. At the end of the iteration Poisson equation is solved to insure error in the resolution of equation (18) do not accumulate over time. According to Denavit the convergence of the scheme is very rapid and a time step can very often be used without iteration. However the only way to ascertain this point is to do at least two iterations. This scheme should be more accurate than the preceding one for the following reasons:

- It uses a third damping order time integration scheme.
- Pressure terms are collected at each level of iteration from particle data (Note that here also the pressure is implicit for stability).

On the side of the drawbacks :

- One has to iterate through particle data which may be expensive.
- Many quantities must be stored at two times levels which is very expensive in term of memory requirement and probably prohibit the use of this scheme in 2D.

3.2 Electromagnetic case.

This topic being developed in another chapter of this book we quote it here only for completeness. The only code that has been built is the one of Brackbill and Forslund use a first order implicit scheme for Maxwell equations which writes

$$B^{n+1} - B^{n} = - c(\nabla X E^{n+\theta})\Delta t$$

$$E^{n+1} - E^{n} = c(\nabla X B^{n+\theta})\Delta t - 4\pi J^{n+1/2}\Delta t$$

$$\nabla \cdot E^{n+\theta} = 4\pi N^{n+\theta}$$

$$\nabla \cdot B^{n} = 0 \qquad (20)$$

$E^{n+\theta}$ is defined by : $E^{n+\theta}=\theta E^{n+1}+(1-\theta)E^{n}$ and particles an advanced using

$$X^{n+1} = X^{n} + V^{n+\theta}\Delta t$$

$$V^{n+1} = V^{n} \pm \frac{q}{m} [E^{n+\theta} + \frac{V^{n+1/2} \times B^{n}}{\Delta t}]\Delta t \qquad (21)$$

Plasma density at level $N^{n+\theta}$ and current density at time $J^{n+1/2}$ are obtained from momentum equation according to the scheme

$$N^{n+\theta} = N^{n} - \nabla \cdot J^{n+\Gamma}(\theta\Delta t)$$

$$J^{n+1/2} = J^{n} - \nabla \cdot \frac{J^{n+1/2}\ J^{n}}{N^{n}} (\Gamma\Delta t)$$

$$+ \frac{q}{m} \quad N^{n+1/2}\ E^{n+\theta} + \frac{J^{n+1/2} \times B^{n}}{c} \quad \Gamma\Delta t$$

$$- q_{s}\ \nabla \cdot \tilde{P}(\Gamma\Delta t) \qquad (22)$$

The density N^{n} and the current J^{n} are collected from particle according to the standard definition

$$N^{(n)}(x) = q_{p}\Sigma_{p}\ h(X - x_{p}^{n})$$

$$J^{(n+1/2)}(x) = q_{p}\Sigma_{p}V^{n+1/2}\ h(X - x_{p}^{n+1/2})$$

where the summation extends over all the particles p and h is the particle shape function.

The pressure term that appears in equation (22) has to be evaluated implicitly for nonlinear stability. The choice made is $P^{n+1} = N^{n+1}T^{n}$. The system of equations defined by (20) and (22) is solved iteratively and once convergence as been obtained particles are advanced using (21). Extensive discussion of this algorithm is given in Brackwill and Forslund [1982].

4 DIRECT METHOD

4.1 Electrostatic case

Another kind of approach has been used in Cohen et al. [1982], Friedman et al.[1981] and Langdon et al. [1981]. The method is called direct because moment equations are not used as auxiliary equations. In particular the kinetic stress tensor is not collected. The potential cost of the algorithm is that it can be necessary to iterate through the particles. The algorithm can be described as follows. One start with the D_1 scheme of section 1 for particle motion

$$V^{n+1/2} - V^{n-1/2} = 1/2\ a^{n+1}\Delta t - 1/2\ (V^{n-1/2} - V^{n-3/2}) \qquad (23)$$

The value X^{n+1} of equation (23) can be written as

$$X^{n+1} = 1/2\ a_{n+1}\Delta t^2 + X^{n+1(0)}$$

$$X^{n+1(0)} = X^{n} + V^{n-1/2}\Delta t - 1/2\ (V^{n-1/2} - V^{n-3/2}) \qquad (24)$$

i.e. as the position obtained from the free streaming motion plus a displacement due to the acceleration. The basis of the direct method is:

1) Compute the charge density $\rho^{(0)}$ obtained from free streaming motion of particle. Since it depends only on quantities defined at t=n or earlier, it is known,

2) Evaluate the change $\delta\rho$ due to the displacement of particle from particle from $X^{n+1(0)}$ to X^{n+1} i.e. $\delta X=(X^{n+1(0)}-X^{n+1})$. As shown in Langdon [1979], this increment $\delta\rho$ is

$$\delta\rho = - \nabla\cdot[\ \rho^{(0)}_{n+1}(x)\ \delta x(x)] \qquad (25)$$

3) Assume that δx is the same for all the particle having the same $X^{n+1(0)}$. This means the electric field is assumed to be sufficiently uniform spatially to insure that dispersion in velocity is unimportant. Then

$$\delta X \sim 1/2\ \frac{q}{m}\ E^{n+1}(x)\Delta t^2 \qquad (26)$$

4) Replace (26) into (25) to obtain $\delta\rho$

$$\delta\rho = -\nabla[\ \frac{q}{2m}\ \rho^{(0),n+1}(x)\Delta t^2 E^{n+1}(x)\ ]$$

$$= -\nabla[\ \omega_p^2(x)^{(0),n+1}\ \frac{\Delta t^2}{2}\ E^{n+1}(x)\ ] \tag{27}$$

5) Solve the Poisson equation at time (n+1)

$$E^{n+1}(x) = \rho^{n+1}(x) = \rho^{(0),n+1} + \delta\rho$$

$$= \rho^{(0),n+1}(x) - \nabla(\ \omega_p^2\ (x)\ \frac{\Delta t^2}{2}\ E^{n+1}(x)$$

$$\nabla[\ 1 + \frac{\omega_p^2\ (x)\Delta t^2}{2}\ ]E^{n+1}(x) = \rho^{(0),n+1} \tag{28}$$

the term $1+(\omega_p^2\Delta t^2)/2$ in equation (28) acts as an effective suceptibility which screens the high frequencies. In principle an iterative refinement of the value of the electric field obtained from equation (28) is possible simply by improving the value of the guess $\rho^{(0)}$ (see Langdon et al., 1982].).

At this stage it becomes easy to explain the relationship between the direct and the moment approaches. This relationship has been observed by Mason by eliminating $\mathcal{J}^{n+1/2}$ between Eq.15b and 15a and replacing E^* by E^n. Poisson equation then becomes

$$\nabla\cdot E_{n+1} + \frac{q}{m}\ \Delta t^2\ \nabla\cdot(\rho_n E_{n+1}) = \rho_n - \Delta t\nabla\cdot(\ J_{n-1/2} - \frac{q\Delta t}{m}\nabla P_n\ ) \tag{29}$$

the right hand side of this equation is obtained by setting $E^* = 0$ in (14) and therefore correspond to the $\rho^{n+1(o)}$ approximation of the direct method.

While the direct approach is conceptually very simple its practical implementation is very sensitive to the details of the difference schemes, so that it is useful to reformulate the basic algorithm in term of finite size particle.

The density in each cell is defined as

$$\rho_J^{n+1} = \frac{q}{\Delta X}\ \Sigma_p S(\ X_p^{n+1} - X_j\ ) \quad . \tag{30}$$

where p is a summation over particles. The free streaming approximation to X_p^{n+1} is

$$X^{n+1(0)} = X_p^n + V_p^n\Delta t\ .$$

then we may write

$$S(\ X_p^{n+1} + X_j\ ) = S(\ X_p^{n+1,(0)} - X_j\ ) + (\ X_p^{n+1} + X_p^{n+1,(0)}\ )\ \frac{\partial S}{\partial \tilde{X}_k} \tag{31}$$

Note that for a linear spline this expansion is exact as long as the particle does not change of cell during the iteration.

Finally one must evaluate $(X_p^{n+1}-X_p^{n+1,(0)})$ which is done using:

$$X_p^{n+1} - X_p^{n+1,(0)} = \frac{q}{m}\Delta t^2 E^{n+1}(X_p^{n+1}) \tag{32a}$$

$$\sim \frac{q}{m}\Delta t^2 \Sigma_i S(\, X_p^{n+1(0)} - X_i \,)\, E_i^{(n+1)} \tag{32b}$$

Note that (32b) is a good approximation of (32a) only if $qE\Delta t^2/mL=\omega_b^2\Delta t^2 \ll 1$, where L is a typical scale length and ω_p is the bounce frequency in the E field for the corresponding scale length. From (30), (31) and (32) one can finally obtain the finite difference form of Poisson equation which writes

$$(\, \phi_{j+1} - 2\phi_j + \phi_{j-1} \,)^{n+1} = \rho_j^{(0)n+1} - \sum_i W_{ij}(\, \phi_{i+1} - \phi_{i-1} \,)/2\Delta X \tag{33}$$

where $$\rho_j^{(0),n+1} = \frac{q}{\Delta X}\sum_p S(\, X_p^{(0),n+1} - X_j \,)$$

and

$$W_{ij} = \frac{q^2\Delta t^2}{m\Delta X} \sum_p S(\, X_p^{(0),n+1} - X_i \,) \frac{\partial S(X_p^{0,n+1} - X_i)}{\partial X_p} \tag{34}$$

For a linear shape function W_{ij} is zero if $|i-j|>1$ and since $\partial S/\partial X_p$ is $\pm(1/\Delta X)W_{ij}$ is identical to $\sum_p S(X_p^{(0),n+1})$ within a numerical factor. This means that $\rho^{(0),n+1}$ and W_{ij} can be collected simultaneously by no more operations than in a standard particle code. However the resolution of Poisson equation becomes more complicated because (33) is no longer a tridiagonal system with constant coefficient.

4.2 Electromagnetic case.

An extension of the direct method to the electromagnetic case has been suggested by A. B. Langdon [1983] at the 10th numerical simulation meeting. It used D1 scheme for both particles and fields. Equations for particles in the electromagnetic case write

$$X^{n+1} - X^n = V^{n+1/2}\Delta t,$$

$$V^{n+1/2} - V^{n-1/2} = \bar{a}^n\Delta t + (\, V^{n+1/2} + V^{n-1/2} \,) \times \frac{qB^2\Delta t}{2mc} \,, \tag{35}$$

$$\bar{a}_n = 1/2\bar{a}^{n-1} + \frac{q}{m} E^{n+1}(X^{n+1}) \,.$$

$$X_{n+1} - X_n = V^{n+1/2}\Delta t$$

$$V^{n+1/2} - V^{n-1/2} = \bar{a}_n^n \Delta t + (V^{n+1/2} + V^{n-1/2}) \times \frac{qB^2 \Delta t}{2mc} \,.$$

$$\bar{a}^n = 1/2\bar{a}^{n-1} + \frac{q}{m} E^{n+1}(X^{n+1}) \,.$$

An approximation $X^{(0),n+1}$ $V^{(0),n+1/2}$ can be obtained for example by

$$X^{(0),n+1} - X^n = V^{(0)n+1/2}\Delta t$$

$$V^{(0),n+1/2} - V^{n-1/2} = 1/2\bar{a}^{n-1}$$

The current at $J^{n+1/2}$ can then be split into two contribution

$$J^{n+1/2} = J^{(0),n+1/2} + \delta J$$

with

$$J^{n+1/2}(X) = \sum_p q_p V_p^{n+1/2} S(X - X_p^{n+1/2})$$

$$J^{(0),n+1/2}(X) = \sum q_p \; V^{(0),n+1/2} S(X - X_p^{(0),n+1/2})$$

δJ can then be evaluated as a function of known quantities and of the value of the electric field at level E^{n+1} by expanding the shape function around $X^{(0),n+1/2}$. The simplest differencing scheme (which assumes again that the displacement $\delta X = (X_p^{n+1/2} - X_p^{(0),n+1/2})$ is the same for all the particle in one cell yield

$$\delta J^{n+1/2} = (1/4\rho^{(0)}\frac{q}{m} \Delta t) \quad (I + R)\cdot E^{n+1}$$

$$+ \frac{1}{8}\frac{q}{m} (\Delta t)^2 \nabla \times J^{(0)n+1/2} \times (I + R)\cdot E^{n+1}$$

I and R will be defined later. Note that as in Brackbill and Forslund scheme the magnetic term has been kept centered. This is essential for good energy conservation. Equation for the fields are:

$$c(\nabla \times B^{n+1/2}) = J^{n+1/2} + \frac{E^{n+1} - E^n}{t}$$

$$c(\nabla \times E^n) = \frac{B^{n+1/2} - B^{n-1/2}}{\Delta t}$$

$$E^n = \frac{1}{2} (E^{n+1} = E^{n-1})$$

B^n necessary to integrate equations (35) is obtained by extrapolation. The spirit is the same as for the electrostatic case one splits the current necessary in (36) into two parts

$$J^{n+1/2} = J^{(0),n+1/2} + \delta J \tag{37}$$

The contribution $J^{(0),n+1/2}$ is obtained by an extrapolation procedure i.e.,

$$J^{(0),n+1/2} = \sum_p q_p \, V^{(0),n+1/2} \, S(X - X_p^{(0),n+1/2})$$

where $X^{(0)}$ and $V^{(0)}$ are obtained from

$$X^{(0),n+1} - X^n = V^{(0),n+1/2}\Delta t$$

$$V^{(0),n+1/2} - V^{n-1/2} = 1/2\bar{a}^{n-1}$$

When $J^{n+1/2}$ is defined by

$$J^{n+1/2} = \Sigma_p \, q_p \, V_p^{n+1/2} \, S(X - X_p^{n+1/2}) \quad .$$

δJ is obtained by including in it every thing that is not in $J^{(0),n+1/2}$. This is done by writing

$$V^{n+1/2} \quad \text{as} \quad V^{(0),n+1/2} + \delta V$$

$$X^{n+1/2} \quad \text{as} \quad X^{(0),n+1/2} + \delta X$$

and carrying out the necessary expansions. The simplest differencing scheme (which again assumes that the displacement $\delta X = (X_p^{n+1/2} - X_p^{(0),n+1/2})$ is uniform for all te particle within one cell yields

$$\delta J^{n+1/2} = (\ 1/4\rho^{(0)}\frac{q}{m} \ \Delta t \ )(\ I + R \ )\cdot E^{n+1} \tag{38}$$

$$- \frac{1}{8}\frac{q}{m} \ (\Delta t)^2 \nabla \times J^{(0),n+1/2} \times (\ I + R \ )\cdot E^{n+1}$$

I is an identity matrix, while R is a rotation matrix by $-qB^n\Delta t/mc$. When (37) is substitute in (36) with this expression for δJ one obtains a closed system for E^{n+1} and $B^{n+1/2}$. The scheme then proceed in the same way as for the electrostatic case.

5. CONCLUSIONS

We hope to have shown that there are lots of possibilities that have been suggested to overcome the time step limitation imposed by the standard particles method. There is still a great deal of effort needed to make some of the approaches practical, but usually when a code is written it becomes difficult to change it, so that the choice of the algorithm deserve some reflexion....

REFERENCES

DENAVIT, J., J. Comput. Phys., 42, 337, 1981.

Cohen, B.,I., LANGDON, A. B., FRIEDMAN, A., J. Comput. Phys., 46, 15, 1982.

BARNES, D. C., KAMIMURA, T., LEBOEUF, J. N., TAJIMA, T., Institute for fusion studies internal report IFSR # 68, 1981.

Gear, C. W., Numerical Initial Value Problems in Ordinary differential equations (Prentice Hall).

MASON, R. J., J. Comput. Phys., 41, 233, 1981.

BRACKBILL, J. V. and FORSLUND, D.,W., J. Comput. Phys., 46, 271, 1982.

FRIEDMAN, A., LANGDON, A. B., COHEN, B. I., Comments on Plasma Physics and control Fusion 6, 225, 1981.

LANGDON, A. B., COHEN, B. I., FRIEDMAN, A., Lawren Livermore Laboratory Univ. of California, Report UCRL 86350, 1982.

LANGDON, A. B., in proceeding of the 10th conference on Numerical of Plasma 1983.

LANGDON, A. B., J. Comput. Phys., 30, 202, 1979.

PART II

MHD SIMULATIONS

PRINCIPLES OF MAGNETOHYDRODYNAMIC SIMULATION IN SPACE PLASMAS

Tetsuya Sato

Institute for Fusion Theory, Hiroshima University, Hiroshima 730, Japan

ABSTRACT

The aim of this article is not to give a tutorial explanation of either the numerical technique or the numerical procedure, but to place a special emphasis on the philosophical as well as physical principles which are essential in the establishment of the magnetohydrodynamic (MHD) simulation study in the solar terrestrial plasma research. Taking the limits of the abilities of the present-day computers for granted, we emphasize the importance of the "local" MHD simulation. Based on the physical insight as well as the observational facts, we divide the solar terrestrial plasma space into several elementary regions where a macroscopic elementary energy conversion process would take place. The local MHD simulation is defined as a self-contained MHD simulation in each elementary region. Differences in the roles between the local and present-day global MHD simulations are briefly discussed. The importance and difficulties of the boundary condition are also discussed in some detail. Finally, a couple of notes on the finite-difference method and the importance of the diagnostics are pointed out.

1 INTRODUCTION

Energy flows unidirectionally and irreversibly with time throughout the Universe. If one wishes to simulate a particular phenomenon which was observed in nature, one must start, in a strict sense, with the Big-Bang of the Universe. As a matter of fact, this is impossible and preposterous.

Physics assumes that any phenomenon has its own cause and that its occurrence is predictable by a universal law. If one could trace back the causality eighteen billions years ago, one would arrive at

H. Matsumoto and T. Sato, Computer Simulation of Space Plasmas,

the Big-Bang. The branches evolving from the Big-Bang are generally so complex and tangled. Closely looking at a particular branch, however, it sometimes happens clearly branched out from the trunk. In such a case, one can regard the branch as being independent and forget the energy feedback to the trunk from which branched out. One thus can assume a unidirectional energy supply from the trunk and regard it just as the input source. The evolution of the branch can then be treated as an initial-boundary value problem. The solar terrestrial plasma physics, or space plasma physics, is one such branch in which the solar wind is assumed as the input source.

The scientific goal of the solar terrestrial plasma physics is to elucidate the physical processes governing the energy flow and conversion, microscopic and macroscopic, in the solar terrestrial plasma environment.

The microscopic, or kinetic, approach deals with the energy conversion process between the particles and (resonant) waves. Local acceleration of electrons or ions, generation of anomalous nature of the medium such as the anomalous resistivity and viscosity, and emission of radio waves are the results of microscopic processes.

On the other hand, the macroscopic, or magnetohydrodynamic (MHD), approach deals with the macroscopic energy conversion governing the global configuration and energy state such as the magnetotail formation and the magnetospheric substorm. The macroscopic energy state consists of the magnetic energy, the bulk flow (kinetic) energy and the thermal energy (pressure). It can be said that the macroscopic approach is to deal with the physical process involved in the energy conversion among the magnetic, kinetic and thermal energies.

When a drastic change appears in the macroscopic energy state, it is usual that any microscopic process participates in it. It is also true, on the other hand, that a drastic microscopic energy conversion is often preceded by a macroscopic energy conversion. For example, magnetic reconnection can sometimes lead to an explosive conversion from the magnetic to bulk flow energy. But for this to happen, an anomalous resistivity is required. For auroral particles to be accelerated, on the other hand, any macroscopic energy conversion, e.g., generation of field-aligned currents, is required. Strictly speaking, therefore, one must take into account simultaneously the microscopic and macroscopic processes for the thorough understanding of the solar terrestrial plasma physics.

The time and spatial scales of the microscopic and macroscopic processes, however, are several orders or more different. For example, let us take n(density)=$5\times10^5 m^{-3}$, B(magnetic field)=2×10^{-8} tesla, T_e(electron temperature)=500 eV and T_p(proton temperature)=5 keV, which are typical in the magnetotail. Then, the typical microscopic time scales and spatial scales are given by

f_{pe}^{-1} (electron plasma period) $\simeq 2\times10^{-4}$ sec

f_{cp}^{-1} (proton gyroperiod) $\simeq 3$ sec

λ_{De} (electron Debye length) $\simeq 500$ m

r_p (proton gyroradius) $\simeq 500$ km

The typical macroscopic scales, on the other hand, are given by

L (characteristic MHD length) $\simeq 30{,}000$ km

τ_A (Alfvén transit time) $\simeq 30$ sec

where $\tau_A = L/V_A$ (V_A is the Alfven speed). Thus, we have $L/\lambda_{De} = 6\times10^4$, $L/r_p \simeq 150$, $\tau_A f_{pe} \simeq 1.5\times10^5$ and $\tau_A f_{cp} \simeq 10$.

This indicates that it is almost impossible to cook the microscopic and macroscopic processes simultaneously in the same oven. The large separation of the time and spatial scales, however, suggests us to cook them separately and mix them when we eat.

The particle simulation, which is described in other parts of this book, is the best cooking tool for the microscopic process, while the magnetohydrodynamic (MHD) simulation, which is the subject in this chapter, is best suited for cooking the macroscopic process.

2. MHD EQUATIONS

Since the majority inhabitants in the space of our interest are plasma particles, their behaviours can be described by the equations of motion

$$m_j \frac{d^2 \underset{\sim}{x}_j}{dt^2} = e_j (\underset{\sim}{E} + \underset{\sim}{v}_j \times \underset{\sim}{B}) \qquad (1)$$

where m, e, $\underset{\sim}{x}$, $\underset{\sim}{v}$, $\underset{\sim}{E}$ and $\underset{\sim}{B}$ are the mass, charge, position, velocity, electric field and magnetic field, respectively and the subscript j specifies each particle.

Eq.(1) states that the behaviours of all particles are solvable if the electric and magnetic fields are known. However, since the plasma particles carry charges, their positions and velocities change the fields so that all particles strongly interact with each other. In other words, our problem is a typical many-body problem and a highly nonlinear problem.

The fields and the particle motions are connected by the following Maxwell equations

$$\underset{\sim}{\nabla} \cdot \underset{\sim}{E} = \frac{1}{\varepsilon_0} \sum_j e_j \, \delta(\underset{\sim}{x} - \underset{\sim}{x}_j) \qquad (2)$$

$$\nabla \times \mathbf{B} = \varepsilon_0 \frac{\partial \mathbf{E}}{\partial t} + \mu_0 \Sigma e_j \mathbf{v}_j \delta(\mathbf{x} - \mathbf{x}_j) \tag{3}$$

where μ_0 and ε_0 are the permeability and the dielectric constant in vacuum, respectively; $\delta(x)$ is the delta function.

Eqs. (1)-(3) constitute the fundamental equations governing the microscopic and macroscopic processes in our solar terrestrial plasma environment. The basic concept of the particle simulation is to solve this set of equations. As we described in the introduction, however, it is not clever to attempt to solve directly these equations for the time scale the magnetospheric body trembles. It is wanted to derive some practical equations which can adequately describe the macroscopic processes.

The shortest way to this end is to smear out the bumpyness of the fields reflecting the particle discreteness. The conventional way relies on a distribution function. The complete statistical description is given in terms of a many-body distribution function which satisfies the Liouville equation. In a plasma where the particle population is very rare compared with a liquid or solid, the particle interactions are long-range and the interparticle force can be approximated by the average force which does not depend on the exact positions and velocities of all the particles. In this case the many-body Liouville equation can be reduced to the one-body Liouville equation which is called the Vlasov or collisionless Boltzmann equation [see, any text book of plasma physics, for instance, Montgomery and Tidman, 1964]. The Vlasov equation is given by

$$\frac{\partial f_\alpha}{\partial t} + \mathbf{v} \cdot \frac{\partial f_\alpha}{\partial \mathbf{x}} + \frac{e_\alpha}{m_\alpha} (\mathbf{E} + \mathbf{v} \times \mathbf{B}) \cdot \frac{\partial f_\alpha}{\partial \mathbf{v}} = 0 \tag{4}$$

where the subscript α represents the species of the particles, and $\mathbf{E}$ and $\mathbf{B}$ are the sum of the external fields and the average fields due to long-range interparticle interactions. these fields must satisfy the Maxwell equations.

Although the Vlasov-Maxwell equations exclude some subtleties due to particle discreteness, they still contain the long-range particle interactions. To remove the particle interactions, it is needed to further reduce the equations so that they can be expressed in terms only of the macroscopic variables.

For this purpose we obtain moment equations by integrating the Vlasov equation over all velocity space. The zero moment equation gives the mass conservation, or continuity, equation and the first moment equation gives the momentum conservation equation, or simply, equation of motion. The second moment equation leads to the energy conservation equation. Since usually the third moment appears in the energy conservation equation, it is required to derive higher and higher moment equations for completeness. However, we usually ignore the third moment appearing in the energy equation and use the equation of state, thus obtaining a closed set of moment equations.

The obtained equations, however, still contain separate equations for electrons and ions. Thus, we call them multi-fluid equations. Unfortunately, the multi-fluid equations are not so useful as expected for solving the macroscopic energy conversion processes. This is because the electron-fluid equation still contains the electron plasma oscillation which is the typical microscopic process. In other words, the multi-fluid equations can get rid of the resonant interactions between particles and plasma waves, notably, the Landau damping, but still contain almost all fundamental particle characteristics such as the plasma frequency and cyclotron frequency. In this sense, the multi-fluid equations appear less practical than the Vlasov equation.

For the description of macroscopic processes, it is needed to remove the particle characteristics contained in the multi-fluid equations. We already have learned that the plasma oscillation is the evil which does not release us from the particle aspect. To remove this evil from the multi-fluid equations, we must rely on the physical nature that the quasi-neutrality be satisfied for the charge. This approximation enables us to remove all particle characteristics and get one-fluid equations which we call the magnetohydrodynamic equations, or simply MHD equations.

Introducing the one-fluid mass density $\rho=\sum_\alpha n_\alpha m_\alpha = n\sum_\alpha m_\alpha$ and the one-fluid velocity $\underset{\sim}{v}=\sum_\alpha m_\alpha \underset{\sim}{v}_\alpha / \sum_\alpha m_\alpha$, the continuity equations of multi-species reduce to

$$\frac{\partial \rho}{\partial t} + \underset{\sim}{\nabla} \cdot \rho \underset{\sim}{v} = 0 \tag{5}$$

The momentum conservation equations reduce to the equation of motion

$$\rho \frac{d\underset{\sim}{v}}{dt} = \underset{\sim}{J} \times \underset{\sim}{B} - \underset{\sim}{\nabla} p \tag{6}$$

and Ohm's law

$$\underset{\sim}{E} + \underset{\sim}{v} \times \underset{\sim}{B} = \eta \underset{\sim}{J} + \frac{\underset{\sim}{J} \times \underset{\sim}{B}}{ne} \tag{7}$$

where $p=\sum_\alpha p_\alpha$, η is the electric resistivity; for simplicity, isotropic pressure and resistivity are assumed.

These one-fluid equations are supplemented by the equation of state

$$\frac{d}{dt} p \rho^{-\gamma} = 0 \tag{8}$$

and the Maxwell equations

$$\frac{\partial \underset{\sim}{B}}{\partial t} = - \underset{\sim}{\nabla} \times \underset{\sim}{E} \quad (9)$$

$$\mu_0 \underset{\sim}{J} = \underset{\sim}{\nabla} \times \underset{\sim}{B} \quad (10)$$

where γ is the ratio of the specific heats. Eqs. (5)-(10) constitute a closed set of one-fluid equations and are called the MHD equations. The Hall term (last term) in ohm's law is negligible if $LV \gg B/ne\mu_0$ where V is the characteristic flow speed. Let us evaluate the order of maginute in the plasma sheet where the Hall term is supposed to be important. Take $B=2\times10^{-8}$ tesla, $n=5\times10^{5}$ m^{-3} and $V=2\times10^{5}$ m/sec. Then $L \gg 1000$ km is the condition for the neglect of the Hall term. Since the validity condition for the MHD approximation is $L \gg$(ion gyroradius)$\simeq$500 km, it can be said that when the Hall term becomes important, the other gyrokinetic effects might also become non-negligible simultaneously. It is usual, therefore, to neglect the Hall term in Ohm's law.

3. GUIDING PRINCIPLES OF MHD SIMULATION

Since we have obtained the basic equations which can adequately describe the evolution of a macroscopic process in the solar terrestrial plasma environment, we shall move on to the basic concept and guiding principles of the MHD simulation.

The most fundamental concept in the MHD simulation is that there is no length scale inherent in the medium.

"Water assumes the Shape of the Vessel"

As this idiom implies, no physical restriction exists on the spatial grid size, Δx, as far as Δx is taken to be small enough compared to the size of the vessel. In principle, therefore, it is possible to do simulations for any large scale phenomena.

The only existing characteristic quantity in the MHD regime is the Alfven velocity V_A. More strictly, the shear Alfven velocity V_A, the fast magnetosonic velocity V_F and the sonic velocity V_S characterize the complete evolution of the system under study, where

$$V_A=B/(\mu_0\rho)^{1/2}, \quad V_S=(\gamma p/\rho)^{1/2}, \quad \text{and} \quad V_F=(V_A^2+V_F^2)^{1/2}$$

Given the length scale L of the vessel in which our MHD fluid fills, then the time scale τ_A is defined by $\tau_A=L/V_A$. This scale is called the Alfven transit time which is the time for the Alfven wave to propagate over the length scale L.

The time step Δt must be chosen sufficiently small that the propagation of the magnetohydrodynamic waves can be adequately described. The explicit condition for this is given by

$$\frac{\Delta x}{\Delta t} > V_A,\ V_F \tag{11}$$

which is called the Courant-Fridrichs-Lewy condition, or simply the Courant condition [Richtmyer and Morton, 1967]. This condition implies that the numerical characteristic velocity be large enough to represent the propagation characteristics of all the necessary physical informations.

We are now ready for making a MHD simulation since we already have the equations to be solved and the characteristic length and time scales. Certainly, one can make a simulation and get a result if one has an available computer and enough computer time. However, it is not guaranteed that the obtained result is physically significant, even though numerically plausible.

In what follows, we shall discuss the role and significance of the MHD simulation. In the fusion plasma it is conceivable that the plasma is confined within a conducting container so that the magnetic flux is totally conserved. In such a case we can definitely define the boundary condition and can rather easily develop a self-contained simulation model.

In the solar terrestrial environment, however, no rigid boundary exists. Many freedoms are left for the choice of the simulation domain and the boundary condition. Besides, our computer capacity is so tiny compared with the vast information the solar terrestrial plasma bears. It may not be exaggerated to say that it is this choice of the simulation domain that influences the scientific contribution of the simulation result. There are two key points in this choice. One is the self-consistency of the chosen model and the other is the temporal and spatial resolutions we use.

Let us start with the discussion of the self-consistency. In order to discuss this matter, we must first touch on the role of the MHD simulation.

Our ultimate goal of the solar terrestrial plasma research is to establish the solar terrestrial plasma physics which can both qualitatively and quantitatively connect any observable phenomenon with its direct cause. The primary role of observations is to acquire data and to induce the cause and effect relationships of the observed phenomena. The primary role of theories, on the other hand, is to generalize the cause and effect relationships induced by observations and to prove their universality based on the physical laws. The simulation approach must be identified as the theoretical approach. Therefore , the primary role of the simulation must be to deduce a universal causality rather than to simply demonstrate what is observed. The only practical difference between the analytical approach and the numerical approach lies in whether the solution is given by a mathematical expression or by a numerical expression. The mathematical treatment is advantageous in that more generality is implied, but disadvantageous in that a highly nonlinear problem is intractable. In contrast, the numerical treatment is powerful for a highly nonlinear problem which is common in space plasmas.

In short, the primary role of the MHD simulation is to deduce a universal causality that governs an observable phenomenon in space. To this end, it is favorable to develop a simulation model in such a way that "minimum" elementary processes can be included in it. A question then arises as to how we can assume "minimum". This is the real key to the MHD simulation. A deep physical insight is required to answer this question. In this regard, one may suspect that the numerical technique is the key to simulation. Certainly, a special technique must sometimes be developed depending upon the problem we wish to attack. However, this is rather minor. Indeed, the necessary technique could eventually be developed by efforts, no matter how it looks hard.

As we have discussed in the introduction, any single phenomenon occurring in nature has a long causal history. Therefore, it is not so easy to extract a reasonable elementary "black box" from the long causal chain. The "self-consistency" must be tested for this extration. Namely, the extracted black-box must be self-contained. The black box has an energy input and output. If an appreciable amount of energy is fed back from the output to the input during the course of evolution in the black box under consideration, then the black box is not self-contained and our model is self-inconsistent. Thus, we ought to extract a larger black box. When the "self-consistency" test is met, we regard the black box as the "minimum" black box. At this moment, the first gate to the MHD simulation can be opened by the key of the self-consistency of the model.

The next key is the resolution of the numerical model. Most interesting macroscopic phenomena such as the magnetospheric substorm are accompanied by some local dissipation, such as the resistivity or viscosity, caused by a microscopic process. The finite difference scheme, which is inevitable in the numerical simulation, generates an unavoidable numerical dissipation. It is essential, therefore, to figure out correctly the contribution of the numerical dissipation to the result obtained. This is because, as we have discussed above, the primary role of the MHD simulation is to obtain a universal quantitative relationship between the input (cause) and the resulting phenomenon (effect).

In general, the self-consistency requirement can be more easily satisfied for larger black box, while the resolution requirement is more favorable for smaller black box. From the primary role of simulation to establish the space plasma physics, a physically sound model is naturally desired. Accordingly, it would be a natural way to start with the minimum black box model and, with the progress of the computer ability, go on to a larger model in which more elementary processes interact with each other.

For convenience we shall hereafter call the minimum black box model the "local" MHD model and the largest black box model involving the whole process of the system concerned the "global" MHD model.

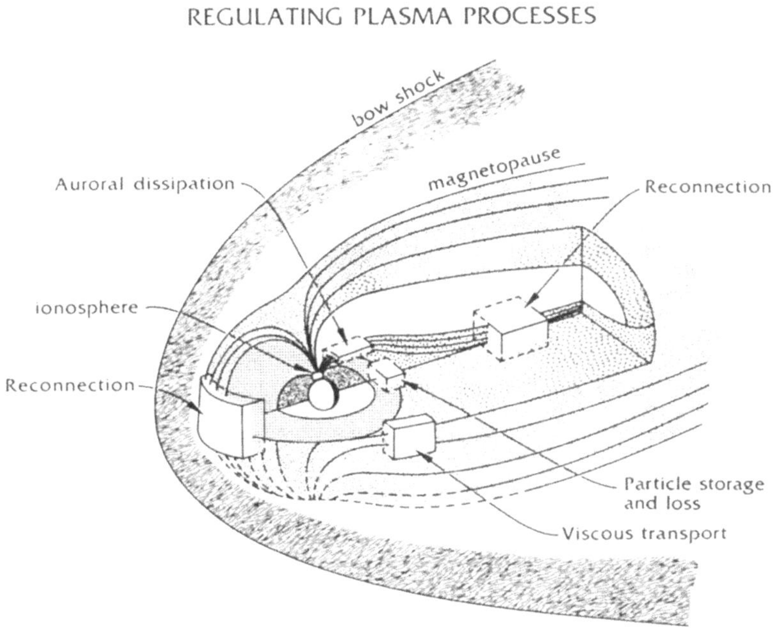

Figure 1. Key regions in the solar wind/magnetosphere system where elementally energy conversion processes would take place. They are candidates for local MHD simulation boxes.

4. LOCAL MHD MODEL

Let us consider the solar wind-magnetosphere system. There are several key regions where remarkable energy conversion processes would take place. Such regions are shown by shades in Figure 1. The first place the solar wind interacts with the magnetosphere is the magnetopause, especially the dayside magnetopause. Some anomalous disipation process, whereby part of the solar wind momentum and energy is introduced into the magnetosphere, must take place there. Otherwise the frozen-in condition is kept and the solar wind flows around the magnetosphere. The solar wind energy can anytime get into the magnetosphere in terms of the magnetohydrodynamic waves, but not in terms of the bulk energy. Thus, the dayside magnetopause becomes the first candidate for the elementary black box where a macroscopic energy conversion process must take place.

Suppose the bulk solar wind energy is transferred into the magnetosphere. The transported energy would at least temporarily be deposited in the magnetosphere, most probably in the tail part of the magnetosphere.

We recall here that there are three macroscopic energy states, i.e., the magnetic energy $B^2/2\mu_0$, the bulk flow energy $\rho v^2/2$, and the thermal energy $p/(\gamma-1)$. In which state the transported solar wind energy is deposited in the magnetosphere may largely depend on the energy transfer mechanism on the magnetopause.

Suppose that the solar wind energy be transferred by a viscous-like interaction [Miura, 1983]. Then, the energy would be deposited primarily in terms of the bulk flow energy, to be specific, in terms of the twin equatorial convective cells. In this case, a steady state can be achieved without having any additional energy conversion inside the magnetosphere. Thus, what we do next is to model the magnetosphere-ionosphere coupling under the condition that an equatorial convection flow is given [Watanabe et al., 1984].

If the energy transfer is caused by magnetic reconnection on the dayside magnetopause, however, a meridional flow would predominantly be generated and sweep the dayside magnetic flux down to the tail. Since the plasma flow is supposed to be sub-Alfvenic and the plasma is low-β, the dominant energy transported into the tail region must be the magnetic energy. Consequently, the magnetotail is more stretched and the neutral sheet (cross tail) current is enhanced.

Thus, the energy is expected to be deposited predominantly in the (thinned) plasma sheet in terms of the tail current, or equivalently free magnetic energy. When the current is locally enhanced or the plasma sheet thinning develops, some microscopic process would act to enhance local dissipation. Accordingly, reconnection would be triggered [Min et al., 1984]. This physical consideration suggests that the current enhanced plasma sheet region can be another elementary black box [Sato et al, 1984]. It should be noted that in contrast to the previous viscous-like interaction, this case is essentially a transient process, since no steady state can be achieved without invoking some energy release mechanism inside the magnetosphere.

The implication of the above argument is very important in the numerical modeling of the energy conversion process in this region. This is because no equilibrium can be assumed as an initial condition. The existence of the dynamic stress is crucial.

The destination of the energy flow after the energy conversion in the thinned plasma sheet does depend on what conversion process has happened. Remember that depending on either the viscous interaction or dayside reconnection, the subsequent energy flow was completely different. This implies that we may not be able to proceed to the next step unless the energy conversion process is elucidated.

Nevertheless, a physical consideration allows us to do a reasonable guess beforehand. For example, if reconnection is driven three dimensionally in the tail, a pair of field-aligned currents (toward the ionosphere on the dawn side and toward the magnetosphere on the dusk) would be generated (see, Figure 2), along with the tailward and earthward plasma jets near the midnight-equatorial region, (see, Figure 3) [Sato, 1982].

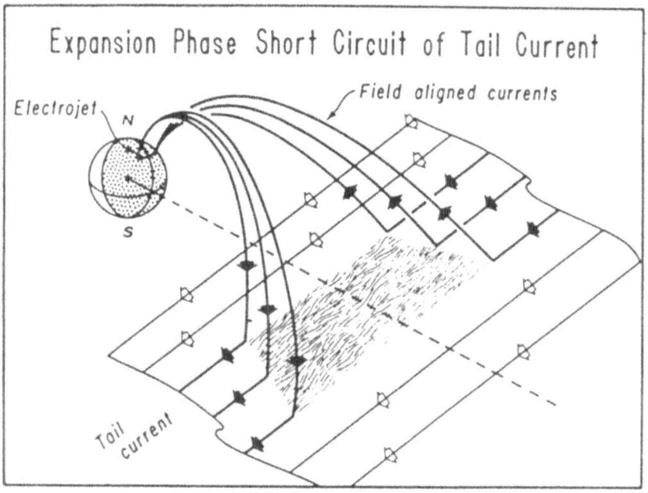

Figure 2. Sketch showing the local interruption of the tail current and generation of the field-aligned current resulting from 3 dimensional reconnection in the tail region.

If this is the case, then the energy flow bifurcates from the reconnection region, one energy carrier being the plasma jets and the other being the field-aligned currents. The tailward jet would simply carry out the eneregy from the reconnection region and eventually join the solar wind further down the tail. The earthward jet would temporarily be deposited in the near earth equatorial region and be eventually dissipated down in the ionosphere. Thus, we can come up with a new black box in which the input energy is the plasma jet and the output is the ionosphere as shown in Figure 3 [Serizawa and Sato, 1984].

The field-aligned currents, though transient, can also provide another energy source which couples to the ionosphere. Incidentally, it is to be noted that a recent 3D local MHD simulation has beautifully confirmed this prediction [Sato et al., 1984].

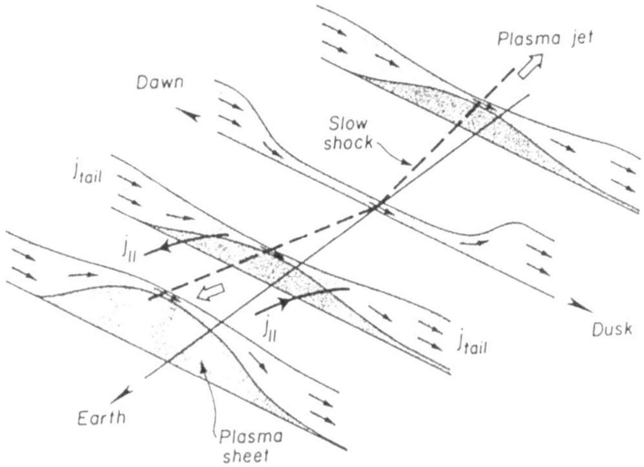

Figure 3. Earthward and tailward plasma jet generations by slow shocks resulting from driven reconnection in the tail.

A special emphasis is placed on the necessity and importance of having a sound philosophy based on a deep physical insight in the simulation study of the solar terrestrial plasma. What has been described here is just an example based on the author's philosophy. There may be different philosophies. What is important in the simulation study, however, is to develop a model persistently and consistently so that all the branches of the energy flow starting from the solar wind can be emerging from the dark. Figure 4 shows one example of the energy flow diagram in the solar-magnetosphere system.

5. GLOBAL MHD MODEL

Our ultimate goal is to have a thorough comprehension of the causal mechanisms governing the overall energy flow in the solar terrestrial plasma environment. There may exist unavoidable semi-global internal feedback loops which we have not been able to predict in the segment analysis for the minimum elementary processes. It is of great value therefore to confirm the validity of our segment analysis of the tangled causal chain based on local MHD simulations

As we have seen in the previous section, it is likely that several independent elementary processes are going on at several different locations in the solar wind-mangnetosphere-ionosphere

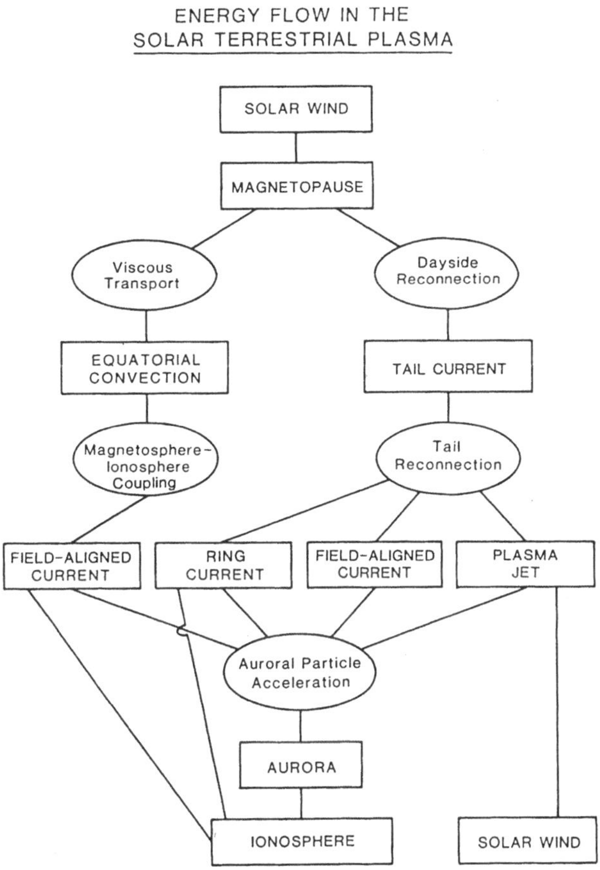

Figure 4. Energy flow diagram in the magnetosphere-ionosphere system as a result of the solar wind-magnetosphere interaction. Depending on the elementaly energy conversion process the output energy states are different, hence, the ensuing energy conversion processes are defferent, as well.

system. Yet, each process appears to involve highly nonlinear interactions in itself. At present, however, even the largest computer at hand cannot afford to give a numerical resolution high enough to correctly follow all the physical processes involved at once. It is not even so sure that the present-day computer is able to totally insure the physical resolution of the local MHD simulation.

As a matter of fact a global simulation is worth while doing. But the scientific objective must be different from that of the local MHD simulation which is to give a "quatitative" understanding of the causal mechanisms governing the macroscopic energy flow in the solar terrestrial plasma environment. The present-day global simulation still stays in the stage to get a rough suggestion to what would happen globally.

The best spatial resolution, Δx, we can have at present would be $2R_E$ (R_E: the earth's radius). This, of course, exceeds the dimension of the elementary region of reconnection. The Alfven transit time, τ_A, would be 300 seconds ($L \simeq 50R_E$ and $V_A \simeq 1000$ km/sec), which, for example, well exceeds even the MHD characteristic time of reconnection. Furthermore, the numerical dissipation is so enormous and numerical error fields are so large that the evaluation of the current distribution, particularly, of the field-aligned current could be erroneous.

The scientific contribution we expect from the present-day global simulation is that it would give us a right direction for the tendency of the overall magnetospheric reaction to the solar wind. For instance, it would tell us at which locations in the magnetosphere reconnection is most likely to take place [Brecht et al., 1982], how the shape of the magnetosphere reacts on the solar wind condition [Wu, 1983] and where field-aligned currents prefer to flow [Ogino, 1984].

6. BOUNDARY CONDITIONS

Suppose that the subject and the simulation segment we wish to study have already been properly determined. The next important step we must take is how to handle the boundary condition.

The set-up of the initial condition does not include any essential difficulty. The boundary condition, however, does. As we have already discussed one place and another, the most essential in the space plasma simulation is to appreciate the implication of

"While the river is constant, the water is always changing."

(from a Japanese classic essay "Hojo-ki" written by Kamo-no-Chomei about 800 years ago). Since we cut out a segment from the ever-lasting causal chain, an elaborate work is needed to numerically express the boundary between the simulation box and the outside region. This is particluarly so for the local MHD simulation in which no clear physical demarcation exists.

Since our main concern is to reveal the energy conversion process, it will be of use to see the energy conversion equations to begin with. From the MHD equations, namely, Eq.(5) through Eq.(10), we can easily derive

$$\frac{\partial K}{\partial t} = - \int \frac{1}{2} \rho v^2 \underset{\sim}{v} \cdot d\underset{\sim}{S} + \int \underset{\sim}{J} \times \underset{\sim}{B} \cdot \underset{\sim}{v} \, dV - \int \nabla p \cdot \underset{\sim}{v} \, dV \qquad (12)$$

$$\frac{\partial T}{\partial t} = - \int \frac{\gamma}{\gamma-1} \, p \, \underset{\sim}{v} \cdot d\underset{\sim}{S} + \int \nabla p \cdot \underset{\sim}{v} \, dV + \int \eta \, J^2 \, dV \tag{13}$$

$$\frac{\partial W}{\partial t} = - \int \frac{\underset{\sim}{E} \times \underset{\sim}{B}}{\mu_0} \cdot d\underset{\sim}{S} - \int \underset{\sim}{J} \times \underset{\sim}{B} \cdot \underset{\sim}{v} \, dV - \int \eta \, J^2 \, dV \tag{14}$$

where $K = \int \frac{1}{2} \rho \, v^2 \, dV$ (bulk flow energy)

$T = \int \frac{p}{\gamma - 1} \, dV$ (thermal energy)

$W = \int \frac{B^2}{2\mu_0} \, dV$ (magnetic energy)

and $d\underset{\sim}{S}$ is the surface element vector and dV is the volume element.

The left hand side of each equation gives the time rate of change of each total energy in the simulation box. The last two terms on the right represent the internal energy exchange, while the first term on the right gives the energy transport through the simulation boundary. It is this first right term that matters to the boundary condition. When this term is negative, an external energy comes in the simulation

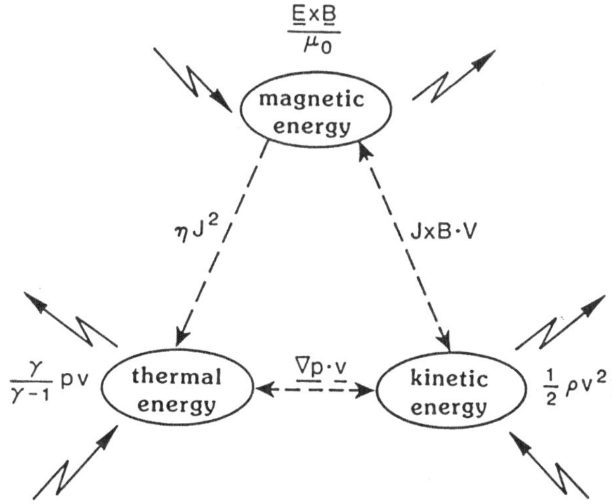

Figure 5. Diagram showing the macroscopic energy conversion relations among the magnetic, kinetic and thermal energies.

box, while it goes out when positive. The diagram of the energy exchange is schematically shown in Figure 5.

Speaking in terms of the energy, the boundary consists basically of the "input" and "output" boundaries. The definition of "output" or "input" here is not the usual one which is defined by the signs of the first terms on the right of Eqs.(12)-(14). We here define the boundary on which we can externally control the energy supply as the "input" boundary, and otherwise as the "output" boundary. According to this definition it may happen that the energy comes in the system through the output boundary.

In the natural system it is usual that the input and output boundaries are not clearly demarcated. In the simulation model, however, it must be clearly discriminated. In other words, the orientation and dimension of the simulation box must be chosen so that the input and output boundaries can be discriminated. This problem may also influence the choice of the coordinates, cartesian, cylindrical or spherical.

Once the location of the input boundary is defined, then we must specify the boundary condition so that all the dependent variables, i.e., ρ, $\underset{\sim}{v}$, $\underset{\sim}{B}$, and p, can be known consistently at each time step. They should not necessarily be specified individually but could be defined by their combinations, e.g., the mass flux, the energy flux, or whatever quantities if they are physically plausible. Usually some freedoms are left for their determination, so that we ought to fill them properly by ourselves. Given all the necessary relations, then it is usually straightforward and no technical difficulty exists any more.

Physically or philosophically speaking, on the other hand, the output boundaries, are rather easy to define, since they are usually supposed to be free boundaries. In some cases, they can be fixed boundaries or reflecting boundaries but this difference does not matter. Technically or numerically, however, they are in general more difficult than the input boundary.

Mathematically, the fixed and reflecting boundary conditions can be well defined. Numerically, as well, they can be defined without any big trouble provided the choice, fixed or reflecting, is physically reasonable for the problem of interest.

When we say "free" boundaries, however, it usually means that we want to treat them as if there were no boundaries. In other words, they are put there because the simulation box is by no means unlimited. There may be no royal way to this treatment. Therefore, depending on the problem and the numerical model under consideration, the best way must be seeked by trial and error.

One example of the treatment of the free output boundary is given here. The boundary values are extrapolated by the condition $\partial f/\partial n=0$

where f stands for any dependent variable such as a component of the magnetic field vector and n is the coordinate variable perpendicular to the boundary plane. Mathematically, this condition is enough to determine the boundary value. However, it is not guaranteed that the assumed condition is physically plausible. Even if it is physically plausible, it is unlikely that it can be rigorous. Therefore, some mismatch would arise between the defined value and the real value which we have no way to predict. This mismatch sometimes accumulates near the boundary and produces an unexpected numerical artifact. One way to get rid of this unexpected artifact is to provide a margin layer inside but adjacent to the output boundary where we add an artificial damping term to the equation which produces the unexpected artifact.

Now we have arrived at a position to be able to construct a physically sound simulation model and define the initial and boundary conditions properly.

What we have to do next is to make a simulation code to solve the MHD equations and then to provide proper diagnostics. We shall see these problems in the following section.

7. SIMULATION CODE AND DIAGNOSTICS

In a word, the MHD simulation code is a numerical solver of the MHD differential equations in a limited spatial domain subject to a boundary condition. To make such a code, the "differential" equations must be aproximated by the "finite difference" equations. Several approximation methods have been developed. Since they are explained in detail in other text books [see, for example, Richtmyer and Morton, 1967; Potter, 1977], they would not be reproduced here but a couple of fundamental points will be mentioned.

The first point is the unavoidable numerical error arising inevitably from the discretization of the spatial derivatives, namely,

$$\frac{\partial f}{\partial x} \cong \frac{\Delta f_3}{\Delta x} = \frac{f_{j+1} - f_{j-1}}{2\Delta x} \tag{15}$$

where j represents j-th discrete position. Let us estimate the numerical error coming from this discreteness by using a Fourier representation, $f=\hat{f}\exp(ikx)$. For simplicity, only one Fourier mode k is considered. The left hand side of Eq.(15) is given by

$$\frac{\partial f}{\partial x} = ikf \tag{16}$$

The left side of Eq.(15), on the other hand, is given by

$$\frac{\Delta f_j}{\Delta x} = if_j \frac{\sin(k\Delta x)}{\Delta x} = ikf_j \{ 1 - \frac{k^2 \Delta x^2}{6} + 0(k^4 \Delta x^4) \} \quad (17)$$

Comparison of Eq.(17) with Eq.(16) indicates that the numerical error due to the finite difference of the first order differenciation is of the order of $(\Delta x/\lambda)^2$, λ being the wavelength, thus indicating that the grid size must be taken to be much smaller than the characteristic length of the phenomenon under study. In the nonlinear state there is a general tendency that the length becomes shorter, so that the error becomes larger. It is also noted that the error takes the form of the diffusion, this indicating the appearance of an unavoidable numerical diffusion in the difference form. Those who wish to do an MHD simulation must remember this fact, particularly when the dissipation plays an essential rule in the evolution. In a 3D global simulation that contains several elementary processes in the same simulation domain, despite of a small grid number, therefore, the elementary processes are unavoidably and largely affected by the numerical diffusion. When one does a global simulation or even a local simulation, the obtained result must be evaluated with this fact deep in his mind.

The second point is that basically there are two ways to approximate the MHD equations. One way is to rewrite the MHD equations in the conservation forms. Eqs.(5)-(10) them take the form of

$$\frac{\partial \rho}{\partial t} = - \nabla \cdot \rho \underset{\sim}{v} \quad (18)$$

$$\frac{\partial \rho \underset{\sim}{v}}{\partial t} = - \nabla \cdot (\rho \underset{\sim}{v}\underset{\sim}{v} + \underset{\approx}{Q}) \quad (19)$$

$$\frac{\partial \underset{\sim}{B}}{\partial t} = - \nabla \times \underset{\sim}{E} \quad (20)$$

$$\frac{\partial U}{\partial t} = - \nabla \cdot \underset{\sim}{S} \quad (21)$$

where

$$\underset{\approx}{Q} = (\frac{B^2}{2\mu_0} + p) \underset{\approx}{I} - \frac{1}{\mu_0} \underset{\sim}{B}\underset{\sim}{B} \quad (22)$$

$$U = \frac{1}{2} \rho v^2 + \frac{p}{\gamma - 1} + \frac{B^2}{2\mu_0} \quad (23)$$

$$\underset{\sim}{S} = (U + p) \underset{\sim}{v} + \underset{\approx}{Q} \cdot \underset{\sim}{v} \quad (24)$$

and

$$\underset{\sim}{E} + \underset{\sim}{v} \times \underset{\sim}{B} = \eta \underset{\sim}{J} \quad (25)$$

$$\mu_0 \underset{\sim}{J} = \nabla \times \underset{\sim}{B} \tag{26}$$

The two-step Lax-Wendroff method is the most conventional and stable method to solve these equations [see, for example, Richtmyer and Morton, 1967]. Usually, however, a simple finite difference expression ends up with a numerical instability even though the Courant-Friedrichs-Lewy condition is satisfied. An artificial numerical smoothing or filtering procedure must be added [see, for example, Sato and Hayashi, 1979].

The two-step Lax-Wendroff scheme is rather favorable to study a high β plasma. This is because the pressure is not explicitly solved but is obtained by subtracting the magnetic and bulk flow energies from the total energy U. In a low β plasma where it is small compared to the magnetic energy, the pressure would be masked by the round-off error and other numerical errors, so that it cannot be evaluated with accuracy. In the space plasma simulation, scientifically interesting regions are usually high β regions such as the plasma sheet. Therefore, the two-step Lax-Wendroff scheme is often useful.

If the plasma of interest is low β everywhere in the simulation box, then the Lax-Wendroff scheme may not be appropriate. It is needed to directly solve the equation of state, Eq.(8). In this case, Eqs.(5)-(10) constitute the equations to be accessed directly.

Depending on the boundary condition, one can take either the finite-difference scheme or the Fourier-expansion scheme. In the space plasma research, as we have already discussed sufficiently, the system is open so that the periodic condition may not be plausible. Therefore, the Fourier scheme may not be so useful. But on some special occasions, it can of course be useful. A mixed usage of the finite-difference and Fourier schemes in different dimensions may also be possible.

We are now at the position to be able to code the MHD simulation program. But there is another very important matter to be seriously considered. That is the diagnostics of the simulation data. The diagnostics is particularly important in 3D simulations which are currently popular.

We must have already had an idea on the diagnostics in our mind when we chose the subject and designed the simulation model. The diagnostics is closely connected to what we really want to disclose by that simulation. We must predict the overall outcome, though not of course precisely, before we make the computer code run. Reasonable prediction requires a deep physical insight and understanding of the space physics. Good simulation or not depends largely on this prediction.

In 3D simulations, enormous amount of data come out. The diagnostics is to seek physically important informations from within the mountain of the data. There may be no royal way to this. But fundamental and minimum quantities to be extracted from the data are the temporal behaviours of the twelve quantities appearing on the left and right hand sides of Eqs.(12)-(14). In addition, depending on the phenomenon we are interested in, we must think out a special diagnostics. For example, when a shock-like structure appears, then the Rankine-Hugoniot relations must be checked.

Before concluding this section, there is one last thing to be noted. As the reader may have already learned, the attitude to examine as carefully as possible the degree of contamination due to the numerical error and to make parameter runs is the minimum requirement for those who wish to do simulations. Since one of the roles of the simulation study is to clarify the causal relationship, it is at least needed to obtain a functional relationship between the input and output or between an internal parameter and the output. For this purpose, several runs where internal and external conditions are differently chosen must be executed. One run simulation gives sometimes a misleading information.

8. CLOSING REMARKS

The numerical technique and procedure of the MHD simulation are given in any text book. In this article, therefore, a special emphasis is placed on the philosophical concepts and physical principles of the MHD simulation in the space plasma research, which are not given in any text book. In the simulation study, to my knowledge, having a sound philosophy based on a profound physical insight and understanding is much more important than the numerical technique. This is particularly the case in space plasma simulation. Frankly, I am very much anxious about the future of the space physics society, particularly, the negative contribution of the simulation study. There appears a tendency, especially among the non-simulation people, to welcome a simulation which gives an apparent similarity to their observations. They do neither care much about the numerical model nor the numerical error. Even if they care, there would be no way to check upon. To have an experimentalist's smile is certainly comfortable to a simulator. However once such a lukewarm atmosphere prevails, it is very dangerous to the space plasma research.

The responsibility for this tendency is solely on the side of the people who are doing simulations. They should not forget that their primary duty and role are not to receive a temporary smile from experimentalists but to collaborate with them in a real scientific sense and contribute steadily to establish the space plasma physics.

The objective of the ISSS is to develop and support a sound simulation society which can contribute to the progress of the solar terrestrial physics.

Acknowledgements

This work is supported by the Grant-in-Aid from the Ministry of Education, Science and Culture in Japan, and also by the STT Grant NAGW-78 and AFGL contract F19628-82-K-0019 in the United States.

REFERENCES

Brecht, S. H., J. Lyon, J. A. Fedder and K. Hain, J. Geophys. Res., 87, 6098, 1982.

Min, K., H. Okuda and T. Sato, 1984, to be published

Miura, A., J. Geophys. Res., 89, 801, 1984.

Montgomery, D. C. and D. A. Tidman, Plasma Kinetic Theory, Mcgraw-Hill, New York, 1964.

Ogino, T. to be published

Potter, D., Computational Physics, John Willey & Sons, New York, 1977.

Richtmyer, R. D., and K. W. Morton, Difference Methods for Initial-Value Problems, 2nd ed., Interscience, New York, 1967.

Sato, T., Auroral Physics, in Magnetospheric Plasma Physics, edited by A. Nishida, center for Academic Publ., Japan and D. Reidel Publ. Co., Tokyo, Japan, 1982.

Sato, T., and T. Hayashi, Phys. Fluids, 22, 1189, 1979,

Sato, T., R. J. Walker and M. Ashour-Abdalla, J. Geophys. Res., in print,1984.

Serizawa, Y., and T. Sato, to be published

Watanabe, K., M. Ashour-Abdalla and T. Sato, to be published

Wu, C. C., Geophys. Res. Lett., 10, 545, 1983.

MHD MODELLING OF THE EARTH'S MAGNETOSPHERE

C. C. WU

Physics Department
University of California, Los Angels
Los Angels, CA 90024

ABSTRACT

A global MHD model of the earth's magnetosphere is defined. Some numerical aspects of the model which include shock capturing technique, nonuniform grid system and multiple time scale problem are discussed. Also presented are some recent results.

1. INTRODUCTION

In a global magnetohydrodynamics (MHD) model, MHD equations are used to describe the solar wind interaction with the magnetosphere. In recent years considerable efforts have been directed towards numerically solving these highly-nonlinear, time-dependent, three-dimensional equations. The first numerical solutions were carried out by Spreiter and his coworkers (Spreiter and Alksne, 1969). They have formulated the model in terms of MHD equations and have provided some justifications for using the MHD descriptions. However, their numerical calculations modeled only flow out side the magnetosphere by using hydrodynamics together with the Chapman-Ferraro model for the magnetosphere.

Recent global MHD models have been studied by many researchers including Leboeuf et al. (1979), Lyon et al. (1980), and Wu et al. (1981). These models include the magnetosphere as well as the flow in the magnetosheath. Bow shock and magnetopause are self-consistently formed in the solution and explicit jump conditions are not required. In principle, the MHD model wil provide information not only about the quiet magnetosphere configurations, but also about the dynamic magnetosphere. Since this is a very large scale numerical work, many numerical problems have to be overcome. The purpose here is to descrbe some numerical aspects of the global MHD model. In Section 2,

H. Matsumoto and T. Sato, Computer Simulation of Space Plasmas,

the global MHD model is formulated. In Section 3, some numerical aspects of the model are discussed. These include shock capturing technique, nonuniform grid system and multiple time scale of the problem. In Secion 4, some recent results are presented for illustrations.

2. GLOBAL MHD MODEL

Our model of the earth's magnetosphere is based on an MHD description of the interaction of the solar wind and the geomagnetic field. The ideal MHD equations are:

$$\frac{\partial \rho}{\partial t} = - \vec{\nabla}\cdot(\rho\vec{v}) , \tag{1}$$

$$\frac{\partial(\rho\vec{v})}{\partial t} = - \vec{\nabla}\cdot[\rho\vec{v}\vec{v} + \vec{I}(p+\frac{B^2}{2}) - \vec{B}\vec{B}] , \tag{2}$$

$$\frac{\partial\vec{B}}{\partial t} = \vec{\nabla}\times(\vec{v}\times\vec{B}) , \tag{3}$$

$$\frac{\partial \varepsilon}{\partial t} = -\vec{\nabla}\cdot[(\frac{1}{2}\rho v^2 + \frac{p}{\gamma-1} + P)\vec{v} - (\vec{v}\times\vec{B}) \times \vec{B}] \tag{4}$$

Here the mass density, pressure, velocity and magnetic field are denoted by ρ, p, $\vec{v}$ and $\vec{B}$ respectively, γ is the ratio of the specific heats, and the energy density (ε) is given by $\varepsilon = \rho v^2/2 + B^2/2 + p/(\gamma-1)$. In addition to these equations, we require $\nabla\cdot\vec{B}=0$, which is satisfied if it is satisfied by the initial data in our initial value problems.

In this model the plasma is assumed to have an equation of state that represents the internal energy by the relation

$$I = \frac{P}{(\gamma-1)p} \tag{5}$$

For simplicity reason we use this equation with constant $\gamma=2$ throughout the whole region. In principle, one could try to use a more complex equation of state to characterize plasma behaviours in various regions.

Since there are discontinuities (bow shock, magnetopause) in te solutions, we should be more precise about the difinition of our model. In the smooth region, we require the solutions to satisfy the MHD eqs. (1-4). Across the discontinuities, we require the solutions to satisfy physical jump conditions. Bow shock and magnetopause are treated in our model as "boundaries" only. In the numerical calculations, the jump conditions and the positions of these boundaries are "captured", it means that they evolve naturally in the solution. (Further discussions about the shock capturing technique are given in the next section). In our calculations the geomagnetic field was approximated by a dipole field,

$$\vec{B}(\vec{r}) = \frac{3\bar{r}(\vec{\mu}\cdot\bar{r}) - \vec{\mu}}{r^3} \tag{6}$$

where $\vec{\mu}$ is the dipole moment and $\vec{r}$ is the position relative to the dipole center.

The model is treated as an initial- and boundary-value problem, and thus both initial conditions and boundary conditions must be specified. There are two kinds of boundary conditions: the ionosphere boundary is a physical one where the ionosphere-magnetosphere interaction is treated; the four boundaries marked by either "inflow" or "outflow" in Figure 1 are numerical boundaries which are introduced to limit the computational domain. The physcal boundaries are located at infinity. Initially, the plasma is in a static equilibrium with the geomagnetic dipole field. At t=0 the solar wind is introduced at the inflow boundary. Subsequent time evolution is obtained by integrating the MHD equations.

Since there are no theorems available concerning the existence or the uniqueness of the solutions, the model is defined to be close to (possible) wind tunnel experiments and hopefully, the results will have physical meaning. Exact specifications of the initial and boundary conditions used in the calculations will be given in Section 4.

Although the ideal MHD equations are used in the model, there are effective viscosity and resistivity in the calculations due to numerical truncation errors. Because of this resistivity term, magnetic field reconnection can take place. Ideally, one would like to reduce these numerical effects and use viscosity and resistivity determined by physical mechanisms.

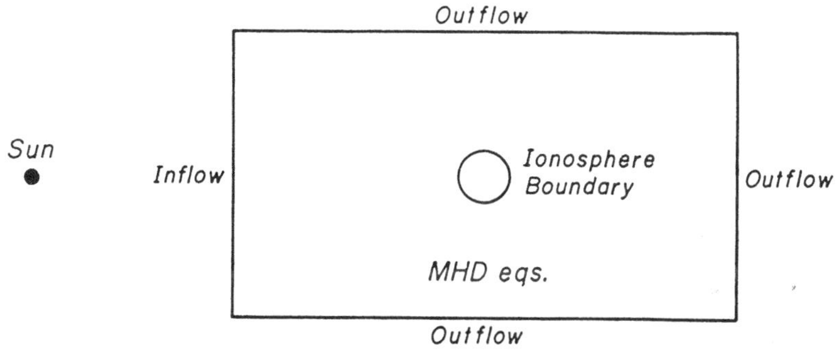

Figure 1. Global MHD model.

3. NUMERICAL METHODS

3.1 Shock Capturing Technique

In our global MHD model, bow shock and magnetopause boundaries are captured in the calculations, and explicit jump conditions are not required. Although the capturing technique has been extensively used in gas dynamics, for instance in studying reentry problems for space vehicles, the technique is relatively new to researchers in magnetospheric physics. Improper use of the technique can lead to incorrect results. Thus we present some discussions of the technique here and for illustration, we shall consider one-dimensional gas flow problem. The governing equations for the one-dimensional gas dynamics are

$$\frac{\partial \rho}{\partial t} = - \frac{\partial}{\partial x}(\rho v_x) \tag{7}$$

$$\frac{\partial (\rho v_x)}{\partial t} = - \frac{\partial}{\partial x}(\rho v_x^2 + p) \tag{8}$$

$$\frac{\partial \varepsilon}{\partial t} = - \frac{\partial}{\partial x} [(\frac{1}{2}\rho v_x^2 + \frac{p}{\gamma-1} + p)v_x] \tag{9}$$

with $\varepsilon=1/2\rho v^2{}_x+p/(\gamma-1)$. This system can be rewritten in the form

$$\frac{\partial u}{\partial t} + \frac{\partial}{\partial x}F = 0 \tag{10}$$

with u=column vector $(\rho,\rho v_x,\varepsilon)$ and F=column vector $(\rho v_x,\rho v^2{}_x+p,(\varepsilon+p)v_x)$.

It is well known that this system of equations can develop discontinuities (Landau and Lifshitz, 1959). When this occurs, the meaning of Eqs. (7-9) must be carefully considered. The set of Eq.(10) represents the conservation laws of mass, momentum and energy. They are derived from the integral equation (in 3D notation)

$$\frac{\partial}{\partial t} \int_v u \, d\tau + \int_\sigma \vec{F}\cdot d\vec{A} = 0 \tag{11}$$

where v represents volume and σ is the boundary surface. These equations indicate that the change of mass, for instance, in the volume is equal to the flux crossing the surrounding boundary.

It is obvious that the solution of the differential Eq. (10) is also the solution of the integral Eq. (11); however the solution of the integral equations need not be the solution of the differential equation. Discontinuous solutions are admissable for integral equations but not allowed for differential equations. Therefore if one is interested in having discontinuos solutions, one must solve the integral equations. However, integral equaitons allow solutions that are not physical. Let us derive the jump conditions for a steady normal shock. From equation (11) we have

$$F_1 = F_2 \tag{12}$$

where subscripts 1 and 2 represent conditions before and after the shock boundary. Particles move from region 1 to region 2 across the shock. From this equation we can relate quantities on one side of the shock to quantities on the other side by

$$\frac{\rho_2}{\rho_1} = \frac{\frac{\gamma+1}{\gamma-1}\frac{p_2}{p_1} + 1}{\frac{\gamma+1}{\gamma-1} + \frac{p_2}{p_1}} = \frac{v_1}{v_2} \tag{13}$$

These relations are called Rankine-Hugoniot equations. In terms of Mach number M, defined as the ratio of the fluid velocity to the fluid sound speed, $M = v/(\gamma p/\rho)^{1/2}$, we have the relations

$$\frac{p_2}{p_1} = 1 + \frac{2\gamma}{\gamma+1}(M_1^2 - 1) \tag{14}$$

and

$$\frac{\rho_2}{\rho_1} = \frac{(\gamma+1)M_1^2}{(\gamma-1)M^2 + 2} \tag{15}$$

According to Eq. (14), p_2 can be either larger or smaller than p_1 depending on the magnitude of M_1. Additional consideration is then required to determine the nature of the shock wave. This additional requirement is based on the second law of thermodynamics, namely, the entropy can only increase following a particle (entropy condition). For an ideal gas the entropy is given by $c_v \log(p/\rho^\gamma)$. It can be shown that the entropy increases across the shock if and only if $p_2 > p_1$. A shock is always compressive. It then follows from Eqs. (13-15) that $M_1 > 1$, $\rho_2 > \rho_1$, $v_2 < v_1$, and $T_2 > T_1$, where $T = p/\rho$. When $M_1 < 1$, the jump conditions given in Eqs. (13-15) refer to a nonphysical rarefaction shock. For $M_1 < 1$ the physical solutions should be smooth. The Rankine-Hugoniot equations which also satisfy the entropy condition are called physical jump conditions. The purpose of a shock capturing technique is to find solutions which in the smooth region will satisfy the Eq. (10), while across the discontinuities, wil satisfy physical jump conditions.

Let us discuss further numerical methods of solving gas flow problems. In one dimension, discretization of Eq. (11) gives, referring to Figure 2,

$$\frac{\partial}{\partial t}(u_i \cdot \Delta \cdot A) + F_{i+\frac{1}{2}} \cdot A - F_{i-\frac{1}{2}} \cdot A = 0 \tag{16}$$

where A is the cross sectional area and Δ is the grid spacing. Dividing by ΔA in Eq.(16) gives

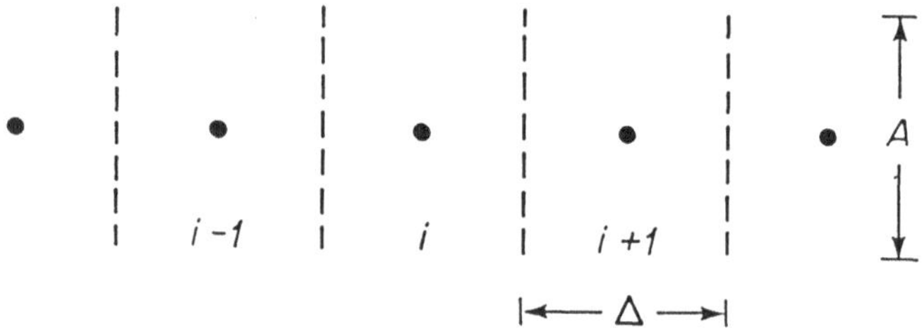

Figure 2. Discrete representation of x variable.

$$\frac{\partial}{\partial t}u + \frac{F_{i+\frac{1}{2}} - F_{i-\frac{1}{2}}}{\Delta} = 0 \tag{17}$$

This is the finite difference form for the differential Eq.(10). Thus solving for differential Eq. (10) in the conservative form is in effect solving for integral Eq. (11).

To ensure an entropy-increase condition, dissipation is required. By using numerical dissipation, we will obtain Rankine-Hugoniot conditions, but not the details of the shock transition layer. The shock transition region will occupy 3 to 6 grid points depending on the numerical scheme (see review by Sod, 1978).

The Lax scheme (Lax, 1954), which is the simplest method of solving the conservation laws, replaces the time derivative in Eq. (17) by a forward difference in which averaging is introduced at the earlier point, i.e.,

$$\frac{\partial u_i}{\partial t} + \frac{u_i^{t+\Delta t} - \frac{1}{2}(u_{i+1}^t + u_{i-1}^t)}{\Delta t} \tag{18}$$

After a slight modification of Eq.(17), the difference equation in the Lax scheme becomes

$$\frac{u_i^{t+\Delta t} - \frac{1}{2}(u_{i+1}^t + u_{i-1}^t)}{\Delta t} + \frac{F_{i+1}^t - F_{i-1}^t}{2\Delta} = 0 \tag{19}$$

The difference equation is equivalent to the equation

$$\frac{\partial u}{\partial t} + \frac{\partial F}{\partial x} = \frac{\Delta^2}{2\Delta t}\frac{\partial^2 u}{\partial^2 x} + \cdots\cdots \tag{20}$$

The dissipation term makes the scheme numerically stable and also provides dissipation for ensuring entropy condition. The integration time step size is limited by the Courant-Friedrichs-Lewy (CFL) condition that

$$\Delta t \leqslant \frac{\Delta x}{v} \tag{21}$$

where v is the maximum characteristic speed. This means that the time step size should be small enough that a disturbance cannot travel more than one grid spacing in that time interval.

Let us emphasize the main points of our discussions in the following. If the solution of the gas flow problem is smooth then one can seek a solution from Eq. (10) or from another set of Eq. (10), such as

$$\frac{\partial \rho}{\partial t} + \frac{\partial}{\partial x}(\rho v_X) = 0 \tag{22}$$

$$\frac{\partial v_X}{\partial t} + v_X\frac{\partial}{\partial x} v_X + \frac{1}{\rho}\frac{\partial \rho}{\partial x} = 0 \tag{23}$$

$$\frac{\partial p}{\partial t} + \frac{\partial}{\partial x}(p v_X) + (\gamma-1)p\frac{\partial}{\partial x} v_X = 0 \tag{24}$$

and one shall obtain the same results. But if there are discontinuities in the solution, one should use the conservation laws of mass, momentum and energy. In addition, one should require entropy conditions across the shock. Although numerical dissipation is used to obtain entropy conditions, sometimes a nonphysical shock can result in some numerical scheme (Osher and Solomon, 1982). It is advisable to check the jump conditons in the solution.

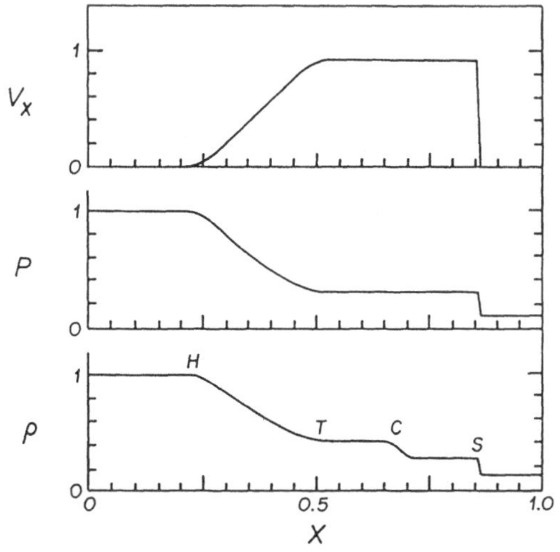

Figure 3. Shock tube problem. S, C, T and H denote the positions of shock, ontact discontinuity, tail and head of a rarefaction wave, respectively.

In the following I will show two examples of gas flow problems in which the Lax scheme was used. The first one is a typical shock tube problem. In this problem the initial conditons consist of two constant states separated by a diaphragm. At time $t = 0$, the diaphragm was abruptly removed. The initial data are $\rho=1$, $p=1$, $v_x=0$ for $x\leqslant 0.5$ and $\rho=0.125$, $p=0.1$, $v_x=0$, for $x>0.5$. In this example $\gamma=1.4$ and 1600 points were used in the calculations. The configuration at $t=0.206$ is shown in Figure 3. (Note the solution is self-similar, i.e., a function of x/t only). It contains five regions separated by a shock front, a contact surface, the tail and the head of a rarefaction wave. The results agree with the analytical solution. The transition of the shock occupied 4 to 6 grid points and the transition of the contact discontinuity occupied a wider region. Further discussions can be found in the review by Sod (1978), where a survey of finite difference schemes was given.

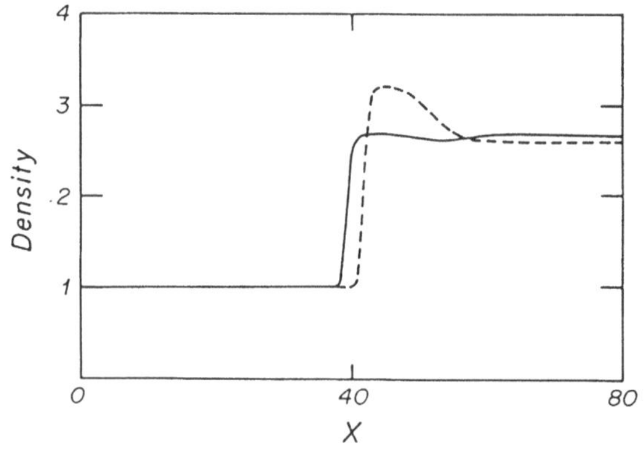

Figure 4. Comparisons between two formulations. The solid line gives the density distribution using conservative form Eqs. (7-9), and the dashed line gives the density distribution using nonconservative form, Eqs. (22-24).

The second example is given to show the importance of using the conservative form of Eqs. (7-9). The initial data are $p = 0.5$, $\rho=1$, $v=4$, for $x\leqslant 40$, $p = 10.5$, $\rho=8/3$, $v=1.5$ for $x>40$, and $\gamma=2$. These two constant states satisfy the Rankine-Hugoniot jump conditions. The calculations were carried out with 400 points. Figure 4 shows the density distributions at $t = 9.6$ in two different calculations, one with Eqs. (7-9) and the other with Eqs.(22-24). The solution from the conservation laws shows consistency with the Rankine-Hugoniot relations. The solution from Eqs. (22-24) however, differs greatly from the Rankine-Hugoniot jump conditions. Since the pressure equation, (24), does not correspond to any conservation law, the jump condition is not defined. As shown in Figure 4, density reaches 3.2 after the discontinuity, while the Rankine-Hugoniot condition gives a limit of $\rho_2/\rho_1 \leqslant 3$ for $\gamma=2$. Similar demonstrations were performed by Weiland (1978).

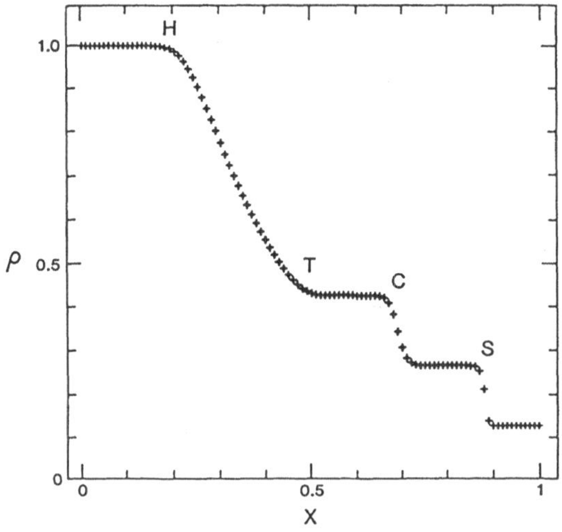

Figure 5. Shock tube problem using Harten's scheme. S, C, T and H denote the positions of shock contact discontinuity, tail and head of a rarefaction wave, respectively.

In the last 30 years, shock capturing technique has been one of the main topics in computational fluid dynamics. Recent developments can be found in the review by Harten et al. (1983). Here we present the results of the shock tube problem using Harten's scheme (Harten, 1982) to show the state of the art. All parameters are the same as in the first example except that now we have 100 grid points and the configuration is at a different time. As clearly seen in Figure 5, the Harten's scheme gives a sharp shock jump condition and a well defined contact discontinuity.

3.2. Computational Requirements and Techniques for Global MHD Modelling

The global MHD modelling requires not only a lot of computer time but also a huge amount of data storage. If it were not for these limitations, the global MHD model might have been attempted long ago. For example, Spreiter and his coworkers (Spreiter and Alksne, 1969) had formulated a similar MHD model in the 1960's, but because of the computational limitations, they had to limit their studies to the flow properties of the magnetosheath by using fluid equations and the

Chapman-Ferraro model. (Spreiter, private communication, 1982). In this section, we discuss the computational requirements of the model and some techniques to alleviate the limitations. The success of the MHD model will strongly depend on our ability in solving these computational problems in addition to having a better and more accurate numerical algorithm. Of course advances in computer technology will play an important role. If the next generation of scientific computers can attain a speed of 10^9 floating point operations per second and have 50 million words of data memory, which represents an order of magnitude faster in CPU speed and larger in memory size from the present Cray-1 computer, some of the limitations imposed in the following may be avoided.

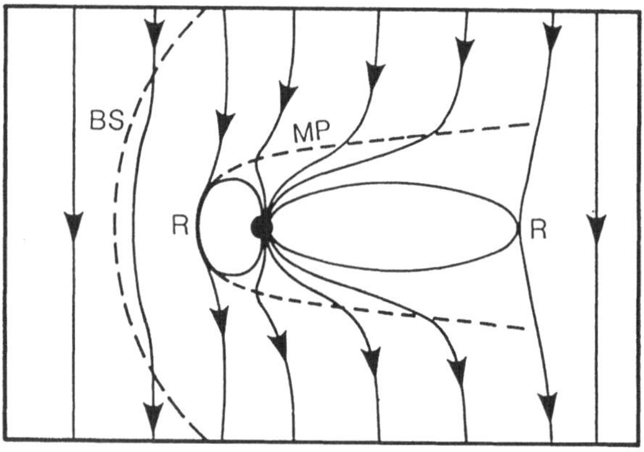

Figure 6. Dungey's model of the magnetosphere.

The solutions of the model have two discontinuities: bow shock and magnetopause, as well as two reconnection regions, one on the dayside and one on the nightside (following Dungey's reconnection model) as sketched in Figure 6. To be self-consistent, calculation domain should include all of these regions. Therefore, in the solar-magnetospheric coordinate system, the domain should extend from 20 R_e on the dayside. On the nightside, in principle, we can have a well defined boundary if it is placed some distance beyond the far-tail reconnection region. However, our knowledge about this reconnection process is limited, and one may even dispute the necessity of this process as long as there is one in the tail at about -20 R_E. Thus in our calculations, we first place our tail boundary at a sufficiently far distance, for example, at -200 R_E. By carrying out computer experiments with different locations, we can study the effects of the distant tail on the dynamics near the earth and eventually the model can give us indications concerning the Dungey's far tail reconnection process. For an estimate, the calculation domain is extended to -200 R_E in the tail. In both the east-west and north-south directions, the calculation domain should be extended far enough that the effects of possible shock reflections from these boundaries on the far tail is minimized. For a magnetotail of 200 R_E long, boundaries need to be placed at 100 R_E in both north-south and east-west directions.

In the solution of the MHD model, serveral areas require very fine spatial resolution, such as at the bow shock, the magnetopause, the ionosphere-magnetosphere boundary and the reconnection regions. For example, in order to model the reconnection processes in the tail which has a current sheet of 2~3 R_E thick, one probably needs about 5 grid points per 1 R_E to model tearing mode instabilities for the magnetic Reynolds number of 1000. For the cases with higher magnetic Reynolds' numbers, much finer spatial resolution is required. Another example is that in the study of supersonic flow past a sphere, it is found that 20 grid points are required between bow shock and body, therefore one may require the same number of grid points between the bow shock and the magnetopause which is about 3 R_E in physical space in the MHD model. One may also require the same spatial resolution from the magnetopause to the earth.

Obviously, we can not have fine spatial resolution throughout the whole calculation region. It is more economical to have a nonuniform grid system with more grid points concentrated in the regions where good spatial resolution is required. To get a feeling about the size of the system, let us assume that a nonuniform grid system of 100 by 100 by 100 points is required for the calculations with 50 grid points on the dayside and 50 grid points on the nightside along the sun-earth line. Since there are 14 data values (ρ, p, $\vec{v}$, $\vec{B}$, $\vec{B}d$, and $\vec{x}$, with $\vec{B}d$ the geomagnetic dipole field, $\vec{x}$ the corrdinates) for each grid points, the total data storage is 14 milion words. In the special case that there is east-west component in IMF, we can carry out the calculations in one quadrant and we need 100 by 50 by 50 grid points, and 3.5 million words of stage. Actual calculations of this size have been performed.

Grid generation is a difficult numerical problem. Most methods of generating systems of coordinates involve the solution of systems of elliptical partial differential equations and are applied in two dimensions. In our magnetospheric problem, we face a large three dimensional system, it is important to find an efficient way of generating grids that satisfy our requirements. Figure 7 show the results of our attempt to generate such a system. As shown in the figure, more grid points are clustered near the bow shock, magnetopause and the earth. In the future when a more realistic magnetosphere-ionosphere boundary condition is used, we plan to overlap the coordinate system with a spherical coordinate system near the earth. The grid is generated by using a equispaced grid in the computational domain. First we prescribe two surfaces that correspond to the bow shock surface and the magnetopause boundary respctively.

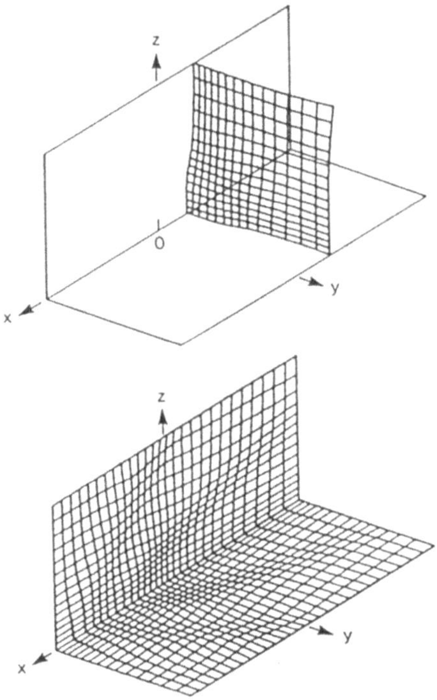

Figure 7. A 3D mesh system used in the global MHD model.

Then we let points move towards these surfaces. This relocation is achieved by inducing a velocity at each grid point, the magnitude and direction depending on its relative distance to the prescribed surfaces. This method is based on the idea given by Rai and Anderson (1981).

Once the nonuniform coordinate system is generated, the MHD equaions are transformed in the computational coordinate system. Since the equations can be written in terms of conservation laws, (10), conservative form for the transformed equations can be obtained in the new coordinate system as follows (Vinekur, 1974). Let us rewrite the conservations laws (10) in the Cartesian system (x,y,z,t),

$$\frac{\partial u}{\partial t} + \frac{\partial E}{\partial x} + \frac{\partial F}{\partial y} + \frac{\partial G}{\partial z} = 0 \tag{25}$$

where E, F and G represent the components of the flux function F in Eq. (10). Let the new coordinates (computational coordinates), η, ξ and

ζ be related to the Cartesian system by the transformation

$$\eta = \eta(x,y,z)$$
$$\xi = \xi(x,y,z)$$

and

$$\zeta = \zeta(x,y,z) \tag{26}$$

The Eq. (25) is then transformed into the conservative form:

$$\frac{\partial \bar{u}}{\partial t} + \frac{\partial \bar{E}}{\partial x} + \frac{\partial \bar{F}}{\partial y} + \frac{\partial \bar{G}}{\partial z} = 0 \tag{27}$$

where u = u/τ

$$\bar{E} = (E\eta_x + F\eta_y + G\eta_z)/\tau$$

$$\bar{F} = (E\xi_x + F\xi_y + G\xi_z)/\tau$$

$$\bar{G} = (E\zeta_x + F\zeta_y + G\zeta_z)/\tau$$

The geometric derivatives are given by

$$\begin{aligned}
\eta_x &= \tau(y_\xi z_\zeta - y_\zeta z_\xi) \\
\xi_x &= \tau(y_\zeta z_\eta - y_\eta z_\zeta) \\
\zeta_x &= \tau(y_\eta z_\xi - y_\xi z_\eta) \\
\eta_y &= \tau(z_\xi x_\zeta - z_\zeta x_\xi) \\
\xi_y &= \tau(z_\zeta x_\eta - z_\eta x_\zeta) \\
\zeta_y &= \tau(z_\eta x_\xi - z_\xi x_\eta) \\
\eta_z &= \tau(x_\xi y_\zeta - x_\zeta y_\xi) \\
\xi_z &= \tau(x_\zeta y_\eta - x_\eta y_\zeta) \\
\zeta_z &= \tau(x_\eta z_\xi - x_\xi z_\eta)
\end{aligned} \tag{28}$$

and

$$\tau = \frac{\partial(\eta,\xi,\zeta)}{\partial(x,t,z)}$$

For an analytical transformation the geometric derivatives can be evaluated analytically. For a numerical transformation these derivatives will have to be evaluated numerically.

In an "adapted" scheme, the coordinate transformation in Eq.(26) is generalized to allow the new coordinates to be functions of time and the transformed equations can also be casted in conservative form. This method will have advantages over the fixed grid system by allowing grid points to move according to the need of the solution; for instance, by having more grid points in the area that has large gradient of some physical quantities such as pressure. However, we have not yet attempted to use the adaptive method in our model for the difficulty of generating a satisfactory grid system.

In our magnetospheric model there are basically three significantly different time scales: the Alfven transit time, the solar wind flow transit time and the magnetic field diffusion time. By the pressure balance principle at the magnetopause, we can estimate the relative time scales of the first two. At the subsolar point, the stagnation pressure Pst is roughly $\rho_{sw}v_{sw}^2$ and the pressure-balance relation is

$$\rho_{sw}v_{sw}^2 = 1/2\ (2\ B_{dipole})^2_{m.p.} \tag{29}$$

where ρ_{sw} and v_{sw} refer to the solar wind density and velocity respectively. Thus the magnitude of the dipole field at the subsolar point is $(0.5\rho_{sw}v_{sw}^2)^{1/2}$. If we set the magnetopause position at 10 Re then the magnetic dipole field at 1 R_E is about $700\times(\rho_{sw}v_{sw}^2)^{1/2}$. Accordingly, the Alfven velocity at 1 R_E is about 700 v_{sw}, if the plasma density at 1 Re is assumed comparable with the solar wind plasma density. (The observed plasma density near the earth is higher and the Alfven speed is lower than the estimated value.) Hence the Alfven transit time is faster than the solar wind flow time by two orders of magnitude.

The magnetic field diffusion time is slower than the solar wind flow transit time. But the estimate depends greatly on our assumptions concerning the mechanism of the reconnection processes. If the reconnection processes at the tail are due to tearing instability, one can expect the time scale to be much slower perhaps by a factor of 10^3 for the magnetic Reynols number of 10^6. If, on the other hand, the reconnection processes are "driven" processes as proposed by Sato et al. (1979), one may expect the time scale to be compatible with the solar wind flow time. Since observationally the magnetospheric dynamics is occuring in the solar wind flow transit time scale, it seems that the reconnection processes should proceed in the same time scale.

In our model we follow the calculations in the solar wind flow time scale. Because of the fast Alfven wave near the earth, the model is more difficult to solve than the problem of hydrodynamics flow past bodies. In the terminology of differential equations, our model may be considered as a stiff problem.

Realizing the fact that the large Alfven speed is localized near the earth since the magnetic dipole field drops off like r^{-3} we can somewhat avoid the stiffness in the problem by devising a numerical scheme which employs two calculation regions: (i) a small region around the earth with time step size limited by the Alfven speed, and (ii) a much larger surrounding region with the time step limited by the solar wind flow velocity. For a given time step in the large region, many time steps are carried out in the small region near the earth. Since the amount of computer work is proportional to the number of time steps times the number of grid points in the region, the amount of computing in the small region is found to be of the same order as in the larger surrounding region. At the interface of the two regions numerical stability is maintained. As shown in Figure 8,

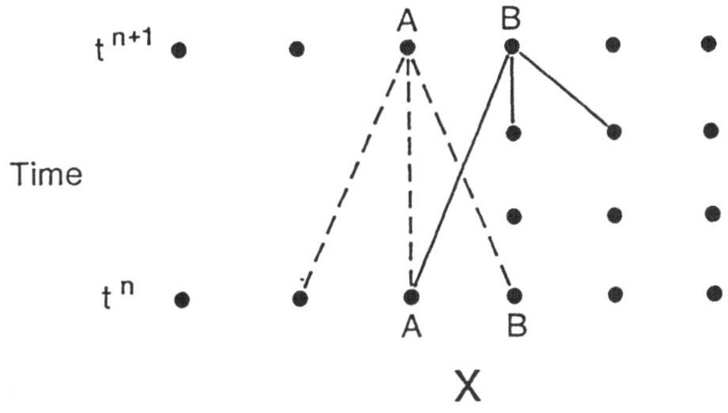

Figure 8. Two region scheme.

calculations at both point A in the large region and point B in the small region satisfy CFL condition.

Using this two region scheme and using a system of 100 by 50 by 50 grid points in one quadrant as described earlier, we can estimate the total amount of computer time for a run. Suppose that the smallest spatial resolution in the system is 1/4 R_E and that the total time in the run is for the solar wind to traverse about 250 R_E, then 1000 time steps are required for the time integration. The processing time is about 10μs per grid for the Rusanov' scheme (Rusanov, 1962) on the Cray-1 computer. Thus, a total of 1.5 hours of CPU time is required. Due to the large data storage requirement, disks are used. During the calculations, data which are arranged plane by plane are transferred between disks and memory at the same time. The I/O charge is an additional important factor.

In this section, we have attempted to state the scale of the problem. Although it is a very large scale numerical work, with some techniques as described above, one can try to carry out the modelling within the power of the present computers. In the next generation of computers, one can certainly carry the task more easily. At that time one will probably go further than the current single fluid MHD model.

4. RESULTS

In the following some results from my calculations will be presented to illustrate the global MHD model.

We used Rusanov's scheme (Rusanov, 1962) in these calculations. Rusanov's method is a variant of Lax's scheme in which the artificial viscosity is maintained at a minimum value except where a large value is needed. Instead of the diffusion term in Eq.(20), the Rusanov scheme used

$$\frac{\Delta^2}{\Delta t}\frac{\partial}{\partial x}\left(\alpha\frac{\partial u}{\partial x}\right) \tag{30}$$

where α in the MHD model is proportional to $(v+c_S+c_A)/(v+c_S+c_A)_{max}$, with sound speed c_S and Alfven speed c_A. The diffusion term is therefore large only in the regions near the bow shock and magnetopause where the change of v is large and in the dipolar region where c_A changes rapidly.

The solar magnetospheric coordinate system is used. The initial data are $\rho=0.2$, $p=0.1$, $\vec{v}=0$, and $\vec{B}=\vec{B}dipole$. At the inflow boundary the values of all variables are fixed to the solar wind parameters: $p=0.5$, $\rho=1$, $\vec{v}=-2.5\ \hat{x}$ and $\vec{B}=\vec{B}_{IMF}$. The boundary conditions in both the y and z directions are treated by linear extrapolation along the direction 45^0 from the x-axis. Linear extrapolation was used at the -x outflow boundary. The boundary conditons at the earth are specified by fixing the values of all variables at their initial values. This does not realistically model the ionospheric effects and is not self-consistent. In our calculations, the numerical diffusion terms are large near the earth and they were used to represent the magnetosphere-ionosphere coupling.

The 3D calculations were performed with 94x53x53 grid points in a nonuniform mesh covering the quadrant $-38\leqslant x\leqslant 16$, $0\leqslant y\leqslant 34$, and $0\leqslant z\leqslant 34$. All quantities were normalized with respect to the solar wind parameters. Thus, the solar wind sound speed equals 1. The solar wind Mach number was chosen to be 2.5. The geometric field is given by Eq.(6) with $\vec{\mu}=-350\hat{z}$. By the pressure-balance principle, we will have the magnetopause at r=5.8 along the sun-earth line, following the relation (29). If this position is identified at 10 R_E then 1 spatial unit is about 1.72 R_E , and 1 unit of magnetic field is about 17.3 gammas.

Figures 9(a) and 9(b) show pressure contours on both the equatorial and noon-midnight planes. The pressure distibution along the earth-sun line is plotted in Figure 9(c). Similar plots for density distributions are given in Figure 10. The accuracy of our results can be evaluated by checking against the Rankine-Hugoniot jump conditions. According to Eqs.(13-14), $p_2/p_1=8$ and $\rho_2/\rho_1=2.27$ for our case. This gives p = 4 and $\gamma=2.27$ just inside the shock along the sun-earth line. The calculated values, see Figures 9 and 10, agree well. The calculated stagnation pressure is 4.8 which gives k = 0.8 for the Newtonian formula $p_{st}=k\rho_{sw}v_{sw}^2$. The bow shock is located at 8.2 and the magnetopause is at 5.8. The ratio of the standoff distance to the magnetopause is then 0.41, which is larger than the observed average value of 0.33. This difference is due to the small Mach number used in our calculation.

On the nightside, the region near the equatorial plane is characterized by a plasma sheet. This is most evident in Figure 9 where the plasma pressure has a local maximum near the equator. The region of low pressure, which contains the dotted contour, corresponds

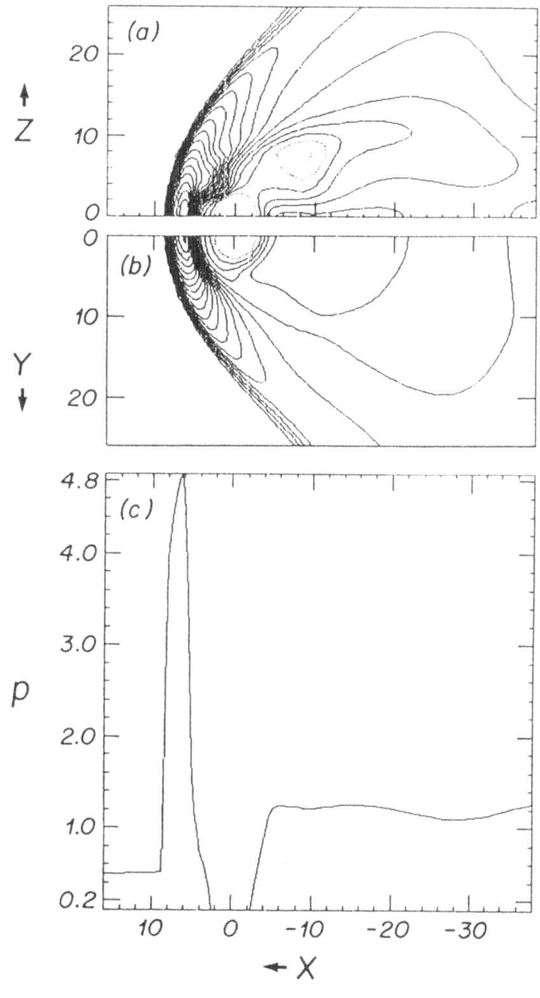

Figure 9. Pressure contours on both the equatorial and the noon-midnight planes in (a) and (b) respectively, and the pressure distribution along the earth-sun line in (c) for the case IMF = 0. The solid lines denote $p \geqslant p_{SW}$ and the dotted lines $p < p_{SW}$.

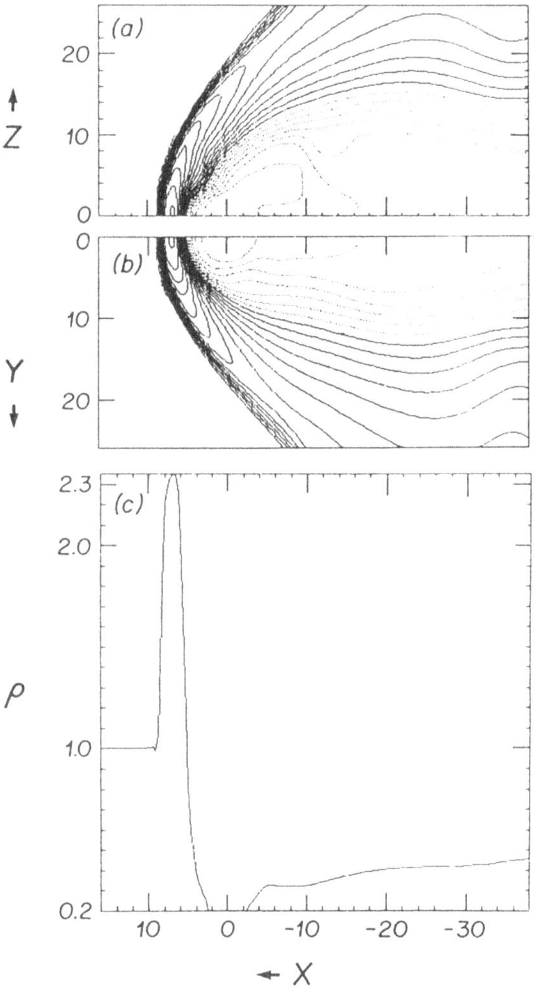

Figure 10. Density contours on both the equatorial and the noon-midnight planes in (a) and (b) respectively, and the density distribution along the earth-sun line in (c) for the case IMF = 0. The solid lines denote $\rho \geqslant \rho_{SW}$ and the dotted lines $\rho < \rho_{SW}$.

to the tail lobes. At x~-13 R_E the presure decreases by about an order of magnitude between z=0 and z=10 R_E. As expected, the region near the equator contains a high beta plasma ($\beta \gg 1$) and that at larger z values contains low beta ($\beta \ll 1$). The model predicts nearly constant density in both the plasma sheet and lobe regions (Figure 10). Thus, most of the change in plasma pressure across the plasma sheet boundary is caused by a change in plasma temperature.

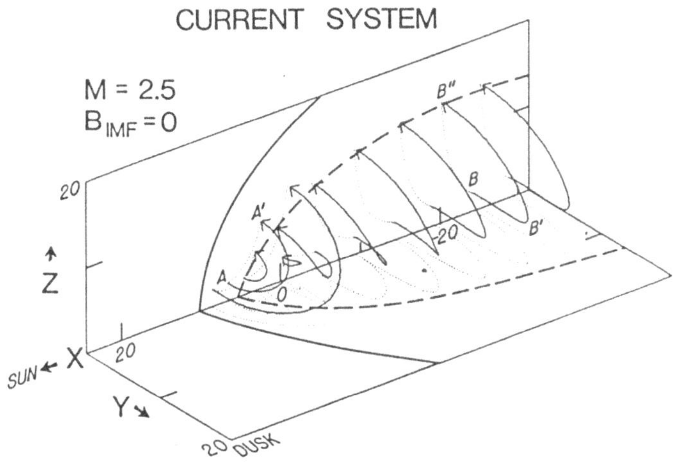

Figure 11. Three dimensional perspective plot of the current system.

The magnetospheric current system (from an earlier run, Wu et al., 1981) is presented in Figure 11. These currents are traced from points at the z=2 line on the noon-midnight meridian plane. Their projections on both the equatorial and noon-midnight meridian planes are shown with dotted lines. A typical magnetopause current element at the nose is indicated by AA'. The magnetopause boundary currents are similar to those obtained from Chapman-Ferraro pressure balance models. But as will be discussed later, the MHD model in fact presents a very different physical picture from that of the Chapman-Ferraro model.

The current loop BB'B" is typical of the cross tail current and its return current on the magnetopause. The cross tail current BB' flows nearly parallel to the equatorial plane and curves towards the earth. The return current flows almost on a plane perpendicular to the earth-sun line from B' to B". Similar currents flow from B' to B" in the other quadrant to complete the current loop and a mirror image current loop is formed in the southern hemisphere to complete the familiar θ shaped current system. The line BB" is tilted from the z axis toward the earth. This tilt is larger closer to the earth. Many features in our tail current system are found in the semi-empirical model of Olson and Pfitzer (1977). Olson and Pfitzer used wire current loops to model the tail field. In their calculation, they found it necessary to tilt the wire current loops towards the sun and to bend them around the earth, in order to reproduce the observed field. The resulting current distribution is very similar to that from our model. The current density distribution along the earth-sun line obtained from our model (solid line) and from the Olson-Pfitzer model are presented in Figure 12. The dotted lines give the current density profile used by Olson and Pfitzer to calculate the field, while the dashed line was generated by calculating $\nabla \times B$ from an analytic version of the mode for B. These distributions are normalized at the peak at $x \sim 15\ R_E$. The agreement is good.

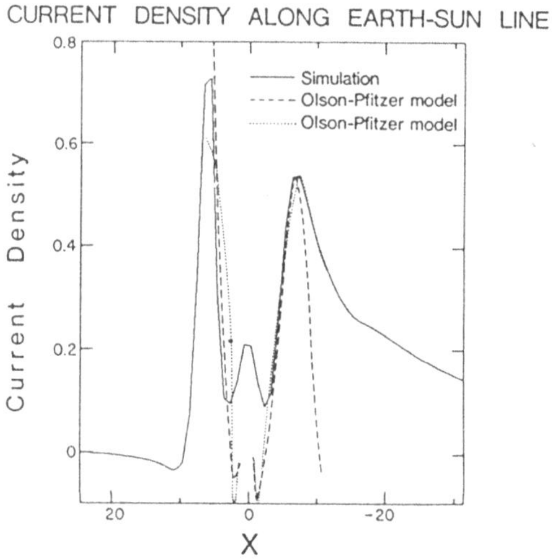

Figure 12. Current density distribution along the earth-sun line. The solid lines are from our model, the dotted lines give the profile used by Olson and Pfitzer (1977) to calculate the field, and the dashed line is from an analytic version of the Olson-Pfitzer model.

One interesting result from the MHD model is about the shape of the magnetosphere (Wu, 1983). The time development of the solar wind interaction with the dipole magnetic field is sketched in Figure 13. Figure 13(a) represents the initial state where plasma of uniform density and uniform pressure is in static equilibrium with the geomagnetic dipole field. The topology of the magnetic dipole field is represented by the 90^0 field lines which close at infinity. Figure 13(b) represents a state at a later time when the bow shock and magnetopause have been formed. Because of the "frozen in" condition, the initial field line topology should be maintained. Figure 13(b) represents one of two possible resulting configurations. The other one has the two neutral points on the dayside. But in our experiment, the dayside magnetosphere was compresesed by the solar wind and the night-side magnetosphere was then compressed by the dayside magnetosphrere, the configuration of Figure 13(b) results.

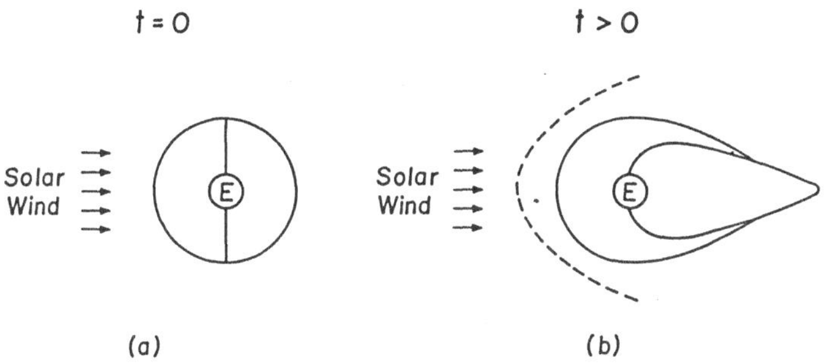

Figure 13. A sketch of the time development of the interaction of the solar wind with a line dipole. The dashed line in (b) represents the bow shock.

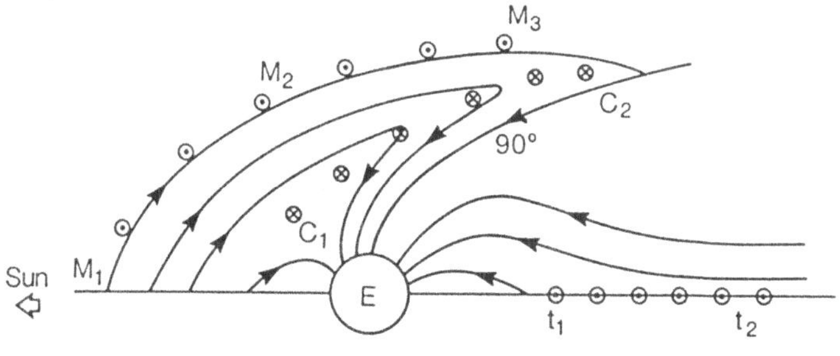

Figure 14. Field line pattern and current structure for the configuration Figure 13(b). $M_1M_2M_3$ denotes the magnetopause surface, C_1C_2 the cusp current sheet and T_1T_2 the tail current sheet. C_1 the denotes the location of the cusp in our MHD model.

The field line pattern and the current structure of the resulting magnetosphere (Figure 13(b)) are sketched in Figure 14. Note the existence of the cusp current sheet, C_1C_2, and that the cusp in this MHD model refers to the region marked by C1 where field lines converge at the earth. This picture of the magnetopause looks quite different from that of the Chapman-Ferraro model, but it can be reconciled in the following way. Curve (M_1M_2 C_1C_2) represents the shape based on the Chapman-Ferraro model, curve (M1M2M3) represents the magnetopause surface and curve (C_1C_2) shows the cusp current sheet from the MHD model. It is clear from this figure that the Chapman-Ferraro model, which is a surface current model, represents the boundary of the magnetopause surface of our model from M1 to M_2 and the cusp current sheet from C_1 to C_2. The magnetopause surface current from M_2 to M_3 is neglected as it is small in comparison with the cusp current sheet. It is not surprising that the current distributions in both the MHD model and the Chapman-Ferraro model should agree. Both models use the presure balance principle, and the magnetopause currents are used to shield the dipole fields. But the MHD model presents a very different physical picture from that of the Chapman-Ferraro model (Wu, 1983). For instance, in the MHD model, solar wind plasma cannot enter the ionosphere directly through the cusp.

Figure 14(a) to 14(c) show the perspective plots of p, $B^2/2$ and $p+B^2/2$ on the noon-midnight plane, respectively from our model. The $p+B^2/2$ plot shows the pressure balance principle (which is true for weak curvature).

5. CONCLUSION

In this paper an introduction to the global MHD model was given. Despite the efforts of many researchers in the last several years, it is fair to say that this mathematical problem has not been solved with complete satisfaction. However, because of the advances in numerical techniques and the availability of more powerful computers, one can expect rapid progress in the field of quantitative MHD modeling. It is hoped that these models will be invaluable in our understanding of magnetospheric physics.

ACKOWLEDGMENTS

This work was supported by National Science Foundation Grant ATM 79-26492, NASA Solar Terrestrial Theory Grant NAGW-78, and Air Force Contract F 196-28-82-K1009 and the MHD code was developed with support by the Department of Energy, DE-AM03-76SF00010 PA26, Task VIB.

References

Harten, A., NYU Report DOE/ER/03077, New York Univ., 1982.

Harten, A., Lax, P. D. and van Leer, B., SIAM Review, 25, 35, 1983.

Landau, L. D. and Lifshitz, E. M., Fluid Machanics, Pergamon Press, 1959.

Lax. P. D., Comm. Pure Appl. Math., 7, 159, 1954.

Leboeuf, J. N., Tajima, T., Kennel, C. F. and Dawson, J. M., Geophys. Res. Lett., 5, 609, 1987.

Lyon, J. S., Brecht, H., Fedder, J. A. and Palmadesso, P. J., Geophys. Res. Lett., 7, 721, 1980.

Olson, W. P. and Pfitzer, K. A., Magnetospheric magnetic field modeling, McDonnell Douglas Astronautics Co., preprint, 1977.

Osher, S. And solomon, F., Math. Comp., 38, 339, 1982.

Rai, M. M. and Anderson, D. A., J. Comput. Phys., 43, 327, 1981.

Rusanov, V., Nat. Res. Council of canada, Translation No. 1027, 1962.

Sato, T. and Hayashi, T., Phys. Fluids, 22, 1189, 1979.

Sod, G. A., J. Comput. Phys., 27, 1, 1978.

Spreiter, J. R. and Alksne, A. Y., Rev. Geophys. Space Phys., 7, 11, 1968.

Vinokur, M., J. Comput. Phys., 14, 105, 1974.

Weiland, C., J. Comput. Phys., 29, 173, 1978.

Wu, C. C., Geophys. Res. Lett., 10, 545, 1983.

Wu, C. C., Walker, R. J. and Dawson, J. M., Geophys. Res. Lett., 8, 523, 1981.

NUMERICALLY-SIMULATED FORMATION AND PROPAGATION OF INTERPLANETARY SHOCKS

S. T. Wu

The University of Alabama in Huntsville
Huntsville, Alabama, 35899, USA

ABSTRACT

Interplanetary shocks caused by solar activity (such as flares, coronal transients, eruptive prominences, etc.) play a significant role in the modulation of solar wind and cosmic rays which affect our earth's environment. With recent advancements in the space program a large data base of in-situ measurements of solar wind plasma parameters has been accumulated by the observers and deposited in the World Data Centers. From these fruitful data, the understanding of interplanetary shock wave physics has been advanced significantly. The numerical simulation method is contributing to this understanding. In this paper, we present several numerical methods for treating the shock propagation problem and use a specific example to demonstrate the capability of the simulation method.

1. INTRODUCTION

Study of the formation and propagation of solar-originated shock waves in heliospheric space has attracted significant attention in the past decade. This attention is appropriate because the propagation of shocks in heliospheric space has been thought as one of the major physical processes for solar wind and cosmic ray modulations and the subsequent, substantial influence on our earth's environment. The early developments of the interplanetary shock generated by solar flares have been given by Dryer (1974; 1975) excellent accounts of the observational and theoretical studies were included in these two review articles. The author discussed the birth of shocks within the flare generation processes; magnetohydrodynamic (MHD) wave propagation through the chromosphere and inner corona; their maturity to fully-developed coronal shock waves; their subsequent propagation into

H. Matsumoto and T. Sato, Computer Simulation of Space Plasmas,

heliospheric space; and finally, their disturbing effects on the solar wind within the context of phenomenological, theoretical, and numerical models.

In this paper, we focus our discussion on recent developments in the study of numerical simulation of shock formation and propagation in heliospheric space. These numerical simulation models are developed in an orderly and systematic fashion: from a simple one-dimensional, time-dependent gasdynamic model (GD) (Wu et al., 1976) to the present development of the quasi-three dimensional time-dependent MHD model (Wu et al., 1983a,b). In the latter case, all the physical vectoral quantities are three dimensional, but the spatial dependence is still in two-dimensions (Wu, 1983; Wu et al., 1983a,b). The numerical methods used for development of these models will be discussed in Section 2; computational procedures and physical examples for the demonstration of the model will be presented in Section 3. Concluding remarks on the simulation models will be given in Section 4.

2. NUMERICAL METHODS AND MATHEMATICAL MODEL

Although numerical methods for dealing with nonlinear fluid mechanisms can be found in many standard text books (i.e., Roache, 1972; Chow, 1980; and Patankar, 1980) we will, for convenience, briefly summarize those numerical methods which have been directly used for the development of the MHD simulation models for the formation and propagation of interplanetary shocks in this section.

In order to illustrate the numerical method, we have chosen a two-dimensional MHD transient planar flow with infinite conductivity as our demonstration model. The governing equations are cast in a quasi-conservation form on the equatorial plan in a spherical coordinate system (r,θ,ϕ) at $\theta=90^0$ plane (Han et al., 1979).

$$\frac{\partial \underset{\sim}{W}}{\partial t} + \frac{\partial}{\partial r}\underset{\sim}{F} + \frac{1}{r}\frac{\partial}{\partial \phi}\underset{\sim}{G} = \underset{\sim}{S} \tag{1}$$

where

$$\underset{\sim}{W} = \left| \begin{array}{l} r^2 \rho \\ r^2 \rho V_r \\ r^2 \rho V_\phi \\ rB_r \\ rB_\phi \\ r^2 \left[\frac{1}{\gamma - 1} p + \frac{1}{2} \rho \, |\vec{V}|^2 + \frac{|\vec{B}|^2}{2\mu_0} \right] \end{array} \right| \tag{2}$$

$$\underset{\sim}{F} = \begin{vmatrix} r^2 \rho V_r \\ r^2 \left[p + \rho V_r^2 + \dfrac{B_\phi^2 - B_r^2}{2\mu_o} \right] \\ r^2 \left[\rho V_r V_\phi - \dfrac{B_r B_\phi}{\mu_o} \right] \\ 0 \\ r \left(V_r B_\phi - V_\phi B_r \right) \\ r^2 \left\{ V_r \left[\dfrac{1}{\gamma - 1} p + \dfrac{1}{2} |\vec{V}|^2 \right] - \dfrac{B_\phi}{\mu_o} \left(V_\phi B_r - V_r B_\phi \right) \right\} \end{vmatrix} \tag{3}$$

$$\underset{\sim}{G} = \begin{vmatrix} r^2 \rho V_\phi \\ r^2 \left[\rho V_r V_\phi - \dfrac{B_r B_\phi}{\mu_o} \right] \\ r^2 \left[p + \rho V_\phi^2 + \dfrac{B_r^2 - B_\phi^2}{2\mu_o} \right] \\ r \left(V_\phi B_r - V_r B_\phi \right) \\ 0 \\ r^2 \left\{ V_r \left[\dfrac{\gamma}{\gamma - 1} p + \dfrac{1}{2} \rho |\vec{V}|^2 \right] + \dfrac{B_r}{\mu_o} \left(V_\phi B_r - V_r B_\phi \right) \right\} \end{vmatrix} \tag{4}$$

and

$$\underset{\sim}{S} = \begin{vmatrix} 0 \\ 2rp + r^2 \rho \dfrac{\partial \psi}{\partial r} + \dfrac{r B_r^2}{\mu_o^2} + r \rho V_\phi^2 \\ \dfrac{r B_r B_\phi}{\mu_o} + r \rho \dfrac{\partial \psi}{\partial \phi} \\ 0 \\ 0 \\ r^2 \rho \left\{ V_r \left[\dfrac{\partial \psi}{\partial r} + \dfrac{1}{r} \dfrac{\partial \psi}{\partial \phi} \right] \right\} \end{vmatrix} \tag{5}$$

where ρ denotes the mass density, $\underset{\sim}{V}=(V_r,0,V_\phi)$ denotes flow velocity vector, T denotes temperature, $\underset{\sim}{B}=(B_r,0,B_\phi)$ denotes the magnetic field, and the other symbols have their usual meaning; $p=\rho RT$ being the gas pressure, ψ the gravity potential, γ the specific heat ratio and finally μ_o the permeability in vacuum. Independent variables are time (t), radial (r) and azimuthal (ϕ) coordinates.

In the following subsections we shall describe the following essential elements: discretization of the governing equations, boundary conditions, shock conditions, grid size, time step, and the relationship of each of these elements with numerical stability of the scheme.

2.1 Discritization of the governing equations.

It is immediately apparent that the set of MHD equations, Eqs. (1) through (5), are non-linear. One cannot therefore, always expect to have a smooth solution in the entire flow field. In fact, this is the most important characteristic in compressible MHD flow, it means that the solution of this set of non-linear, partial differential equations allows a discontinuity to exist; this discontinuity represents the formation of MHD shocks (Richtmyer, 1963). One of the most formidable difficulties associated with the numerical computation of compressible MHD flow is to include the discontinuity (i.e., MHD shocks) within the flow field because the apperance of MHD shocks implies the existance of large gradients in the flow field. In the event that strong gradients develop in the flow field, the convential difference scheme of numerical computation usually breaks down. Therefore, it is necessary to select those methods which are able to handle these strong gradients. A class of finite difference methods known as the shock-capturing-difference schemes (Roache; 1972; Lax, 1954; Lax and Wendroff, 1960) exists. We now discuss some essential features that are relevant to our computations in this paper.

For convenience, we shall consider only one component of the conservation Equation (1) for this demonstration; that is:

$$\frac{\partial W}{\partial t} + \frac{\partial F}{\partial r} + \frac{1}{r}\frac{\partial G}{\partial \phi} - S = 0 \tag{6}$$

where W, F, G and S are single elements of the $\underset{\sim}{W}, \underset{\sim}{F}, \underset{\sim}{G}$, and $\underset{\sim}{S}$ vectors given by Equations (2)-(5).

Eq. (6) allows one to obtain a physically-acceptable discontinuity within the solution subject to a given initial condition; i.e., $W(r,\phi,0) = \Phi(r,\phi)$. Then, W will satisfy the divergence-free character of the vector field (W, F'), which is expressed by an integral equation as discussed by Lax (1957):

$$\int_0^\infty \int_{-\infty}^\infty \int_{-\infty}^\infty (\omega_t W - \nabla\omega \cdot F') \, dr d\phi dt + \int_{-\infty}^\infty \int \omega(r,\phi,0) \, \phi(r,\phi) \, dr d\phi = 0 \tag{7}$$

where F' is defined as follows:

$$\nabla \cdot F' = \frac{\partial F}{\partial r} + \frac{1}{r}\frac{\partial G}{\partial \phi}$$

For a smooth test, vector $\omega(r,\phi,t)$ must vanish at large values of $|\vec{r}|$ and t. The solution W is then called a weak solution of Eq. (6).

The immediate consequence of the divergence-free character of the vector field (W, $\underset{\sim}{F}'$) is the requirement that the conservation laws of flow quantities must be inviolate throughout the entire flow region. This will give us the generalized Rankine-Hugoniot relationships (Jeffrey and Taniuti, 1964) when we apply this conservation principle to a narrow region which contains the discontinuity.

For the particular form of Eq. (6), the generalized Rankine-Hugoniot relation can be expressed by

$$[- \sigma^* W + \underset{\sim}{F}' \cdot \underset{\sim}{n}] = 0 \tag{8}$$

where σ^* is the propagation speed of the discontinuity surface and $\underset{\sim}{n}$ is the unit normal vector to this discontinuity surface usually designated by []. The symbol [] implies that the quantity in the bracket is conserved across the discontinuity. Therefore, a discontinuity which satisfies the generalized Rankine-Hugoniot relation is part of a weak solution of the governing equation (6). Thus, the numerical solution contains discontinuities as part of its solution; i.e., the discontinuity does not arise from any boundary but is, rather, an integral part of the interior region of the flow field. Under these circumstances, we have chosen the shock-capturing finite difference method of the Rubin and Burstein (1967) scheme for our system of governing equations. The result is outlined by the following difference equations according to the mesh systems and coordinate systems as shown in Figure 1. The intermediate and final steps are given as follows:

Intermediate Step

$$\begin{aligned} \bar{\underset{\sim}{W}}^{n+1}_{i+1/2,j} = \bar{\underset{\sim}{W}}^{n+1}_{A} &= \frac{1}{2} (\underset{\sim}{W}^n_{i,j} + \underset{\sim}{W}^n_{i+1,j}) - \frac{\Delta t}{\Delta r} (\underset{\sim}{F}^n_{i+1,j} - \underset{\sim}{F}^n_{i,j}) \\ &- \frac{\Delta t}{4\Delta\phi} [\frac{1}{r_{i+1}} (\underset{\sim}{G}^n_{i+1,j+1} - \underset{\sim}{G}^n_{i+1,j-1}) + \frac{1}{r_i} (\underset{\sim}{G}^n_{i,j+1} - \underset{\sim}{G}^n_{i,j-1})] \\ &+ \frac{\Delta t}{2} (\underset{\sim}{S}^n_{i+1,j} + \underset{\sim}{S}^n_{i,j}) \end{aligned} \tag{9}$$

and

$$\begin{aligned} \bar{\underset{\sim}{W}}^{n+1}_{i,j+1/2} = \bar{\underset{\sim}{W}}^{n+1}_{B} &= \frac{1}{2} (\underset{\sim}{W}^n_{i,j+1} + \underset{\sim}{W}^n_{i,j}) \\ &- \frac{\Delta t}{4\Delta r} [(\underset{\sim}{F}^n_{i+1,j+1} - \underset{\sim}{F}^n_{i-1,j+1}) + (F^n_{i+1,j} - F^n_{i-1,j})] \\ &+ \frac{\Delta t}{2} (\underset{\sim}{S}^n_{i+1,j} + \underset{\sim}{S}^n_{i,j}) - \frac{\Delta t}{\Delta\phi r_i} (\underset{\sim}{G}^n_{i,j+1} - \underset{\sim}{G}^n_{i,j}) \end{aligned} \tag{10}$$

where the subscripts i and j refer to spatial grid coordinates shown in Figs. 1 (a) and 1 (b); the superscript refers to the time step. Similarily the intermediate variables $\bar{W}_C^{n+1}$ and $\underset{\sim}{W}_D^{n+1}$ can be found at C(i=(1/2),j) and D(i,j-(1/2)). Using the intermediate variables $\underset{\sim}{W}^{n+1}_{i,j}$ and the old variables $\underset{\sim}{W}^{n+1}_{i,j}$, the final value of $\underset{\sim}{W}^{n+1}_{i,j}$ is found from the following equation:

Final Step

$$\underset{\sim}{W}^{n+1}_{i,j} = \underset{\sim}{W}^{n}_{i,j} - \frac{\Delta t}{2}\left[\frac{1}{2\Delta r}\left(\underset{\sim}{F}^{n}_{i+1,j} - \underset{\sim}{F}^{n}_{i-1,j}\right) + \frac{1}{\Delta r}\left(\bar{\underset{\sim}{F}}^{n+1}_{A} - \bar{\underset{\sim}{F}}^{n+1}_{C}\right)\right]$$
$$- \frac{\Delta t}{2r_i}\left[\frac{1}{2\Delta\phi}\left(\underset{\sim}{G}^{n}_{i,j+1} - \underset{\sim}{G}^{n}_{i,j-1}\right) + \frac{1}{\Delta\phi}\left(\bar{\underset{\sim}{G}}^{n+1}_{B} - \bar{\underset{\sim}{G}}^{n+1}_{D}\right)\right] \tag{11}$$
$$+ \frac{\Delta t}{2}\left[\underset{\sim}{S}^{n}_{i,j} + \frac{1}{4}\left(\bar{\underset{\sim}{S}}^{n+1}_{A} + \bar{\underset{\sim}{S}}^{n+1}_{B} + \bar{\underset{\sim}{S}}^{n+1}_{C} + \bar{\underset{\sim}{S}}^{n+1}_{D}\right)\right]$$

All flow variables at the interior mesh points are advanced with these equations to the $(n+1)\Delta t$ time step except at four boundaries.

The reflection rule is applied at the axis of the symmetry, and non-reflexive conditions are applied at the boundary; i.e. $(\partial^2 \underset{\sim}{W}/\partial r^2) = (1/r^2)(\partial^2 \underset{\sim}{W}/\partial\phi^2) = 0$. At the solid boundary, a similar treatment due to Lapidus (1967) is applied. This method follows the concept of the conservational principle. New grids are constructed as shown in Fig. 1 (b). The flux " B'_A " into a rectangle which contains point A during time $\Delta t/2$ is

$$B'_A = -\frac{t}{2\Delta r}\left[\frac{1}{2}\left(\underset{\sim}{F}^{n}_{2,j} + \underset{\sim}{F}^{n}_{2,j-1}\right) - \frac{1}{2}\left(\underset{\sim}{F}^{n}_{i,j-1} + \underset{\sim}{F}^{n}_{i,j}\right)\right]$$
$$- \frac{\Delta t}{2\Delta\phi}\left[\frac{1}{2}\left(\underset{\sim}{G}^{n}_{i,j}/r_1 + \underset{\sim}{G}^{n}_{2,j}/r_1\right) - \frac{1}{2}\left(\underset{\sim}{G}^{n}_{2,j-1}/r_2 + \underset{\sim}{G}^{n}_{1,j}/r_2\right)\right] \tag{12}$$
$$+ \frac{\Delta t}{2}\left[\frac{1}{4}\left(\underset{\sim}{S}^{n}_{1,j} + \underset{\sim}{S}^{n}_{2,j} + \underset{\sim}{S}^{n}_{1,j-1} + \underset{\sim}{S}^{n}_{2,j-1}\right)\right]$$

using these fluxes B'_A, the intermediate conservative variable vector $\bar{\underset{\sim}{W}}^{n+1/2}$ at the points A and B are given by

$$\bar{\underset{\sim}{W}}^{n+1/2}_{A} = \frac{1}{4}\left(\underset{\sim}{W}^{n}_{1,j} + \underset{\sim}{W}^{n}_{1,j-1} + \underset{\sim}{W}^{n}_{2,j-1} + \underset{\sim}{W}^{n}_{2,j}\right) + \underset{\sim}{B}'_A \tag{13}$$

and

$$\bar{\underset{\sim}{W}}^{n+1/2}_{B} = \frac{1}{2}\left(\underset{\sim}{W}^{n}_{1,j} + \underset{\sim}{W}^{n}_{1,j-1}\right) + \underset{\sim}{B}'_A \tag{14}$$

Similar expressions for points C and D can be written. The final variables at the point (1,j) are then found by applying the conservational principle to rectangle ABCD.

$$\underset{\sim}{W}^{n+1}_{i,j} = \underset{\sim}{W}^{n}_{1,j} - \frac{\Delta t}{\left(\frac{\Delta r}{2}\right)}\left[\frac{1}{2}\left(\bar{\underset{\sim}{F}}^{n+1/2}_{C} + \bar{\underset{\sim}{F}}^{n+1/2}_{A}\right) - \frac{1}{2}\left(\bar{\underset{\sim}{F}}^{n+1/2}_{D} + \bar{\underset{\sim}{F}}^{n+1/2}_{B}\right)\right]$$
$$- \frac{\Delta t}{\Delta\phi}\left[\frac{1}{2}\left(\frac{\bar{\underset{\sim}{G}}^{n+1/2}_{A}}{r_1 + \frac{\Delta r}{2}} + \frac{\bar{\underset{\sim}{G}}^{n+1/2}_{B}}{r_1}\right) - \frac{1}{2}\left(\frac{\bar{\underset{\sim}{G}}^{n+1/2}_{C}}{r_1 + \frac{\Delta r}{2}} + \frac{\bar{\underset{\sim}{G}}^{n+1/2}_{D}}{r_1}\right)\right] \tag{15}$$
$$+ \frac{\Delta t}{4}\left(\bar{\underset{\sim}{S}}^{n+1/2}_{A} + \bar{\underset{\sim}{S}}^{n+1/2}_{B} + \bar{\underset{\sim}{S}}^{n+1/2}_{C} + \bar{\underset{\sim}{S}}^{n+1/2}_{D}\right)$$

In Eq. (15), the intermediate flux vectors $\underset{\sim}{F}^{n+1/2}$, $\underset{\sim}{G}^{n+1/2}$ and source vector $\underset{\sim}{S}^{n+1/2}$ are evaluated by using the intermediate conservative flow variables $\tilde{W}^{n+1/2}$. Computations of conservative variables at the solid boundary are performed by using Eqs. (12)-(15). One critical assumption needed for this method is that the rectangle ABCD is so small that the variation of the flow quantities at the boundary point (1,j) is equal to the variation of the flow quantities at the point E. (see Fig. 1.) The radial momentum at point E found by eqn. (15) may not be zero. Since the radial momentum at the solid boundary should be zero, this radial momentum at point E is artificiality converted to thermal energy of the gas P'_E, i.e.

$$P'_E = (\gamma - 1)\left(\rho \frac{|V_r|^2}{2}\right)\left(-\frac{V_r}{|V_r|}\right) \tag{16}$$

Therefore, the pressure and the radial velocity at the solid boundary becomes

$$P^{n+1}_{1,j} = P^{n+1}_{E} + P'_E$$

and

$$V^{n+1}_{1,j} = 0 \tag{17}$$

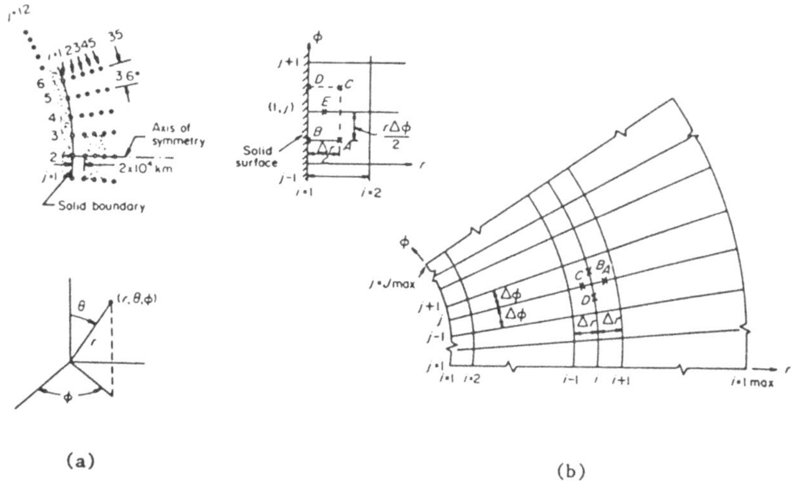

Figure 1. The schematic representations of (a) Grid point arrangement and coordinate system, and (b) discretized mesh system for interior domain and solid boundary.

2.2 Numerical Stability Considerations

The set of finite difference equations; (i.e., Equations (9)-(15)), subject to proper initial and boundary conditions, do not always generate a stable numerical solution. The possibility is admissible because, in a non-linear interaction process, the small amplitude pertubation, superimposed on a mean smooth solution, can grown in amplitude without bound as time progresses. When the amplitude of the numerical oscillation becomes too large, the numerical solution becomes meaningless. In general, the numerical instability can be classified into two categories; static instability and dynamical instability. The static instability is caused by using an improper difference method. For instance, if one uses the centered difference analog in time for an advection equation, the resulting finite difference equation is unconditionally unstable. On the other hand, the dynamic instability is caused by one or more of the following reasons: (i) too large gradient of physical quantities in the flow (such as shocks); (ii) too large time step Δt ; and (iii) improper handling of the boundary conditions. For further details of these discussions, we refer the reader to the literature (i.e., Richtmyer and Morton, 1967; Rubin and Burstein, 1967; Lindemuth, 1971; and Roache, 1972). We merely discuss the specific stability criteria which we employed for this example.

It is understood that the Von Neumann-Richtmeyer's artificial diffusion scheme is not effective in two-dimensional MHD flows (Lindemuth, 1971). The Lapidus difference scheme has been used in many multi-dimensional flow problems with good, simple and effective results (Lapidus, 1967; Lindemuth, 1971; Han, 1977). The essence of this scheme is to add the artificially viscosity terms to all partial difference equations, thus, the governing equations to be solved for the physical system become

$$\frac{\partial \underset{\sim}{W}}{\partial t} + \underset{\sim}{T} + \underset{\sim}{T}_r + \underset{\sim}{T}_\phi = 0 \tag{18}$$

where

$$\underset{\sim}{T} = \frac{\partial \underset{\sim}{F}}{\partial r} + \frac{1}{r}\frac{\partial \underset{\sim}{G}}{\partial \phi} - \underset{\sim}{S}$$

and $\underset{\sim}{T}_r$ and $\underset{\sim}{T}_\phi$ are artifical diffusion terms expressed by

$$\underset{\sim}{T}_r = - b_r^2 \, (\Delta r)^2 \frac{\partial}{\partial r} \left[\left| \frac{\partial V_r}{\partial r} \right| \frac{\partial \underset{\sim}{W}}{\partial r} \right] \tag{19}$$

$$\underset{\sim}{T}_\phi = - b_\phi^2 \, (r\Delta\phi)^2 \frac{1}{r}\frac{\partial}{\partial \phi} \left[\left| \frac{\partial V_\phi}{\partial r} \right| \frac{\partial \underset{\sim}{W}}{\partial \phi} \right] \tag{20}$$

with b_r and b_ϕ being constants of order of unity. Eq. (18) is then solved by a fractional time step (Han, 1977). Effects of artificial viscosity on the overall flows field have been carefully investigated to meet requirements specified by Richtmyer and Morton (1967).

The time step is an important factor for control of the numerical instability and efficiency of the computation. Therefore, we need to determine the maximum allowable time step for the computation. According to the method given by Richtmyer (1963) the maximum allowable time step was determined for two dimensional transient MHD flow. In this method, the maximum allowable time step has to satisfy following three inequalities simultaneously:

$$\frac{\Delta t}{\Delta t} \; , \; \frac{\Delta t}{r\Delta\phi} \leqslant \frac{1}{\sqrt{2}(|\underset{\sim}{V}| + C_A + C_S)}$$

$$2b_r^2 \; \Delta t \; \left| \frac{\partial V_r}{\partial r} \right| \leqslant 1$$

and

$$2b_\phi^2 \; \Delta t \; \left| \frac{\partial V_\phi}{r\partial\phi} \right| \leqslant 1 \tag{21}$$

where

$$|V|^2 = V_r^2 + V_\phi^2, \quad C_A^2 = C_\phi^2 + C_S^2, \quad C_S^2 = \frac{\gamma P}{\rho}$$

and

$$C_{r,\phi} = \frac{|B_{r,\phi}|}{\sqrt{\rho\mu_o}} \tag{22}$$

3. COMPUTATION PROCEDURES AND PHYSICAL RESULTS

3.1 Computation Procedures

In order to demonstrate the ability of the numerical method to solve MHD flow problems associated with the formation and propagation of interplanetary shocks that are initiated by solar activity, we have chosen the non-planar, two-dimensional, time-dependent MHD model (Han et al., 1982; Wu et al., 1983b) for illustration in this paper.

For the benefit of the reader, we briefly summarize the actual computation in a flow diagram as shown in Figure 2. The essential computing steps are: (i) Definition of the grid size will depend upon the time step. An optimal condition for choosing the grid size is used on the basis of the maximum allowable time step discussed above. (ii) choice of the initial conditions for all physical parameters is based upon either a relaxation technique or by using a solution of the governing equations by setting $\partial/\partial t=0$ (i.e., the steady state solution of the system). For example, the initial conditions chosen here are those for a isothermal hydrodynamic atmosphere. Using these values, the initial conditions of the conservative flow variables defined in Eq. (2) can be found, i.e., $(W_1)^n_{i,j}=r_i^2 \; \rho^n_{i,j}$, $(W_2)^n_{i,j}=r_i^2 \; \rho^n_{i,j} \; V^n_{r\,i,j}$....., etc. (iii) Spatial and temporal perturbations in one variable (or a combination of physical variables) are introduced in this step. Each particular physical case may suggest the choice of variable (s), such as velocity, etc., and its (their) temporal variation and spatial location. For example, the physical situation

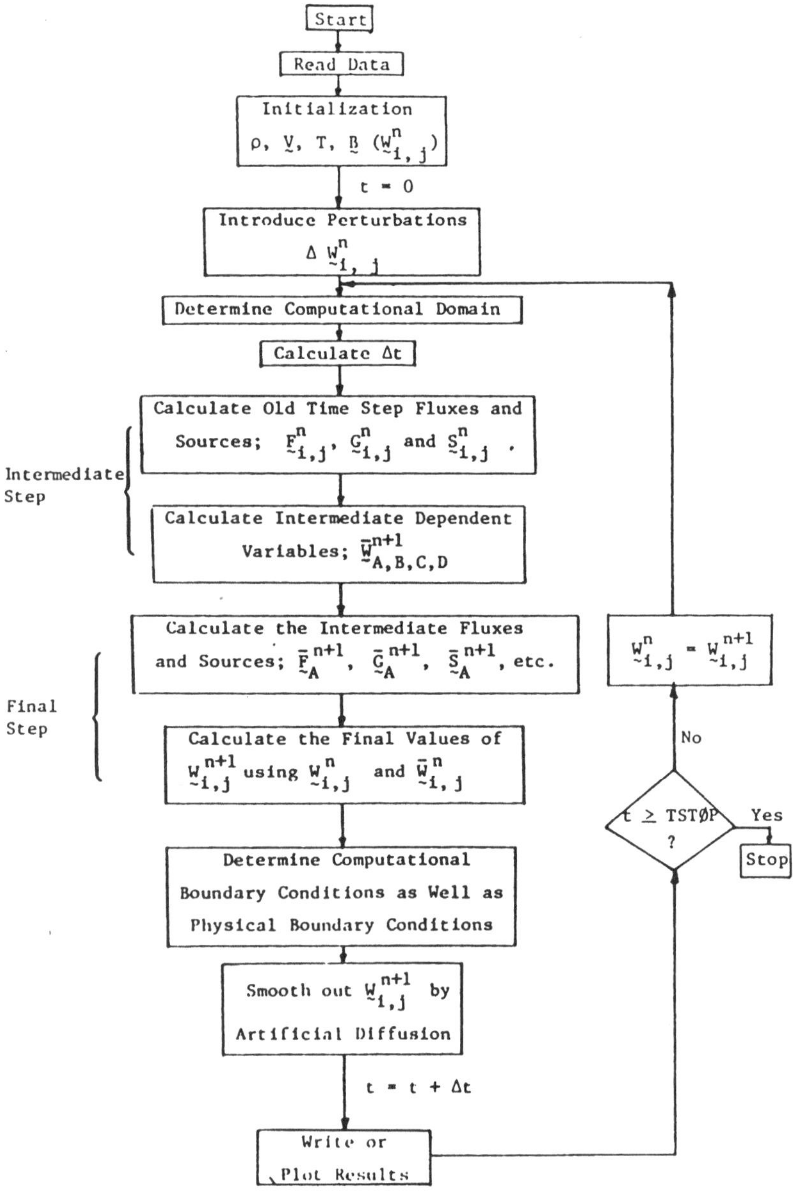

Figure 2. Computation Flow Chart

of a solar flare suggests that a shock-type perturbation may be introduced at the lower boundary for the study of shock propagation in interplanetary space (c.f., Wu et al., 1983). For this case, the Rankine-Hugoniot equations are solved in a separate routine for desired input shock velocities. Other cases (simple velocity, density, and temperatures pulses) were performed by Wu et al. (1976). (iv) The stability conditions (i.e. Equations (21) and (22)) are used to determine the time step. (v) Find $\underset{\sim}{F}^n_{i,j}$, $\underset{\sim}{G}^n_{i,j}$, and $\underset{\sim}{S}^n_{i,j}$ by using Equations (3), (4) and (5) according to the initial conditions given in step (ii). (vi) Using Equations (9) and (10) determine $\overline{W}^{n+1}_{\sim i,j}$ (vii) Returning to the original set of governing Equations (1) through (5) with the aid of $\underset{\sim}{W}^{n+1}_{i,j}$ found in step (iv), determine the values of $F^{n+1}_{\sim i,j}$ $G^{n+1}_{\sim i,j}$and $S^{n+1}_{\sim i,j}$(viii) Calculate the final variables $\underset{\sim}{W}^{n+1}_{i,j}$ by using Eq. (11). (ix) Determine the boundary conditions by using compatibility relationships (see, Nakagowa, 1981 a,b). (x) The variables $\underset{\sim}{W}^{n+1}_{i,j}$ found in step (viii) are smoothed by using artificial diffussion (Eq. (18)). (xi) Write out all the necessary information, then complete one cycle of computation.

3.3 Physical Results

To illustrate the numerical code, we have chosen (as noted earlier) a non-planar, two-dimensional, time-dependent MHD model (Wu et al., 1983b) for presentation here. The initial conditions for this example includes an initial magnetic field configuration which has a canonical Archimedean topology in the equational plane; the solar wind plasma includes a small meridional velocity. The three magnetic field components at the inner boundary (taken at 18 solar radii) are found via an iteration process by using the relaxation method (i.e., with $\partial/\partial t \neq 0$, an initially-assumed profile of all physical parameters, but with <u>no</u> pulse), until the desired representative values of the various plasma and field parameters are obtained at 1 A.U. The final representative values are within the bounds of the observed values. The parameters at 1.10 A.U. (and 18 solar radii) are as follows, respectively; V_r=355 km sec^{-1} (250 km sec^{-1}), V_θ=4/4x10^{-3} km sec^{-1} (0.2 km sec^{-1}); V_ϕ=0.5 km sec^{-1} (4.0 km sec^{-1}); T=3.3x10^4 K (1.1x10^6 K); n=10 cm^{-3} (1.8x10^3 cm^{-3}); B_r=2.3 gama (300 gama) B_θ=4.4x10^{-2} gama (1.0 gama); and B^Φ=-2.0 gama (-300 gama).

Figure 3 shows the model calculations of the disturbed solar wind plasma parameters due to the propagation of an interplanetary shock generated by a flare event, this flare event is characterized by an initial shock velocity being 3000 km sec^{-1} which corresponds to a local mach number of ~23. This perturbation is introduced at computational procedure step (iii) with a time duration being 5,400 sec. Also, the maximum strength (3000 km sec^{-1}) is centered at =30^0 and is taken to decay sinusoidally over an included in angle of 24^0 in order to simulate the shock from a solar flare. In these results, the total pressure contours (i.e. P_{tot}=2nkT+B^2/8π) which are superimposed on the deformed magnetic field lines ~20.2 hrs after introduction of the perturbation are shown in Fig. 3a.; the vectoral velocity enhancement is exhibited in Fig. 3b; the plasma β(i.e. 16πnkT/B^2) is shown in Fig. 3c; and finally the three-dimensional representation of the disturbed solar wind configuration is depicted in Fig. 3d. These numerical results shown in the figures clearly demonstate that there is interaction between the quiet solar wind and a flare-generated shock with the IMF (i.e., Interplanetary-Magnetic-Field). For example, the MHD fast forward shock arrives at 0.75 A.U. and the longitudinally more limited MHD fast reverse shock develops at ~0.5 A.U., as shown in Figures 3a and 3d. A blob of compressed plasma propagates out between 0.2 to 0.65 A.U. as seen in Figure 3c. These results also demonstrate the basic MHD shock structures, where both the quasi-parallel and perpendicular shock-structure variation along its periphery are exhibited in Figures 3a and 3c. The plasma β contour map shown in Figure 3b may be used as an indicator of the dynamic behavior chracteristics; i.e., whether the region is dominated by magnetic field or by plasma kinetic energy.

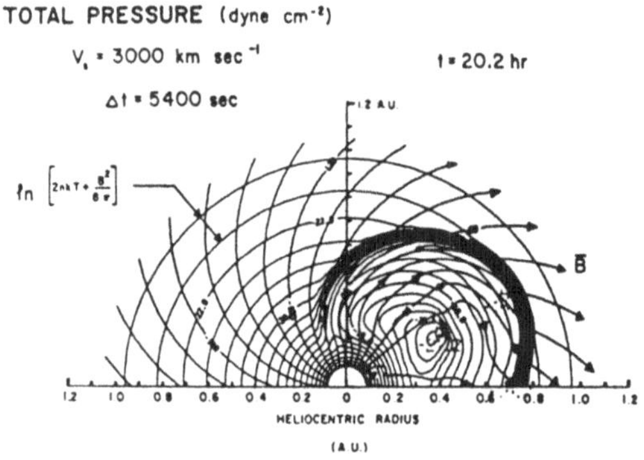

(a)

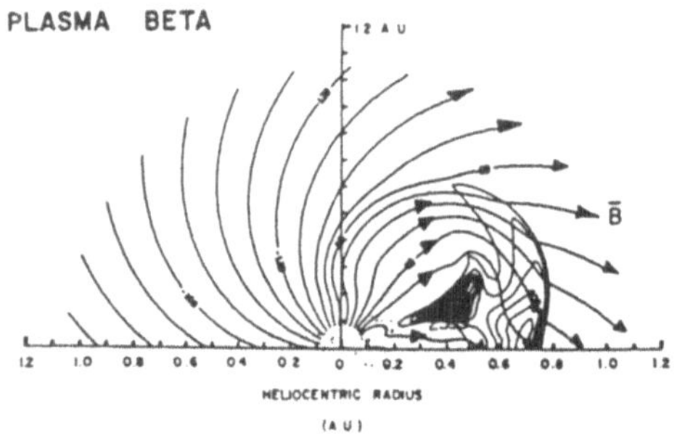

(b)

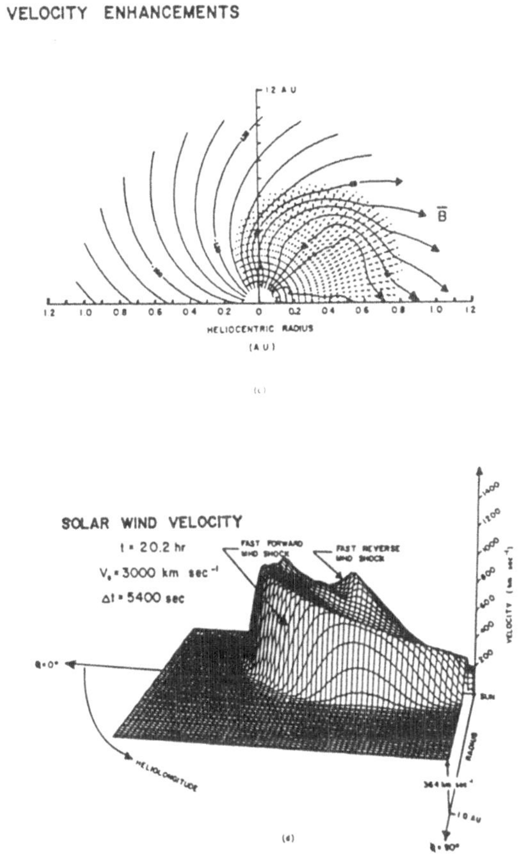

Figure 3. Disturbed solar wind plasma parameters and magnetic field configurations in the equatorial plane at 20 hrs after a 3,000 km sec^{1} shock being introduced for a period of 1 1/2 hours (5,400 sec.); (a) Total pressure, (b) plasma Beta, (c) Velocity enhancements and (d) three-dimensional representation of the shocked solar wind velocity.

In order to illustrate, the formation of the interplanetary shocks, we have chosen a rather weak disturbance (i.e. initial shock velocity being 1,000 km sec^{-1}) with a duration of 3,500 sec. The three-dimensional representative of the disturbed solar wind configuration at various times is shown in Figure 4. Because the maximum strength of the distrubance (introduced at $\phi=30^0$) is supersonic, the fast MHD shock propagates radially outward with

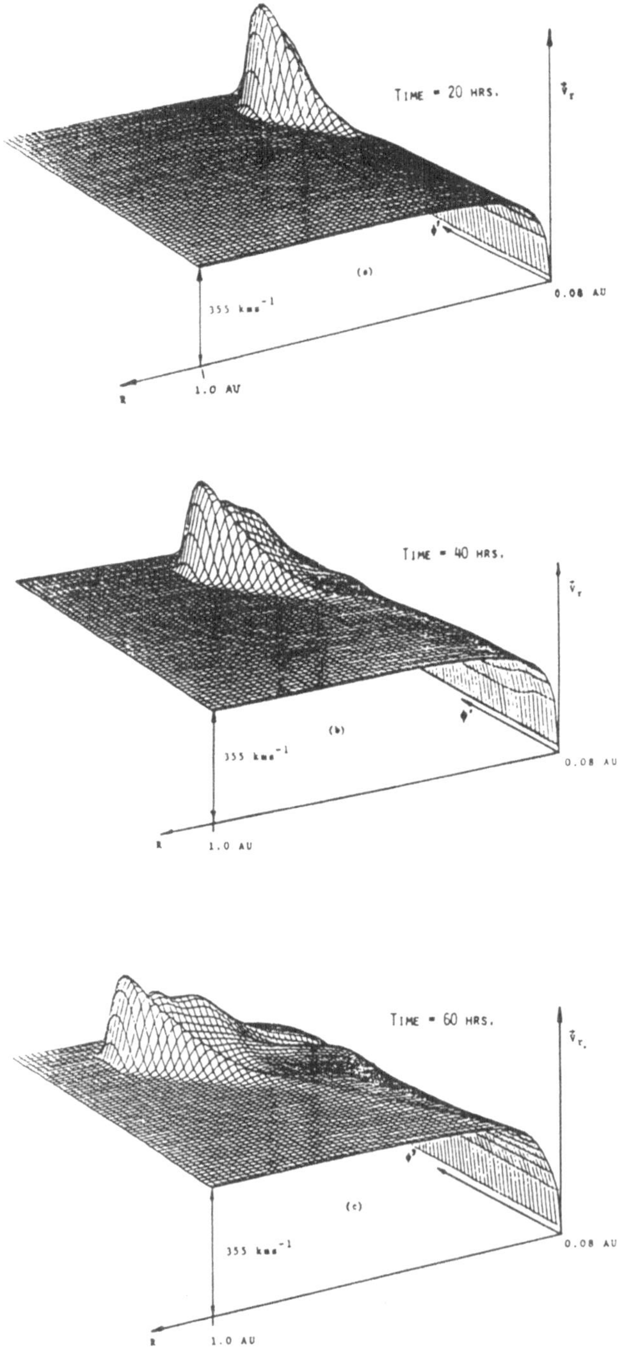
TIME = 20 HRS.
$\vec{V}_r$
ϕ'
(a)
355 kms^{-1}
0.08 AU
R
1.0 AU
TIME = 40 HRS.
$\vec{V}_r$
ϕ'
(b)
355 kms^{-1}
0.08 AU
R
1.0 AU
TIME = 60 HRS.
$\vec{V}_r$
ϕ'
(c)
355 kms^{-1}
0.08 AU
R
1.0 AU

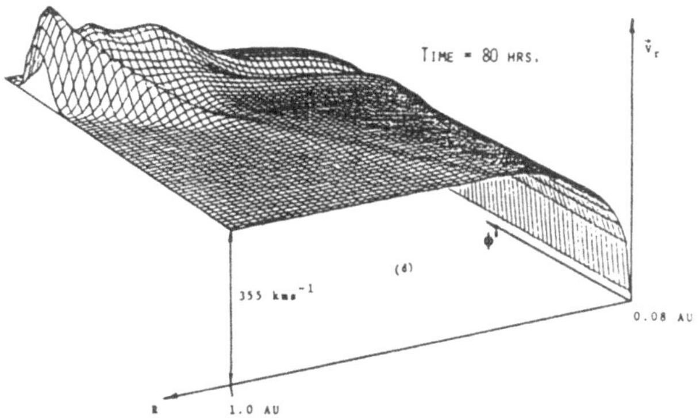

Figure 4. Evolution of disturbed solar wind velocity, which shows the development of shocks, i.e., fast mode MHD forward shock and reverse shock in both radial and heliolongitudinal directions, ϕ' the angle ϕ' in this presentation, is the angle measured from veiwing plane, thus, the central meridian ($\phi=30^0$) is corresponding to ϕ' being 150^0.

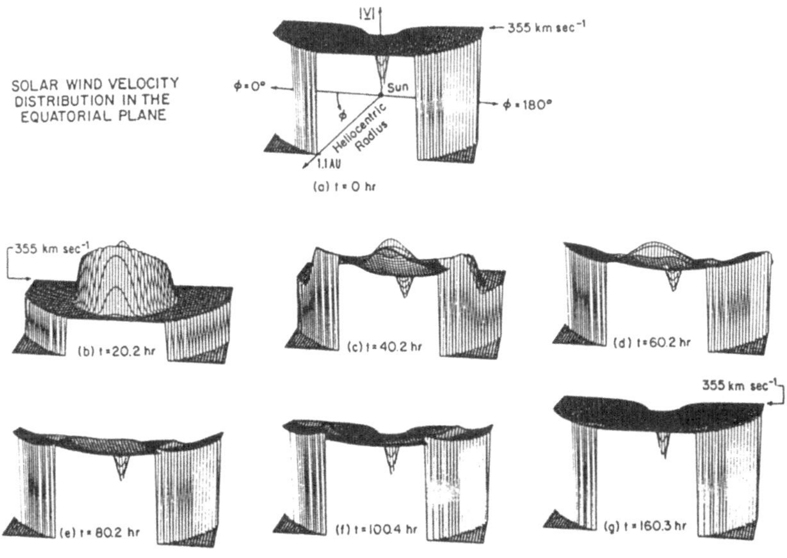

Figure 5. Three-dimensional representation of the solar wind velocity distribution in the equatorial plane. (Dryer et al., 1983)

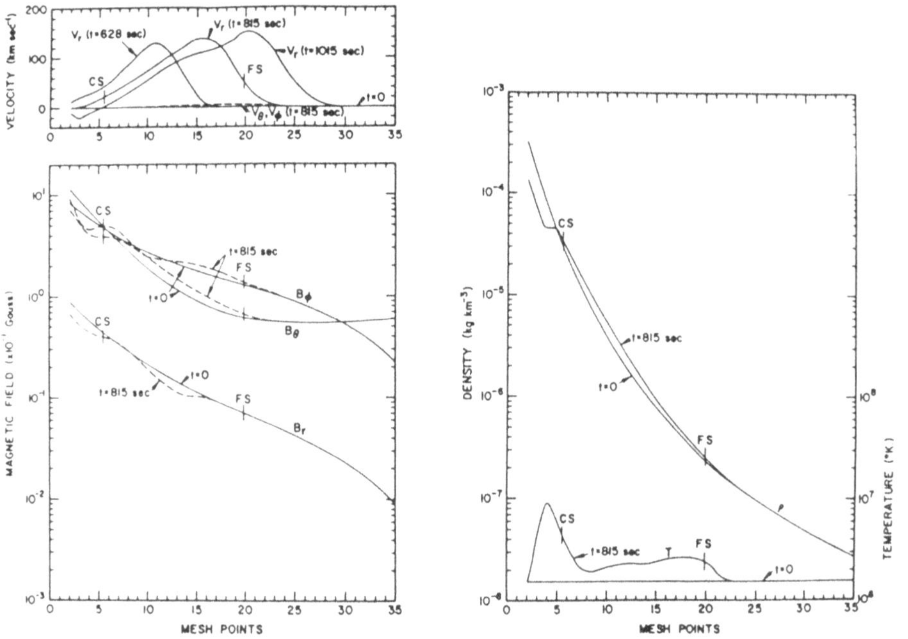

Figure 6. For the case of the initial locally closed magnetic field configuration; (a) Velocity profiles along the radial grid position (j=-2) at time t=600, 815 and 1015 sec. (b) Magnetic field profiles along the radial grid position (j=2) at time t=815 sec. (c) Density and temperature profile along the radial grid position (j=2) at time t=815 sec, where tic markes indicate the contact surface (CS) and fast shock (FS) respectively (Han et al., 1982).

heliolongitudinal spreading. At t=20 hr, the early phase of heliolongitudinal spreading of the fast forward shock is beginning to appear. Formation of the reverse MHD shock (caused by the finite pulse duration) is about to be seen from our choice of viewing angle in Figure 4a. At t=40 hr, the fast reverse MHD begins to appear in addition to the heliolongitudinally-spreading fast forward MHD shock. Development of quasi-parallel and quasi-perpendicular portions of the MHD shock is seen more clearly in Figure 3a and 3b. The spreading phenomenon caused by nonlinear wave coupling becomes more apparent as time progresses as shown in Figure 4c and 4d. In Figure 5, a complete evolution of the three-dimensional representation of the solar wind velocity distribution in the equatorial plane is plotted from another viewing angle and for another initial shock velocity (3,000 km sec^{-1}). It is worthwhile to note that the first (top) and last (lower right corner) display (i.e. Figure 5a and 5g, detail see Dryer et al., 1983) are identical, thereby confirming the excellent numerical accuracy of this model. Physically, this result demonstrates that the

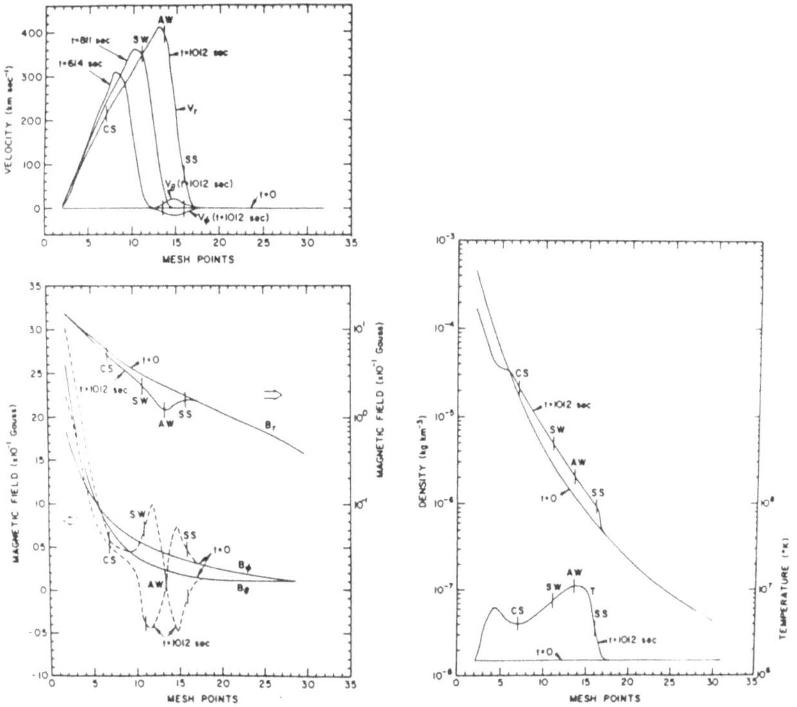

Figure 7. For the case of the initial locally open magnetic field configuration, (a) Velocity profiles along the radial grid position (j=2) at time t=600,811, and 1011 sec. (b) Magnetic field profiles along the radial grid position (j=2) at time t= 1011 sec. (c) Density and temperature profiles along the radial grid position (j=2) at time t=1011 sec, where the tic marks indicate the contact surface (CS), slow wave (SW), slow shock (SS) and Alfvén Wave (AW) respectively (Han et al., 1983).

solar wind has returned to its original undisturbed state after the complex disturbances passes through the entire domain ($0 \leqslant \phi \leqslant 180^{0}$`; 0.08 A.U. $\leqslant R \leqslant 1.1$ A.U.)

To continue our discussion on the accuracy of our numerical results, we now analyze the wave propagation along a fixed radial direction. Figures 6 and 7 show disturbed solar wind plasma properties due to a thermal disturbance ($T/T_0=10$) at several time steps as a function of radial position (i.e., the indices "i" as shown in Figure 1) which lies next to the axis of symmetry (j=2) in the canonical closed and open magnetic field configurations (Wu et al., 1982). Figure 6a exhibits the radial velocity profiles at t=628, 815 and 1015 sec. Measuring the distance between the wave front at t=628 sec and 815 sec, we obtain the absolute disturbed speed V_S=855 km sec^{-1}. This speed is greater than a local fast wave speed C_f=843 km

sec^{-1} (at i=20), implying that the disturbance is a fast MHD shock wave. Thus, the complete post shock condition can be computed by using the analytical MHD shock jump conditions in one-dimension (See Appendix). The results given by these analytical MHD shock jump conditions are $\rho_2/\rho_1=1.22$; $T_2/T_1=1.36$; $B_{\theta 2}/B_{\theta 1}= B_{\phi 2}/B_{\phi 1}=1.21$; $B_{r2}/B_{r1}=1.0$; $V_{r2}=151$ km sec^{-1}; $V_{\theta 2}=3.7$km sec^{-1}; and $V_{\phi 2}=7.41$ km sec^{-1}. Using the numerical results at i=20 (t=815 sec) and considering the numerical diffusion effects, we have $\rho_2/\rho_1=1.11$; $T_2/T_1=1.76$; $B_{\theta 2}/B_{\theta 1}=1.15$; $B_{\phi 2}/B_{\phi 1}=1.18$; $B_{r2}/B_{r1}=1.01$; $V_{r2}=14$ km sec^{-1}, $V_{\theta 2}=2$ km sec^{-1} and $V_{\phi 2}=2.2$ km sec^{-1}. Comparison of the corresponding ratios shows that, indeed, there are discrepancies between the analytical and numerical results. The largest differences between the tangential velocity components, V_θ and V_ϕ, are probably due to the two-dimensional nature of the wave propagation (i.e., the waves are not strictly one-dimensional in the numerical computation) and to the smallness of the radial magnetic field strength compared with tangential components at this location (i=20). Figure 6 (b, c) shows magnetic fields, density and temperature profile at t=815 sec. The Contact Surface (CS) often called the entropy wave, and fast MHD shock (FS) waves are indicated by the tic marks. The Alfven wave mode is not excited along this direction since the magnetic field is perpendicular to the wave front.

In order to demonstrate the excitation of Alfven wave (AW) and slow, shock (SS), we have chosen a computation within an open magnetic field configuration, Figure 7a shows the velocity profiles at t=614,811 and 1011 sec. Similarly, as in the previous case, we estimate the disturbance speed: $V_S=600$ km sec^{-1}. The Alfven wave speed at i=16 (t=1011 sec) is $C_A=610$ km sec^{-1}. Since the disturbance speed, V_S is faster than the slow wave speed $C_S=145$ km sec^{-1}, and slower than the fast wave speed, $C_f=843$ km sec^{-1}. Note, further, that the disturbance occurs within an environment in which the flow velocity is nearly parallel to the magnetic field line; thus, the disturbance is a slow MHD shock. Using the MHD shock jump conditions we obtain, $\rho_2/\rho_1=3.43$, $T_2/T_1=6.47$ and $V_{r2}=437$ km sec^{-1}. Computational results show that $\rho_2/\rho_1=2.25$, $T_2/T_1=7.33$ and $V_{r2}=420$ km sec^{-1}. Large differences of ~50% are show in the density jump. This is due to the two-dimensional nature of the slow shock front propagation in this direction. If the propagation were strictly one-dimensional, the radial magnetic field B_r would remain constant. Computational results show that B_r decreases up to ~50% during the slow shock passage, as shown in Figure 7b. Immediately behind the slow shock front, disturbed solar wind plasma expands rapidly into the θ-direction thereby decreasing the density and the radial component of the magnetic field B_r. The strength of B_r increases again as the disturbed solar wind plasma undergoes expansion, as seen in Figure 7b. This change of B_r in the opposite sense with density introduces the change of V_ϕ(and B_ϕ as well) in the opposite sense. In other words, the Alfven waves are excited in this region (i=12-15). Figure 7 shows no abrupt changes in V_r, ρ and T where the Alfven wave is located. The tangential component of the magnetic field shows a 180^0 rotation while the total strength remained almost the same. The

Contact Surface (CS), Slow Shock (SS), Alfven Wave (AW) and Slow Wave (SW) are indicated by tic marks in Figure 7. In the case of parallel propagation, if the magnetic field strength (B_r) is extremely large compared to the disturbed solar wind plasma gas pressure B_r does not change and consequently, no complex MHD waves are excited. This case thus becomes the one-dimensional ordinary gasdynamic flow case. On the other-hand, the Alfven wave will always be present in multi-dimensional, transient, MHD flows.

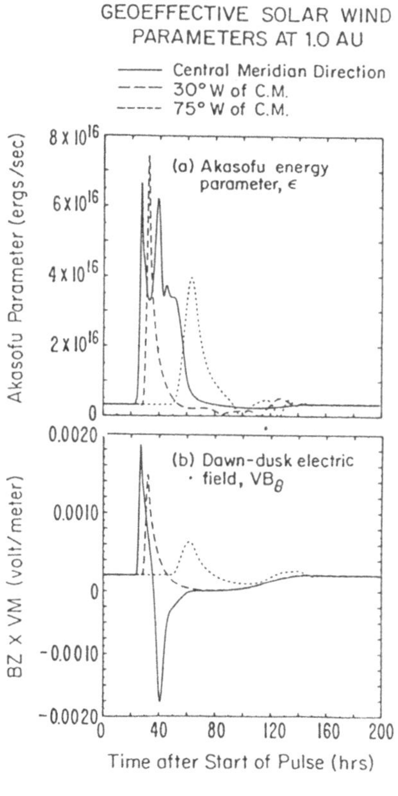

Figure 8. Time evolution of the geoeffective solar wind parameters at 1.0 A.U. (Dryer et al., 1984).

As a final remark, we show a most recent application of this numerical model for the prediction of geoeffective solar wind parameters given by Dryer et al., (1984). Figure 8 shows the energy parameter ε(Akasofu, 1981) and Dawn-dusk electric field (i.e., VB_θ; Claver et al., 1983). It can be easily noted that this kind of calculation can be used to estimate the arrival time of a solar disturbance at 1 A.U. (in particular, at earth's environment). Further one can predict, in principle, how long the disturbance and how severe this disturbance will be. Since the model output gives all solar wind plasma parameters other geoeffective empirical parameters (Akasofu, 1981) can also be computed.

4. CONCLUDING REMARKS

The currently used numerical method for simulation of the formation and propagation of interplanetary shocks has been presented. This method is based on the shock-capturing finite difference scheme (Lax, 1950; Lax and Wendroff, 1960) together with the newly-developed method of NEAR characteristics (Nakagawa, 1980; 1981 a,b). Some specific examples are also discussed in this paper. These examples strongly suggest that all the shocked solar wind plasma parameters due to given physical perturbations (such as flare-generated shocks) can be predicted by using this method. However, the numerical method presented here is limited to the supersonic and super-alfvénic flow. Because of interest in the study of solar flare energy build up, the sub-alfvénic flow has attracted some attention. Therefore, a parallel numerical simulation model was developed (Wu et al., 1983a). This model is based on a full-implicit difference scheme (Wang et al., 1982; Hu and Wu, 1984) which has not been discussed here. It is important to notice that observable solar disturbances begin at the solar surface; thus, it is interesting to construct a model which can accomodate both sub-and super-alfvénic flows. Such a model would then enable us to investigate the progress of a solar disturbance from its birth to its maturity and demise as it propagates throughout heliospheric space as interplanetary shocks. This model was described by Han et al. (1982a).

On the other hand, all of these numerical models have ignored the dissipative mechanisms except at the shock itself. In order to include the dissipative mechanisms on a self-consistent physical base, the hybrid numerical model (Cuperman, 1983; Wu, 1983) needs to be developed. To include dissipative mechanisms, the implicit difference scheme must be used because the criteria of the time-step for satisfying the numerical stability must be significantly relaxed. In the past decade, it has been proved that numerical simulation for a physical system is an essential method to reveal and interpret physical processes whose outward manifestations have been observed by a variety of instruments. The numerical method is a necessary tool that is essential to improve our understanding of the physics of interplanetary shocks as well as other subjects.

ACKNOWLEDGEMENT

I am very much indebted to Dr. M. Dryer, who leads me to this field, read this manuscript, and has given many valuable comments. This work is supported by a NASA Grant (NAGW-9), NASA/MSFC contract (NAS8-33526) and AFGL contract F19428-83-K-0019.

REFERENCES

Akasofu. S. -I., Solar Phys. 71. 175, 1981.

Chow, C. Y., An Introduction to Computational Fluid Mechanics, John Wiley & Sons, Inc., New York, 1979.

Clauer, C. R., R. L. McPherron and C. Searls, J. Geophys. Res., 88, 2123, 1983.

Cuperman, S., Space Sci. Rev., 34, 63, 1983.

Dryer, M., Space Sci. Rev., 15, 403, 1974.

Dryer, M., Space Sci. Rev., 17, 277, 1975.

Dryer, M., S. T. Wu, G. Gislason, S. M. Han, Z. K, Smith, D. F. Smart and M. A. Shea, Astro. Phys. & Space Sci., 1984 (in press).

Han, S. M., Ph. D. Thesis, University of Alabama in Huntsville, Huntsville, Alabama, 35899, USA, 1977.

Han, S. M., S. T. Wu and Y. Nakagawa, Int'l. J. of Computers and Fluids, 7, 97, 1979.

Han, S. M., S. T. Wu and Y. Nakagawa, Int'l. J. of Computers and Fluids, 10, No. 2, 127, 1982.

Hu, Y. Q. and S. T. Wu, J. Comp. Phys, 55, #1, 33, 1983.

Jeffrey, A. and T. Taniuti, Non-Linear Wave Propagation, Academic Press, New York, 1964.

Lapidus, A., J. Comput. Phys., 2, 154, 1967.

Lax, P. D., Comm. Pure Appl. Math., 7, 159, 1954.

Lax, P. D. and B. Wendroff, Comm. Pure and Appl. Math., 8, 217, 1960.

Lindemuth, I. R., Ph. D. Thesis, University of California/Livermore, USA, 1971.

Nakagawa, Y., Ap. J., 240, 275, 1980.

Nakagawa, Y., Ap. J., 247, 707, 1981a.

Nakagawa, Y., Ap. J., 247, 719, 1981b.

Patankar, Suhas, V., Numerical Heat Transfer and Fluid Flow, Hemisphere Publishing Corporation Washington/New York/London, 1980.

Richtmyer, R. D., A Survery of Difference Methods for non-steady Fluid Dynamics, NCAR Tech Note 63-2. Boulder, Colorado. USA. 1963.

Richtmyer, R. D. and K. W. Morton, Difference Methods for Initial Value Problem, 2nd edition, Interscience Publishers, John-Wiley and Sons, New York, 1967.

Roache, P. J., Computational Fluid Dynamics, Hermosa Publishers, Alburquerque, N.M. USA, 1972.

Rubin, E. L. and S. F. Burstein, J. Comput. Phys., 2, 1978, 1967.

Wang, Shui, Youqiu Hu and S. T. Wu, Scientia Sinica, XXV, 1305, 1982.

Wu, S. T., Space Sci. Rev., 32, 115, 1982.

Wu, S. T., Space Sci. Rev., 34, 73, 1983.

Wu, S. T., M. Dryer and S. M. Han, Solar Phys., 49, 187, 1976.

Wu, S. T., Y. Q. Hu, Y. Nakagawa and E. Tandberg-Hanssen, Ap. J., 266, 866, 1983a.

Wu, S. T., M. Dryer and S. M. Han, Solar Physics, 84, 395, 1983.

APPENDIX

MHD SHOCK JUMP CONDITIONS

According to Equation (8), the generalized Rankine-Hugoniot shock-jump conditions for one-dimensional MHD flow can be expressed by

$$V_S[\underset{\sim}{W}] = [\underset{\sim}{F}] \tag{A-1}$$

where V_S is the absolute shock speed, $\underset{\sim}{W}$ is the column vector for solar wind plasma variables and $\underset{\sim}{F}$ is the radial flux vector. [] indicates that quantities inside the brackets are conserved across a shock. A detailed derivation is given by Han et al., (1982), we only present the results here; By solving Equations (A-1), the density jump is determined by a fourth-order polynominal algebraic equation as follows:

$$\begin{aligned}
&\sigma^4 \left[\frac{1}{\gamma M_S^2} + \frac{\gamma-1}{2\gamma} + \frac{3(\gamma-1)}{2} \left\{ \left(\frac{M_r}{M_\theta}\right)^2 + \left(\frac{M_r}{M_\phi}\right)^2 \right\} \right] \\
&- \sigma^3 \left[1 + \frac{1}{\gamma M_S^2} + \frac{2}{\gamma}\left(\frac{M_r}{M_S}\right)^2 + \frac{\gamma-1}{\gamma} M_r^2 + \left\{ \left(\frac{M_r}{M_\theta}\right)^2 + \left(\frac{M_r}{M_\phi}\right)^2 \right\} \right. \\
&\qquad \left. \times \left(1 + \frac{2(\gamma-1)}{\gamma} + \frac{\gamma-2}{2\gamma} M_r^2 \right) \right] \\
&+ \sigma^2 \left[1 + 2M_r^2 + \frac{2}{\gamma}\left(\frac{M_r}{M_S}\right)^2 + \frac{1}{\gamma}\left(\frac{M_r}{M_S}\right)^2 M_r^2 + \left(\frac{\gamma-1}{2\gamma}\right) M_r^4 \right. \\
&\qquad \left. - \frac{\gamma-1}{2\gamma} + \left\{ \left(\frac{M_r}{M_\theta}\right)^2 + \left(\frac{M_r}{M_\phi}\right)^2 \right\} \left(1 + \frac{\gamma-1}{\gamma} M_r^2 + \frac{\gamma-1}{2} \right) \right] \\
&- \sigma \left[\frac{\gamma+1}{\gamma} M_r^2 + M_r^4 + \frac{1}{\gamma}\left(\frac{M_r}{M_S}\right)^2 M_r^2 + \frac{M_r^2}{2} \left\{ \left(\frac{M_r}{M_\theta}\right)^2 + \left(\frac{M_r}{M_\phi}\right)^2 \right\} \right] \\
&+ \left(\frac{\gamma+1}{2\gamma}\right) M_r^4 = 0
\end{aligned} \tag{A-2}$$

where

$$\sigma = \rho_2/\rho_1$$
$$M_S = \frac{V_S - V_{r_1}}{a_1} \tag{A-3}$$

and

$$M_{r,\theta,\phi} = \frac{V_S - V_{r_1}}{\dfrac{B_{r_1,\theta_1,\phi_1}}{\sqrt{\rho_1 \mu_o}}} \tag{A-4}$$

Equation (A-2) holds for all orientations of the magnetic field.

For the non-magnetic field case, Equation (A-2) is reduced to the ordinary gas-dynamic shock jump conditions. The remainder of the jump conditions follow in terms of σ and other known parameters being given below.

$$\frac{P_2}{P_1} = \sigma \; [1+(\frac{\gamma-1}{2})M_S^2[1-\frac{1}{\sigma^2}+2\{(\frac{M_r}{M_\theta})^2+(\frac{M_r}{M_\phi})^2\}\frac{(1-\sigma)}{(M_r^2-\sigma)^2}(M^2+\frac{1}{2}-\frac{3}{2}\sigma)]] \qquad \text{(A-5)}$$

$$\frac{B_{\theta_2}}{B_{\theta_1}} = \frac{\sigma(1-1/M_r)}{1-\sigma/M_r^2} \qquad (\sigma \neq M_r^2) \qquad \text{(A-6)}$$

$$\frac{B_{\phi_2}}{B_{\phi_1}} = \frac{\sigma(1-1/M_r^2)}{1-\sigma/M_r^2} \qquad (\sigma \neq M_r^2) \qquad \text{(A-7)}$$

$$V_{r_2} = \frac{V_{r_1}}{\sigma} + V_S \; (\; 1 \; - \; \frac{1}{\sigma} \;) \qquad \text{(A-8)}$$

$$V_{\theta_2} = V_{\theta_1} + (\frac{B_{\theta_1}}{B_{r_1}}) \; (V_S - V_{r_1}) \; (\frac{1-\sigma}{M_r^2-\sigma}) \qquad (\sigma \neq M_r^2) \qquad \text{(A-9)}$$

and

$$V_{\phi_2} = V_{\phi_1} + (\frac{B_{\phi_1}}{B_{r_1}}) \; (V_S - V_{r_1}) \; (\frac{1-\sigma}{M_r^2-\sigma}) \qquad (\sigma \neq M_r^2) \qquad \text{(A-10)}$$

ANOMALOUS TRANSPORT BY KELVIN-HELMHOLTZ INSTABILITIES

Akira Miura

Geophysics Research Laboratory, University of Tokyo
Bunkyo-ku, Tokyo, 113, Japan

ABSTRACT

Simulation of magnetohydrodynamic Kelvin-Helmholtz instabilities has been performed for parallel ($B_0 \parallel v_0$) and transverse ($B_0 \perp v_0$) configurations, modeling high latitude (or downstream flanks) and dayside low latitude magnetospheric boundaries. In the parallel configuration, a super-Alfvénic and trans-sonic shear flow develops into small eddies, which strongly compresses, twists, and hence amplifies the magnetic field by the dynamo action with an amplification factor $M_A/2$. In the nonlinear stage, however large the initial Alfvén mach number M_A may be, the magnetic field amplified and twisted by the hydromagnetic flow vortices reacts back upon the flow evolution, and the flow vortices cascade into smaller structures. In the transverse configuration the instability leads to the formation of a fast shock discontinuity from an initially sub-fast shear flow. Anomalous tangential stress by the instability in the transverse configuration reaches 1% of the magnetosheath momentum flux, but for the parallel configuration, the anomalous transport is 2-3 times larger than the anomalous transport in the transverse configuration. The anomalous transport for both configurations satisfies the requirement of the viscous-like interaciton at the mangetospheric boundary.

1. INTRODUCTION

The Kelvin-Helmholtz (K-H) instability is important in understanding a variety of space and astrophysical phenomena involving sheared plasma flow. Of particular interest in the space plasma is the consequence of the Kelvin-Helmholtz instability in the hydromagnetic interaction between solar wind and magnetosphere at the magnetosphere boundary (Figure 1): It has long been suggested that at the magnetospheric boundary the Kelvin-Helmholtz instability is excited by velocity shears and leads to a "viscous-like" interaciton

H. Matsumoto and T. Sato, Computer Simulation of Space Plasmas,

(Axford and Hines, 1961) at the boundary, i.e., the net transport of solar wind momentum and energy into the magnetosphre, in order to drive the large-scale plasma convection inside the magnetosphere. Most of theoretical attempts in the past on the Kelvin-Helmholtz instability have been directed to the linear analysis and have been successful in showing that the magnetospheric boundary is linearly unstable for the Kelvin-Helmholtz instability; however, there has been no self-consistent nonlinear treatment of the instability, which could answer what nonlinear state is realized and how much momentum and energy is tranfered by the instability into the magnetosphere, a question being particularly of interest in understanding energetics of the solar wind-magnetosphere interaction.

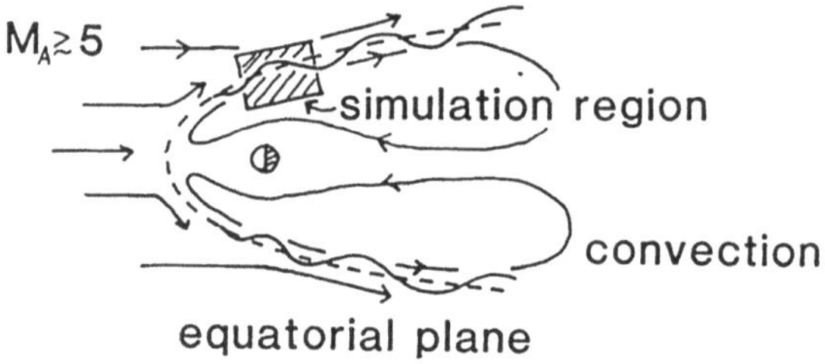

Figure 1. Solar wind-magnetosphere interaction in the equatorial plane.

In this paper we study by means of magnetohydrodynamic (MHD) simulation the MHD Kelvin-Helmholtz instability in a compressible plasma as an initial value problem. By such a self-consistent nonlinear treatment, we will be able to elucidate the basic nonliear dynamics of the instabilities (Miura, 1982) in the basic configurations of sheared plasma flow and magnetic field and answer how much momentum and enrgy of the solar wind is transfered into the magnetosphere by the Kelvin-Helmholtz instabilities at the magnetospheric boundaries.

2. BASIC EQUATIONS AND MODELS

The conservative form of the ideal MHD equations, which describe the hydromagnetic Kelvin-Helmholtz instability are:

$$\frac{\partial \rho}{\partial t} = - \nabla \cdot (\rho \mathbf{v}) \tag{1}$$

$$\frac{\partial}{\partial t} (\rho \mathbf{v}) = - \nabla \cdot \left(\rho \mathbf{v}\mathbf{v} - \frac{1}{\mu_0} \mathbf{B}\mathbf{B} \right) - \nabla \left(p + \frac{B^2}{2\mu_0} \right) \tag{2}$$

$$\frac{\partial \mathbf{B}}{\partial t} = \nabla \times (\mathbf{v} \times \mathbf{B}) \tag{3}$$

$$\frac{\partial \varepsilon}{\partial t} = - \nabla \cdot \left[\left(\varepsilon + p + \frac{B^2}{2\mu_0} \right) \mathbf{v} - \frac{1}{\mu_0} (\mathbf{B}\cdot\mathbf{v}) \mathbf{B} \right] \tag{4}$$

here, ρ, $\mathbf{v}$, $\mathbf{B}$, and p are the plasma mass density, bulk velocity of the plasma, magnetic field and plasma pressure, and ε is the energy density defined by

$$\varepsilon = \frac{1}{2} \rho v^2 + \frac{1}{2\mu_0} B^2 + \frac{p}{\gamma - 1} \tag{5}$$

We show in Figure 2 the two basic configurations of the instability in a uniform magnetic field B_0 investigated in the present simulation. The parallel configuration models the magnetospheric boundary at high latitudes in the noon-midnight meridian plane or the downstream flanks, where the magnetic field has a large component parallel to the sheared plasma flow. The transverse configuration models the magnetospheric boundary at the dayside low latitude near the equatorial plane, where the magnetic field is almost transverse to the sheared plasma flow. The parameters which appear in the following simulation are the sound mach number $M_S=V_0/c_S$ and the Alfvén mach number $M_A=V_0/v_A$, where V_0 is the total jump of the velocity across the shear layer, c_S and v_A being the sound speed and Alfvén speed, respectively. Regarding stability of the velocity shear layer in the parallel configuration we should notice that the tension of the magnetic field lines sets the following instability condition for K-H instability

$$M_A > 2 \tag{6}$$

By using these simple configurations and parameters, we will be able to elucidate the basic nonlinear dynamics of the instability involved in the above two basic configurations and their parametric dependence, which are applicable not only to magnetospheric boundary but also to wide regions of space and astrophysical problems. For the velocity profile we assume a hyperbolic tangent form,

$$v_{0y}(x) = - (V_0/2) \tanh(x/a) \tag{7}$$

characterized by a total velocity jump V_0 and velocity shear scale length a.

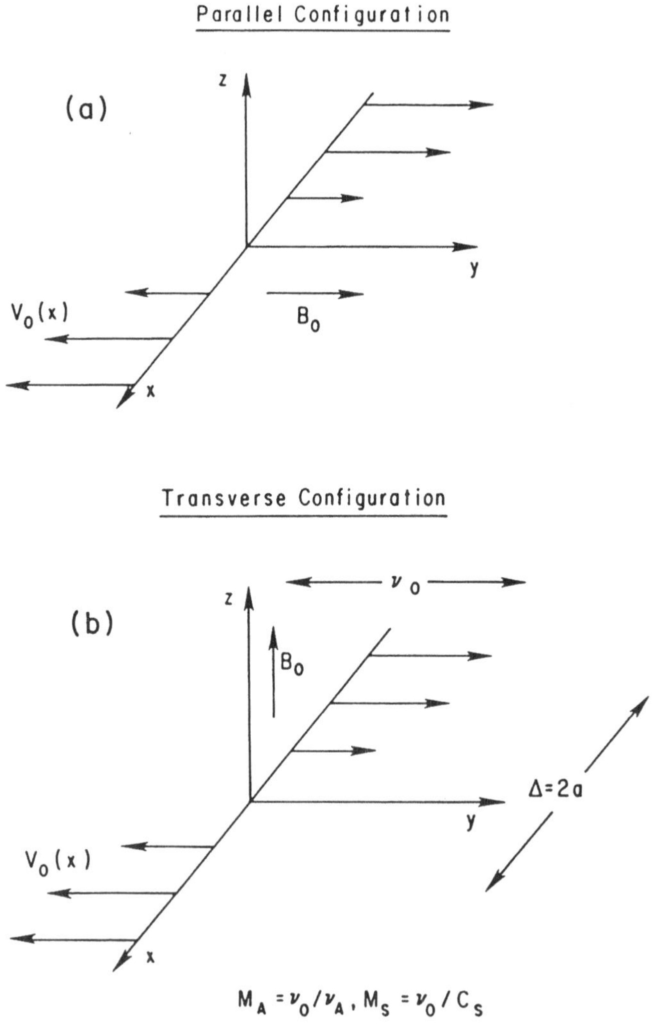

Figure 2. Two basic configurations of the MHD Kelvin-Helmholtz instability. The magnetic field is either parallel to the flow (parallel con-figuration) or perpendicular to the flow (transverse configuration). The velocity profile is characterized by a total velocity jump V_0 and a scale length 2a.

3. NUMERICS

Simulations are performed in the x-y plane where the initial flow velocity v_y has a shear profile in the x direction. We impose a boundary condition such that there is no mass flow ($v_x=0$) across boundaries at $x=\pm x_b$ and all quantities are periodic in the y direction. It then follows from (1)-(4) that B_x and derivatives with respect to x of the remaining quantities ($\rho, v_y, v_z, B_y, B_z, p$) must vanish at the boundaries ($x=\pm x_b$). This boundary condition means that the flow kinetic energy flux and the poynting flux across the boundaries vanish. Therefore, there is no inflow and outflow of energy across the boundaries. For the present simulation we have placed the boundaries at $x=\pm 10a$, which is far enough from the shear region to make boundary effects negligible. In the following time is normalized by $\tau=2a/V_0$, and we use a mesh system with a (100,100) mesh.

In the present simulation we have given a linearly unstable perturbation at the initial stage (t=0) as an initial seed of the growing perturbation. Such a linearly unstable perturbation was obtained by linear eigenmode analysis by the initial value code (Miura and Pritchett, 1982), which follows the linearized MHD equation numerically to produce the linear eigenmode. We have used the two-step Lax-Wendroff scheme (Richtmyer and Morton, 1967) to solve equations (1)-(4) and a mesh system with a (100,100) mesh. Although the Lax-Wendroff scheme includeds an artificial viscosity term implicitly, it turned out during the computation that the artificial viscosity only by the Lax-Wendroff scheme was not enough to smooth the final MHD solution and a large mesh oscillation was actually observed. Therefore, we have added additional artifical viscosity term following Lapidus (1967) in order to smooth the solution and provide a dissipation mechanism, which is necessary to resolve a fast shock discontinuity generated by the K-H instability. Namely, following Lapidus (1967), we have added artificial viscosity term to the MHD variables first by smoothing those variables in x direction and then smoothing them in y direction as follows:

$$\underset{\sim}{V}^1 = \frac{\Delta t}{\Delta x}\frac{\kappa}{4}\Big[\,|u^1_{x+2\Delta x,y} - u^1_{x,y}|\,(\underset{\sim}{U}^1_{x+2\Delta x,y} - \underset{\sim}{U}^1_{x,y}) - |u^1_{x-2\Delta x,y} - u^1_{x,y}|\,(\underset{\sim}{U}^1_{x,y} - \underset{\sim}{U}^1_{x-2\Delta x,y})\,\Big] \tag{8}$$

$$\underset{\sim}{U}^2 = \underset{\sim}{U}^1 + \underset{\sim}{V}^1 \tag{9}$$

$$\underset{\sim}{V}^2 = \frac{\Delta t}{\Delta y}\frac{\kappa}{4}\Big[\,|v^2_{x,y+2\Delta y} - v^2_{x,y}|\,(\underset{\sim}{U}^2_{x,y+2\Delta y} - \underset{\sim}{U}^2_{x,y}) - |v^2_{x,y-2\Delta y} - v^2_{x,y}|\,(\underset{\sim}{U}^2_{x,y} - \underset{\sim}{U}^2_{x,y-2\Delta y})\,\Big] \tag{10}$$

$$\underset{\sim}{U}^3 = \underset{\sim}{U}^2 + V^2 \tag{11}$$

where $u=v_x$, $v=v_y$, $\underset{\sim}{U}=(\rho,v_x,v_y,v_z,B_x,B_y,B_z,p)$, $\underset{\sim}{V}^1$ and $\underset{\sim}{V}^2$ are added artificial viscosity terms, suffixes 1-3 represent one cycle of each time step, and κ is a constant of order of unity. We have to be specially careful in such a smoothing process, however, since we try to elucidate the anomalous momentum transport or the anomalous viscosity induced by the Kelvin-Helmholtz instabilities, which is due to the finite large amplitude K-H mode. Therefore, by trial and errors, we have made those artificial viscosity terms as small as possible.

4. NUMERICAL RESULTS

4.1 Parallel configuration

First, we show results of a simulaiton run performd for the parallel configuration. Solid and dashed curves in Figure 3 marked by the symbol ∥ show the time evolution of normalized peak amplitudes

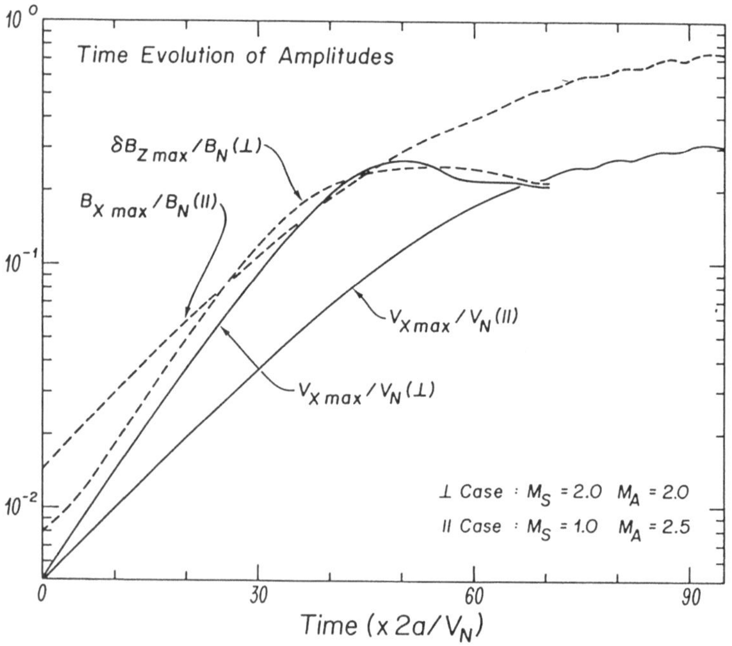

Figure 3. Time evolution of maximum normalized transverse velocity (solid curves) and maximum normalized compressional component of the magnetic field (dashed curves) for the parallel configuration with $M_S=1.0$, $M_A=2.5$ and the transverse configuration with $M_A=M_S=2.0$. Time is normalized by $2a/V_0$.

$V_{x\ max}/V_N$ and $B_{x\ max}/B_N$, where $B_N=B_o$ and $V_N=V_o$, for the parallel configuration with $M_s=1.0$, $M_A=2.5$. In the early stage, the amplitude grow linearly with the predicted linear growth rate. At $t/\tau=90$ the instability saturates, $V_{x\ max}/V_N$ reaches 30 %, and the magnetic field perturbation (transverse component) becomes comparable to the initial beckground magnetic field intensity.

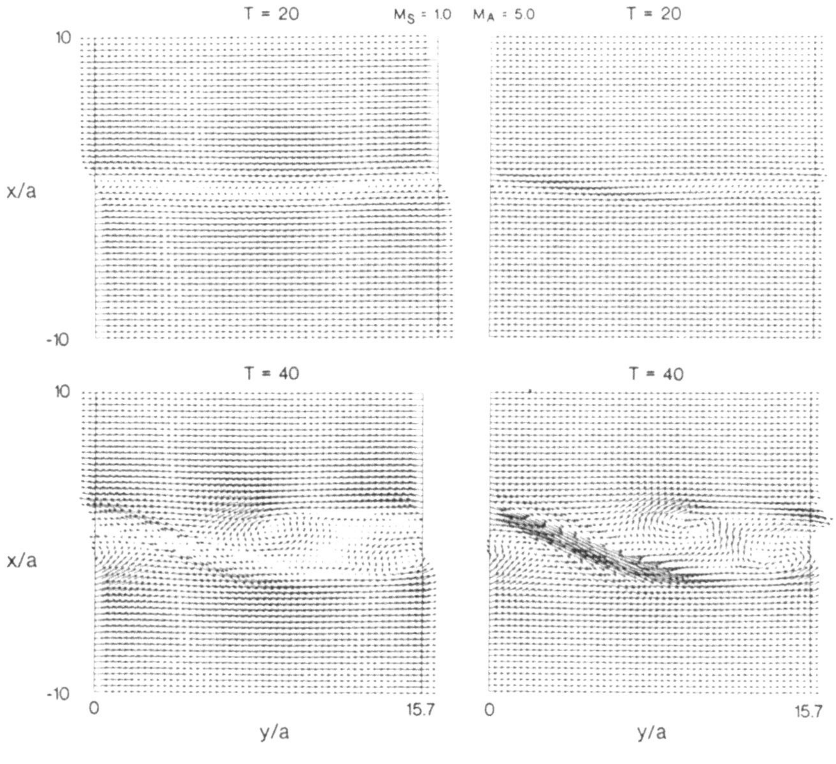

Figure 4. Flow velocity (left) and magnetic field (right) at $t/\tau=20$ and 40 for the parallel configuration ($M_s=1.0$, $M_A=5.0$).

Figure 4 shows flow vectors (left panels) and magnetic field vectors (right panels) at $t/\tau=20$ and 40 for a simulaiton run with $M_s=1.0$ and $M_A=5.0$. System length in the y direction L_y in this case is equal to $L_y=15.7a$, which corresponds to the wavelength of the fastest growing mode. In the early state ($t/\tau=20$), the sheared plasma flow undulates slightly with the development of the instability (top left panel). Since the magnetic field is frozen into the plasma, the above undulaiton of the plasma flow leads to a slight bending of the magnetic field lines as seen in the top right panel. The shear flow is extremely disturbed, however, by the saturation stage $t/\tau=40$, and a pair of eddies is formed inside a large vortex at $7a<y<15.7a$. In the center of the large vortex, the flow is almost stagnant. A stagnation region also appears in between the large vortices, and the plasma flow toward this stagnation region induced by vortices is forced to diverge along a layer formed tangent to the vortices. The initially uniform magnetic field is slightly sheared and compressed inside the velocity shear layer at $t/\tau=20$ and eventually at $t/\tau=40$, a strong compression of the magnetic field occurs at $0<y<7a$ along the layer formed tangent to vortices. On the other hand at $7a<y<15a$ the magnetic field line is stretched and twisted strongly as a consequence of the wrapping-up of the field lines by the differential rotation associated with each of the twin vortices. In this case, the total magnetic energy in the whole calculation domain increased by 26 % of the initial total magnetic energy at the expense of the initial flow kinetic energy.

Shown in the upper panels of Figure 5 are 3-D views of the pressure distributions for this case. Initially the pressure was uniform, but with time, the pressure decreases along the region where the mangnetic flux tube is compressed on both sides. A substantial depletion of the plasma pressure is seen for later stage along the compressed flux tube formed tangent to vortices. This is because the flux tube is compressed on both sides by the incoming plasma flow induced by the vortex motion causing the plasma inside to be squeezed out of the flux tube.

In the bottom panel of Figure 5, we summarize plasma dynamics involved in the parallel configuration for the present case of $M_A=5.0$, which has led to a strong compression and twisting of magnetic field lines and the formation of a slow rarefaction layer. Illustrated also in this figure is an amplification of the magnetic field at the site of small eddies by the dynamo action, $\underset{\sim}{E}\cdot\underset{\sim}{J}=\underset{\sim}{v}\cdot(\underset{\sim}{J}\times\underset{\sim}{B})<0$, where $\underset{\sim}{E}$ and $\underset{\sim}{J}$ are the electric field and current induced by eddy motion and field line twisting. At the site of the slow rarefaction layer, the magnetic field is also amplified by slow magnetosonic rarefaction due to accumulation of magnetic field lines by flux transfer associated with frozen-in vortex motion. Both of these processes, i.e., twisting and compression of magnetic field lines, contribute to a dynamo mechanism, whereby the flow kinetic energy is converted into magnetic energy by a deceleration of the flow by the $\underset{\sim}{J}\times\underset{\sim}{B}$ magnetic force.

In order to see the dependence of the instability consequences on the Alfvén mach number M_A, we have performed simulation runs for $M_A=2.5$ and 10.0 and for a fixed sound mach number $M_S=1.0$. We show in Figure 6, flow velocity vectors at the saturation stage of the instability for $M_A=2.5, 5.0, 10.0$ from the top. For $M_A=2.5$, the tension of the magnetic field line is large in comparison with the inertial term and therefore the flow is only slightly undulated with twin eddy-like circulations developed in a large vortex. With the increase of the mach number M_A, the flow is disturbed more and more and for M_A=10.0, a large vortex initially formed cascade into smaller eddies and the flow is more "turbulent" in the sense that the flow perturbation is now of much smaller scale size. The three panels in Figure 7 show magnetic field vectors corresponding to those flow vectors in Figure 6. For M_A=2.5, the magnetic field lines oscillate

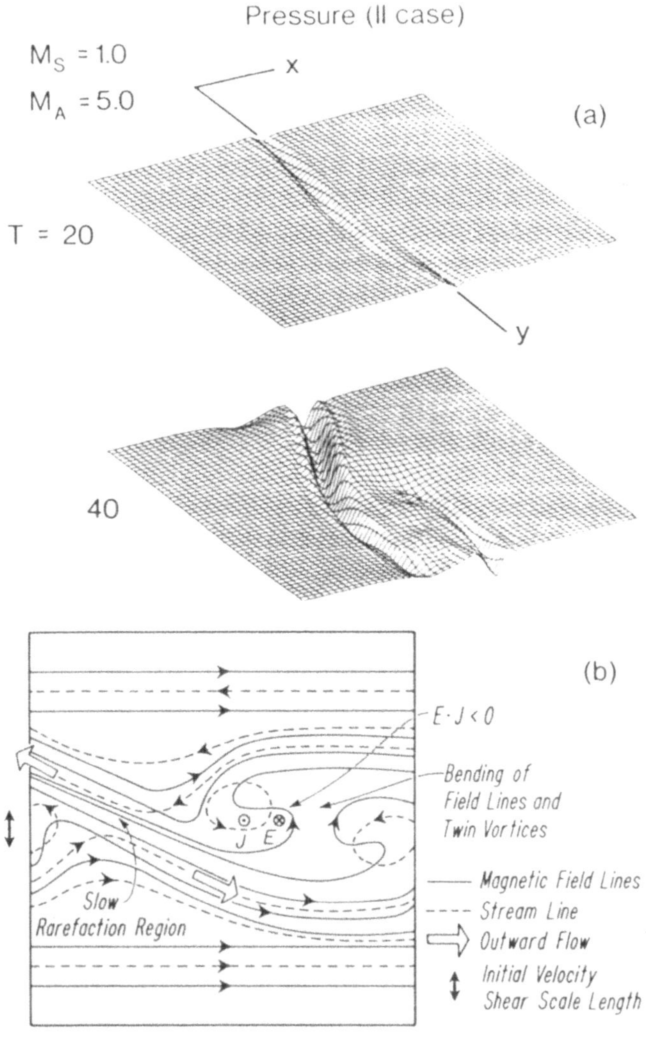

Figure 5. Upper panels; three-dimensional plots of the pressure (top surface) at t/τ=20 and 40 for the parallel configuration. Lower panel; plasma dynamics in the parallel configuration.

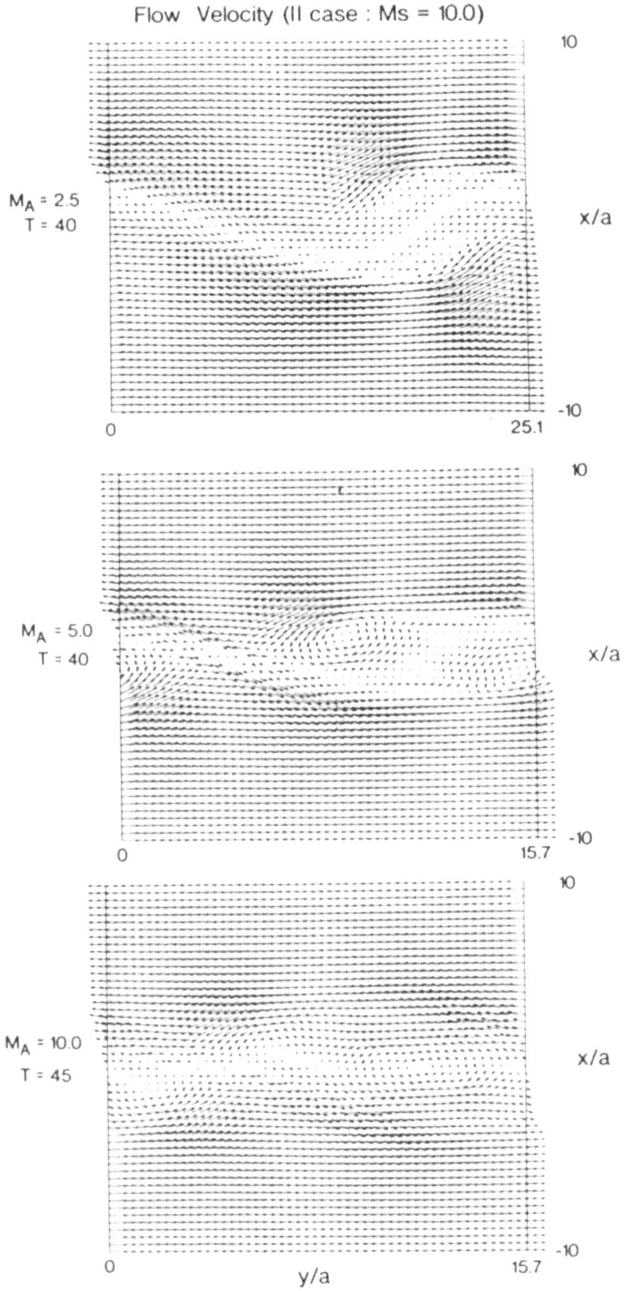

Figure 6. Flow velocities for three different Alfvén mach numbers (M_A=2.5,5.0,10.0, and M_S=1.0) in the parallel configuration.

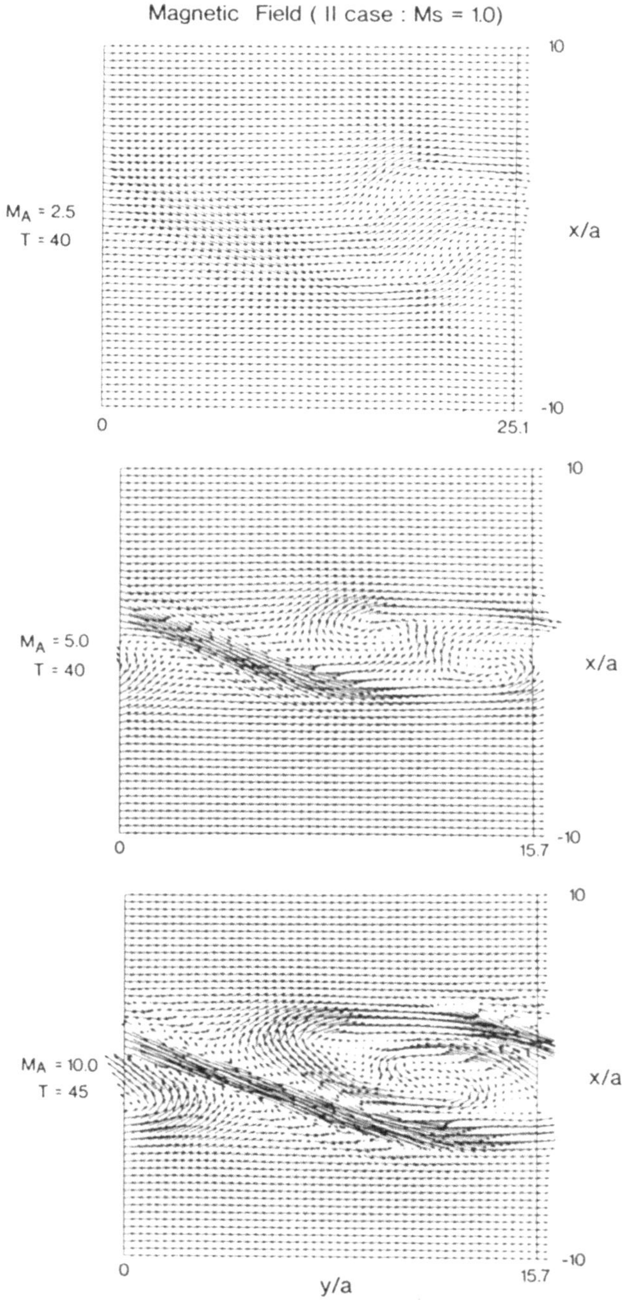

Figure 7. Magnetic fields for three different Alfvén mach numbers (M_A= 2.5,5.0,10.0, and M_S=1.0) in the parallel configuration.

back and forth slightly owing to the frozen-in vortex motion inducedby the instability. However, with the increase of M_A, the magnetic field line is more strongly compressed and twisted by the flow vortices and for $M_A=10.0$, the magnetic field is amplified six times of the initial magnetic field in the slow rarefaction layer, and twisted several times. Notice regarding the present numerical scheme that in the middle and bottom panels, regions of magnetic field reversal are formed, but they are not subject to reconnection by the numerical resistivity. This fact provides us with a proof of the soundness of the present numerical scheme for the ideal MHD plasma.

By a compression and twisting of the magnetic field line, the magnetic energy in the whole calculation domain has increased (dynamo action). In order to see this increase of the magnetic energy, we have plotted in Figure 8 the time evolution of the relative magnetic energy increase $[W_B-W_B(t=0)]/W_B(t=0)$ and the relative decrease of the flow kinetic energy $[W_K-W_K(t=0)]/W_K(t=0)$. It is obviously seen in these plots that the total magnetic energy in the whole calculation domain increases exponentially with time at the expense of the flow kinetic energy. This provides a solid evidence of the dynamo action by the K-H instability in the parallel configuration. Both compression and twisting of magnetic field lines contribute to a dynamo mechanism, whereby the flow kinetic energy is converted into magnetic energy. This dynamo process is caused by a deceleration of the flow by the $\underset{\sim}{J}\times\underset{\sim}{B}$ magnetic force; that is, the magnetic field gains its energy from the vortex flow by $\underset{\sim}{v}\cdot(\underset{\sim}{J}\times\underset{\sim}{B})<0$. For the present parameter ($M_S=1.0$), it is obvious from Figure 7 that the magnetic field amplification by slow rarefaction process contributes most to the increase in the magnetic field. The efficiency of the magnetic field amplification using this process depends on the Alfvén mach number M_A or the plasma $\beta(\beta\simeq M_A^2/M_S^2)$. If we assume a quasi stationaly state, the amplification factor of the magnetic field, i.e., B_f/B_i, where B_i is the initial uniform magnetic field $(=B_0)$ and B_f is the final magnetic field strength inside the slow rarefaction layer, may be calculated as follows: The lower panel in Figure 5 shows that the slow rarefaction layer is formed as a consequence of the compression of the magnetic flux tube by the incoming plasma flow induced by vortices. The pressure balance between the inside and outside of the slow rarefaction layer becomes

$$p_{out} + \rho_{out} v_{out}^2 + \frac{B_{out}^2}{2\mu_0} \sim p_{in} + \frac{B_{in}^2}{2\mu_0} \tag{12}$$

where v_{out} is the velocity of the incoming plamsa flow, which is almost normal to the slow rarefaction layer. From simulation results, we found that $p_{out}-p_{in}<<\rho_{out}v_{out}^2$. Therefore, the above pressure balance condition gives simply

$$\rho_{out} v_{out}^2 \sim \frac{B_{in}^2}{2\mu_0} \tag{13}$$

This relation means that the dynamic pressure by the incoming plasma flow is nearly balanced by the magnetic pressure inside the rarefaction layer. If we use the empirical fact obtained from simulation results

$$v_{out} \sim v_{x\,max} \sim \frac{V_0}{3} \tag{14}$$

(13) can be rewritten as

$$\rho_{out} \left(\frac{V_0}{3}\right)^2 \sim \frac{B_{in}^2}{2\mu_0} \tag{15}$$

Thus, we can conclude that the amplification factor of the magnetic field is

$$\frac{B_f}{B_i} = \left(\frac{B_f^2/2\mu_0}{B_i^2/2\mu_0} \right)^{1/2} \sim \left(\frac{\rho_0 (V_0/3)^2}{B_i^2/2\mu_0} \right)^{1/2} \sim \frac{1}{2} M_A \tag{16}$$

We show in Table 1 the amplification factors, obtained from simulations and calculated from (16), of the magnetic field for three

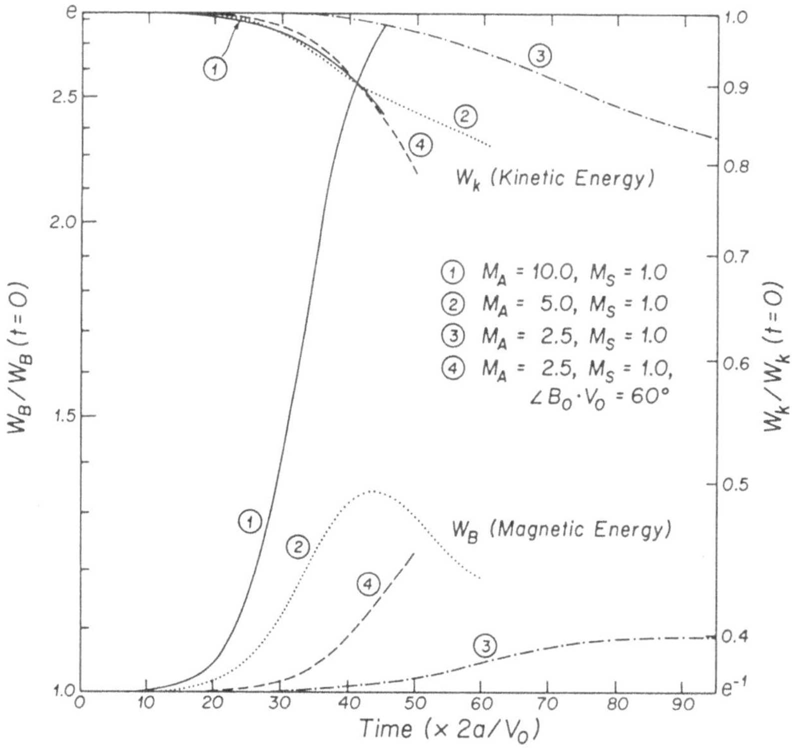

Figure 8. Time evolutions of total magnetic and kinetic energies in the whole calculation domain.

different values of M_A in the saturation stage. Although the above calculations are based on a crude argument, the calculated amplification factor has agreed very well with the simulation results, suggesting that a simple dynamo relation (13) by slow magnetosonic rerefaction is well satisfied in the actual simulation results. Summarized also in Table 1 are the value of the initial plasma β and that of the plasma β within the slow rarefaction layer at the saturation stage for three different values of M_A. Although the initial plasma β is much larger than unity for all cases, the plasma β at its saturation stages becomes 1-2 for all cases, owing to the squeezing process. Because of this large decrease of the plasma β within the rarefaction layer, ∇p and $\tilde{J}\times\tilde{B}$ forces become almost comparable, and hence the flow is strongly affected by the magnetic field, even though the initial (seed) magnetic field satisfying $\beta \gg 1$ is too weak to affect the plasma motion.

4.2 Transverse configuration

Solid and dashed curves in Figure 8 marked by the sign ⊥ show the time evolution of normalized peak amplitudes $v_{x\ max}/V_N$, $B_{z\ max}/B_N$ for the transverse configuration with $M_S=M_A=2.0$ ($M_f=1.44$), where M_f is the fast magnetosonic mach number defined by $M_f=V_0/(c_S^2+v_A^2)^{1/2}$. Both amplitudes grow linearly with the predicted linear growth rate $\gamma= 0.09\cdot 2a/V_0$. In the saturation stage, both normalized amplitudes reach 25%. In this case, the total magnetic field energy increased only slightly, by 1.6 % of the initial total magnetic field energy by the fast compression.

We show in Figure 9 simulation results of the transverse configuration with its time evolution shown in Figure 3. The system length L_y in this case is equal to 17.9a, which is equal to the wavelength of the fastest growing mode λ. Left panels show flow pattern at t/τ=25,55. The initially laminar sheared plasma flow is disturbed slightly at t/τ=25, and it develops into a flat vortex cell at t/τ=55. This time evolution may be regarded in the magnetospheric inertial frame as a spatial evolution along the magnetospheric boundary from upstream to downstream over a distance of $55\tau\times 1/2V_0 \sim 3\lambda$ where λ is the wavelength of the fastest growing mode. The transverse vortex size at t/τ=55 becomes much larger than the initial thickness of the velocity shear layer (2a), and therefore a large scale mixing of plasma is accomplished by a vortex motion. It is seen that the initial laminar shear flow is accelerated and decelerated periodically in the y-direction, since the perturbed vortex motion periodically changes the direction of rotation. Notice that the ultimate energy for this acceleration is provided by the inertial force $\rho_0(\partial v_y/\partial x)v_x$ due to the velocity shear. An interesting consequence of the instability found for this case is the formation of a pair of fast shock structures aligned side by side across the velocity shear layer, even though the initial maximum flow speed was less than the magnetosonic speed ($v_{y\ max}=V_0/2=0.71\ V_f$). The right panels in Figure 9 show a 3-D view of the pressure distribution. It is seen that at t/τ=55, the pressure gradient presents a clear-cut discontinuity, which appears at the strongly accelerated flow region in the bottom left panel. The physical picture leading to this fast shock formation is as follows: Initially, the plasma was uniform and the maximum flow

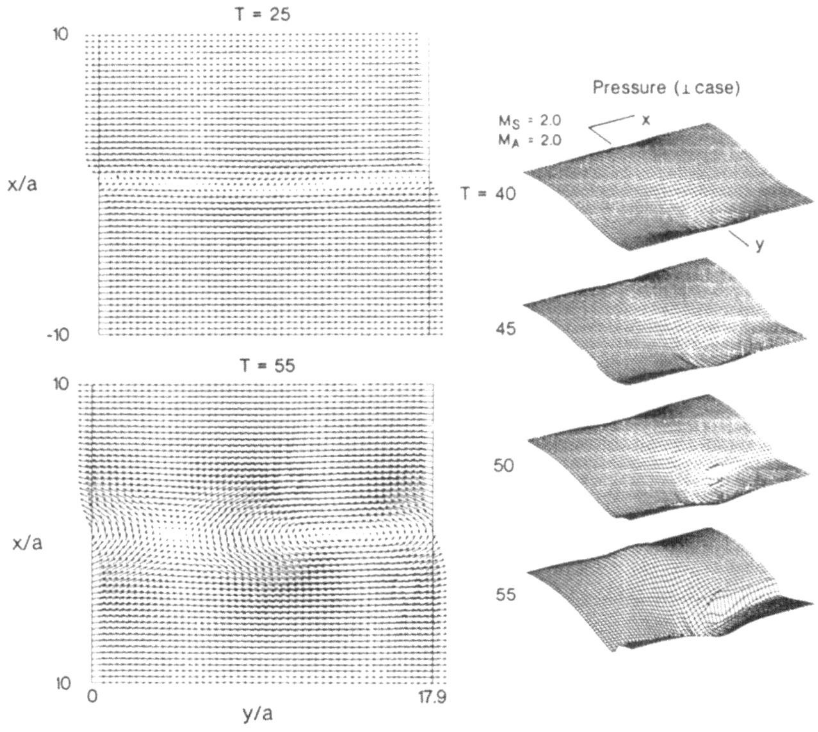

Figure 9. Flow velocity at $t/\tau=25$ and 55 (left panels) and three-dimensional plots of the pressure distribution (right panels) at $t/\tau=40,45,50,55$ for the transverse configuration ($M_S=M_A=2.0$).

speed was below the magnetosonic speed. As the instability grows, however, the vortices are excited and the flow is accelerated and decelerated periodically in the y direction by the perturbed vortex motion. Therefore, the decelerated flow is overtaken by the accelerated flow causing the pressure gradient to steepen more and more with time. Eventually, the accelerated flow speed exceeds the local magnetosonic speed, and a fast shock discontinuity is formed.

In order to resolve the fast shock structure, we show in Figure 10 profiles in the y direction of pressure (p), density (ρ), and temperature (T) normalized by their initial values $p_N=p_0, \rho_N=\rho_0, T_N=T_0$, and profiles of V_y and the magnetosonic speed $(c_S^2+v_A^2)^{1/2}$ normalized by V_0 at x=3.0a in the saturation stage ($t/\tau=55$). At y=10a, those quantities present clear-cut discontinuities, across which the flow speed changes from super-fast ($M_f=1.09$) in the upstream side to sub-fast ($M_f=0.7$) in the downstream side, consistent with the shock condition (Landau and Lifshitz, 1959). Notice that in the present MHD scheme the dissipation mechanism necessary for the formation of the fast shock discontinuity is provided by the artificial viscosity implicitly included in the Lax-Wendroff scheme and that introduced following Lapidus (1967).

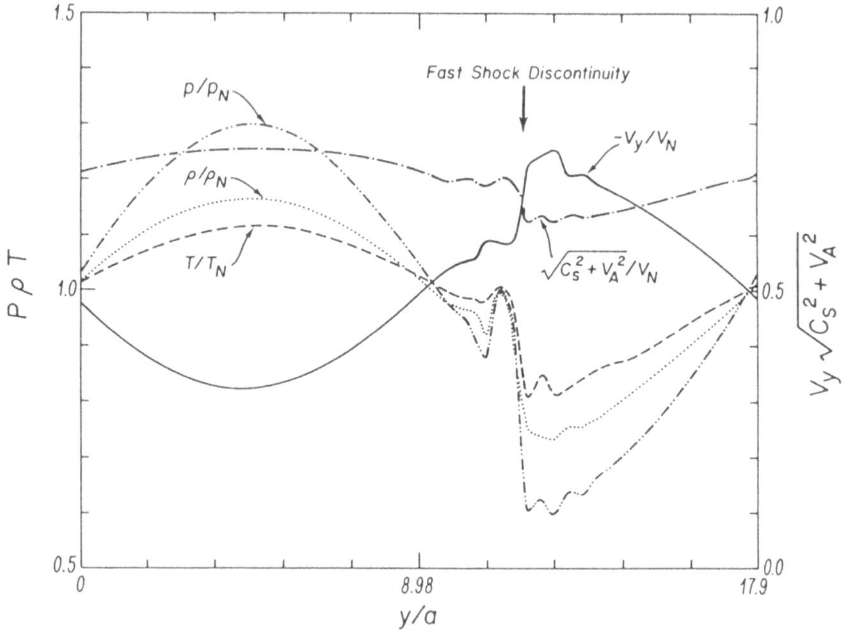

Figure 10. Profiles in the y-direction of pressure (p), density (ρ), temperature (T) normalized by their initial values p_N, ρ_N, T_N, and profiles of v_y and the magnetosonic speed normalized by V_0 for the transverse configuration ($M_S=M_A=2.0$) at x=30a and $T/\tau=55$.

It is obvious in Figure 10 that the initial uniform flow velocity (v_y) is perturbed and steepened to form a fast shock discontinuity where this velocity exceeds the local magnetosonic speed. The maximum perturbation of v_y becomes about 20 % of V_0, which is almost comparable to the maximum velocity in the x-direction of the vortex motion. Therefore, if we assume $(\Delta v_y)_{max} \sim v_{x\ max}$ in the saturation stage, the condition for the fast shock formation can be simply written as

$$\frac{1}{2} V_0 + (\Delta v_y)_{max} \sim \frac{1}{2} V_0 + v_{x\ max} > (c_s^2 + v_a^2)^{1/2} \qquad (17)$$

If we use the fact that $v_{x\ max} \lesssim 0.25\ V_0$, we obtain from the above equation

$$M_f > 1.3 \qquad (18)$$

as a rough condition for the fast shock formation. Therefore, for a fast magnetosonic Mach number less than some critical value, say 1.3, the acceleration of the initial flow is not expected to be strong enough to form the fast shock discontinuity.

4.3 Anomalous transport

Let us now investigate anomalous momentum and energy transport by the instabilities. If we take a spatial average of the y-component of (2) in the y-direction, we obtain for the two-dimensional case ($\partial/\partial z=0$) using the periodicity of perturbations in the y-direction

$$\frac{\partial}{\partial t} \langle \rho v_y \rangle = - \frac{\partial}{\partial x} \langle \rho v_x v_y - \frac{B_x B_y}{\mu_0} \rangle \qquad (19)$$

where the brackets denote the spatial average over the wave period. From (19) we find that the instability can exert an anomalous tangential stress $\langle \rho v_x v_y - B_x B_y/\mu_0 \rangle$ on plasma, where the first term is the hydrodynamic Reynolds stress and the second term is the hydromagnetic Maxwell (magnetic) stress. In order to calculate the change of momentum flux in a rectangular volume extending from x=0 to $x=\infty$ and surrounded by a unit surface at x=0, we integrate equation (19) from $x=\infty$ to x=0 to obtain

$$\frac{\partial}{\partial t} \int_\infty^0 \langle \rho v_y \rangle\ dx = - \langle \rho v_x v_y - B_x B_y/\mu_0 \rangle_{x=0} \qquad (20)$$

This indicates that across the surfact at x=0 (magnetospheric boundary) there is a net transfer of momentum in the y-direction by the instability, which is equal to the anomalous tangential stress at x=0. In the magnetospheric inertial frame, where the velocity changes from zero to V_0, the net energy flux across velocity shear layer is given by the tangential stress multiplied by V_0.

We show in Figure 11 anomalous stresses (upper panel) normalized by $\rho_0 V_0^2$ and velocity shear profiles (lower panel) for the two basic configurations ; the time evolutions for these cases are shown in Figure 3 and the velocity shear profiles are those at their saturation stages. For the transverse configuration ($\underset{\sim}{B}_0 \perp \underset{\sim}{V}_0$) assuming two-dimensionality where $\partial/\partial z=0$, i.e., the field line is not allowed to bend, the Maxwell stress vanishes and only the Reynolds stress at $t/\tau=30$ is plotted (dot-dash curve). The anomalous Reynolds stress peaks at x=0, and the peak stress becomes $0.006\ \rho_0 V_0^2$, which is 0.6 %

of the flow momentum flux far from the shear layer. This anomalous momentum transport by the Reynolds stress leads to a finite diffusion of momentum shown as a relaxation of the velocity profile from dashed curve to dot-dash curve in the lower panel. In the parallel

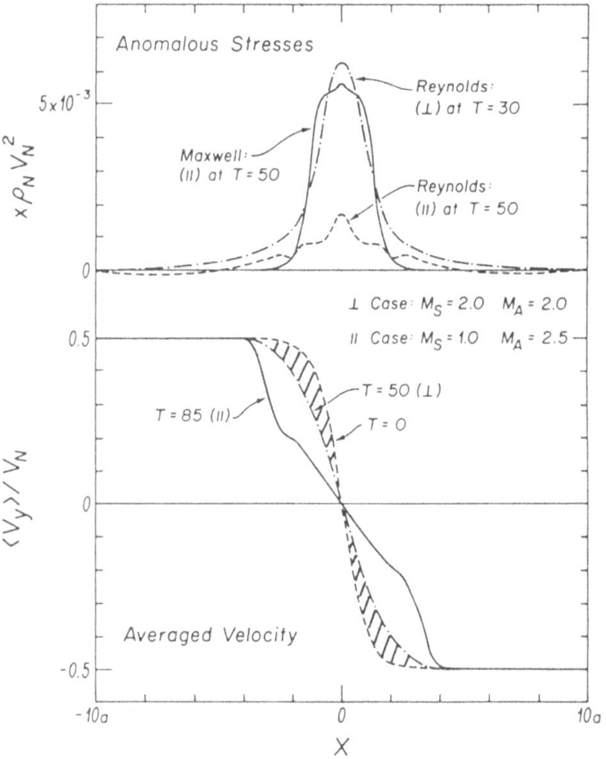

Figure 11. Spatial averages of v_y at t=0 and in the growing phases (lower panel) for the parallel (M_S=1.0, M_A=2.5) and transverse (M_S=M_A=2.0) configurations. Spatial averages of anomalous stresses for the two configurations (upper panel). The hatched area corresponds to the momentum transport for the transverse case.

configuration, the Maxwell stress (solid curve) at t/τ=50 is much larger than the Reynolds stress (dashed curve), and the Maxwell stress reaches ~0.5 % of the flow momentum flux at this time far from the velocity shear layer. Note that the anomalous Maxwell stress is strongly confined within the region of the velocity shear where the magnetic field line is bent most strongly; this causes a very strong relaxation and widening of the initial velocity shear (solid curve in the lower panel), which in turn leads to dynamo amplification of magnetic field. Since the net transfered momentum is proportional to the area between the initial velocity shear profile and the velocity shear profile at the saturation stage (shown by the hatched area for the transverse case), it is seen from the lower panel that the parallel configuration has a larger (about 3 times) momentum transport than the transverse configuration. This means that the hydromagnetic Maxwell stress is more efficient than the hydrodynamic Reynolds stress in the momentum transport. If we define the anomalous viscosity by

$$\nu_{ano} = \rho_N^{-1} \langle B_x B_y/\mu_0 - \rho v_x v_y \rangle (d\langle v_y \rangle/dx)^{-1} \qquad (21)$$

we obtain for the parallel case (M_S=1.0, M_A=2.5) at t/τ=50, $\nu_{ano}=2.6\times10^{-2}aV_0$ at x=0, which is mainly due to magnetic stress (magnetic viscosity; Eardley and Lightman, 1978). For the transverse case ($M_S=M_A$=2.0) we obtain $\nu_{ano}=1.2\times10^{-2}aV_0$ at x=0, t/τ=80.

These results suggest that the momentum diffusion process is an intrinsic feature of the Kelvin-Helmholtz instabilities which in turn leads to the saturation of the instability. It is interesting to note that the importance of the MHD wave, the Alfvén wave or slow magnetosonic wave in the compressible case, in increasing the diffusion rate of the magnetic field has been recognized by Petchek (1964) and Levy et al., (1964). In the present case, the slow rarefaction wave contributes to the strong diffusion of momentum or dissipation of vorticity as was seen in Figure 11.

5. SUMMARY AND DISCUSSION

We have demonstrated by a MHD simulation that the MHD Kelvin-Helmholtz instabilities leads to finite transport of momentum and energy across the magnetospheric boundary. For both parallel and transverse configurations, important results revealed by the present simulation may be summarized as follows:

<u>Parallel configuration</u>

(1) For super-Alfvénic and transonic shear flow (M_S=1, 2.5<M_A<4), the instability leads to the oscillation of the velocity shear layer, which bends the initially uniform magnetic field.

(2) For hyper-Alfvénic shear flow (M_A>4, M_S=1), the instability leads to formation of eddies trapped in a large vortex, and the initially uniform magnetic field is strongly compressed, twisted, and hence amplified with the amplification factor $\sim M_A/2$ by flow vortices and the total magnetic energy in the whole calculation domain increases by the dynamo action $\tilde{E}\cdot\tilde{J}=\tilde{v}\cdot(\tilde{J}\times\tilde{B})<0$.

(3) The anomalous momentum and energy fluxes across the velocity shear layer reach to 2% and 4% of those of the background flow far from the shear layer and the initial velocity shear is strongly relaxed by those finite transport.

Transverse configuration

(1) For a magnetosonic mach number larger than a critical value, the instability leads to the formation of the fast shock discontinuity from an initially subfast shear flow due to the acceleration of the flow velocity by a vortex.

(2) The anomalous momentum and energy fluxes across the velocity shear layer reach to 1% and 2% of those of the background flow far from the shear layer.

With regard to the transport of solar wind momentum and energy into the magnetosphere, it has been said that in order to satisfy the energy consumption in the magnetosphere, the necessary momentum flux is 1-2% of the solar wind momentum flux. Therefore, the anomalous momentum flux by the Kelvin-Helmholtz instabilities seems to well satisfy the requirement of the viscous-like interaction hypothesis (Axford and Hines, 1961). For a typical condition at the magnetospheric boundary, $a \gtrsim \rho_{Li}$ and $V_O \sim v_{ti}$, where ρ_{Li} and v_{ti} are Larmor radius and thermal speed of typical ions at the boundary, the observed anomalous viscosities $\nu_{ano} \simeq 0.02a\ V_O$ for the parallel configuration and $\nu_{ano} \simeq 0.01a\ V_O$ for the transverse configuration become comparable to the Bohm diffusion, which is usually regarded as the upper bound for the anomalous particle transport in the low-β plasma. For typical parameters at the magnetospheric boundary a=250 km, V_O=400 km/sec, these anomalous viscosities also become equal to or larger than the viscosity calculated by Axford and Hines (1961) $\nu_{ano} \simeq 10^{13}$ cm^2/sec to account for the magnetospheric convection with a reasonable intensity. These results based on a MHD simulation of the Kelivn-Helmholtz instabilities strongly suggest that the Kelvin-Helmholtz instabilities play an important role in the transfer of momentum and energy across the magnetospheric boundary, particularly when the magnetic stress (viscosity) by the tangled magnetic field lines is involved in the interaction.

Acknowledgments

This work was performed while the author was visiting I.G.P.P. of University of California, Los Angeles. The author would like to thank M. Ashour-Abdalla and all members of the institute for useful discussions and comments. Thanks are also due to F. V. Coroniti, J. W. Dawson, J. W. Dungey, R. M. Kulsrud, H. Okuda, P. L. Pritchett, C. T. Russell, and T. Sato for helpful disscusions and D. D. Sentman for providing the auther with plotting routines. This work was supported by NASA STTP grant NAGW-78, and the computing money was party supplied by National Science Foundation grant ATM82-18746.

References

Axford, W.I., and C.O. Hines, A unifying theory of high-latitude geophysical phenomena and geomagnetic storms, Can. J. Physics., 39, 1433, 1961.

Eardley, D.M., and A.P. Lightman, Magnetic viscosity in relativistic acceretion disks, Ap.J., 200, 187, 1975.

Landau, L.D., and E.M. Lifshitz, Fluid Mechanics, Pergamon Press, Oxford, p310, 1959.

Lapidus, A., A detached shock calculation by second-order finite differences J. Comput. Phys., 2, 154, 1967.

Levy, R.H., H.E. Petschek, and G.L. Siscoe, Aerodynamic aspects of the magnetospheric flow, AIAAJ., 2, 2065, 1964.

Miura, A., and P.L. Pritchett, Nonlocal stability analysis of the MHD Kelvin-Helmholtz instability in a compressible plasma, J. Geophys. Res., 87, 7431, 1982.

Miura, A., Nonlinear evolution of the magnetohydrodynamic Kelvin-Helmholtz instability, Phys. Rev. Lett., 49, 779, 1982.

Petcheck, H.E., Magnetic field annihilation, NASA Spec. Publ., SP-50, 425, 1964.

Richtmyer, R.D., and K.W. Morton, Difference methods for initial-value problems,2nd ed., Interscience Publishers, New York, 360, 1967.

Table 1. The plasma β at the initial stage, the plasma β at the saturation stage of the instability, and the amplification of the magnetic field for three different Alfvén mach numbers.

Table 1

M_A	Initial β	β at the saturation stage	B_f/B_i	$\frac{W_B-W_B(t=0)}{W_B(t=0)}$
2.5	7.5	1.6	1.9	8.7 %
5.0	30.0	1.7	3.6	26 %
10.0	120.0	2.3	6.0	140 %

PART III

OTHER-METHOD SIMULATIONS

PARTICLE BEHAVIOR IN THE MAGNETOSPHERE

R. A. Wolf and R. W. Spiro

Department of Space Physics and Astronomy
Rice University
Houston, TX 77251, U. S. A.

ABSTRACT

The Rice Convection Model deals with large-scale processes in the Earth's inner and middle magnetosphere, including coupling to the ionosphere. Starting from appropriate initial and boundary conditions, the model computes the following physical parameters: ionospheric electric fields and currents; magnetospheric particle distributions, electric fields, and electric currents; and magnetic-field-aligned (Birkeland) currents connecting the two regions. This paper reviews work on the model, with emphasis on the assumptions made, the basic equations, and the numerical methods. The theoretical basis of the model is compared and contrasted with standard magnetohydrodynamics. The limitations imposed by the major assumptions are discussed. Model inputs and boundary conditions are listed, and the methods of specifying them discussed. Some physical conclusions and insights that have been gained from the model are listed and described very briefly. References are given to published discussions of the major points of physics.

1. INTRODUCTION

The Rice Convection Model (RCM) deals with the closed-magnetic-field-line region of the inner magnetosphere, including the coupling of the inner magnetosphere to the ionosphere. The model is based on the "quasi-static" or "slow-flow" approximation, in which magnetospheric particles are assumed to be drifting slowly relative to their thermal velocities. The requirement that the electric current density be divergence-free leads to the flow of Birkeland currents along magnetic field lines, connecting magnetospheric and ionospheric current patterns. The divergence of magnetospheric drift current is balanced by the divergence of ionospheric conduction current. The

H. Matsumoto and T. Sato, Computer Simulation of Space Plasmas,

model keeps track of electric fields and currents in both the ionosphere and magnetosphere, as well as the time evolution of the magnetospheric particle population. The magnetic field model is taken as input, not computed self-consistently with the model currents. Use of an input magnetic field model reduces a three-dimensional problem to two coupled two-dimensional problems.

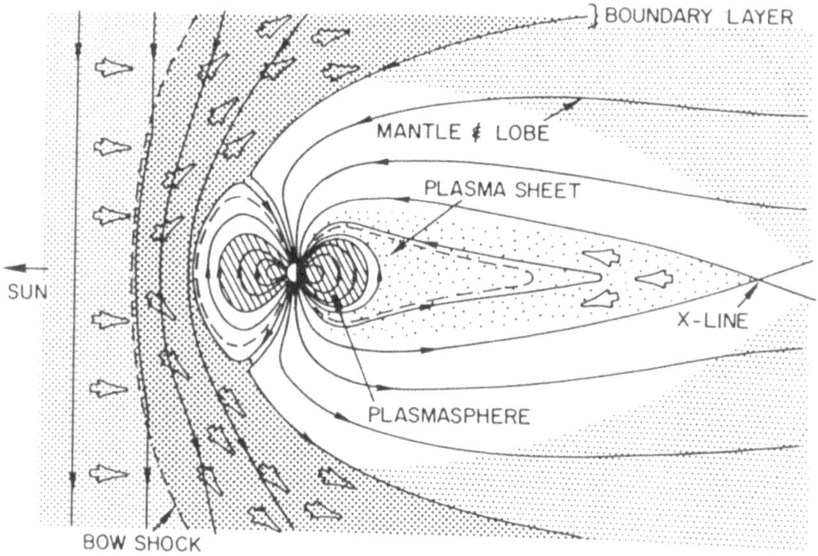

Figure 1. View of the Earth's magnetosphere in the noon-midnight meridian plane. The Sun is to the left. Hollow arrows indicated flow velocities.

The RCM is a quantitative formulation of standard magnetospheric-convection theory. The basic idea of magnetospheric convection is illustrated in Figure 1. The flow of the solar wind past the magnetosphere causes plasma in the outermost part of the magnetosphere to flow systematically away from the Sun. There is a sunward return flow in the interior of the magnetosphere. Because magnetic field lines generally tend to be good conductors, this systematic magnetospheric circulation pattern causes a related flow pattern in the ionosphere: antisunward over the polar caps, which connect to the lobes of the magnetotail, and sunward through the lower-latitude part of the auroral zone, which maps to the plasma sheet.

Of the various types of numerical simulations now being applied in space plasma physics, the global MHD simulations discussed in this volume by and Wu (1984), and in various earlier papers (e.g., LeBoeuf et al., 1978; Lyon et al., 1980; Brecht et al., 1981; Wu et al., 1981) most resemble the RCM. Like the global MHD simulations, the RCM is used to investigate large-scale phenomena in the Earth's magnetosphere. However, there are many differences between these global MHD models and the RCM, in origin, in approach, and in physical approximations used.

The RCM does not represent a conversion to space physics of a code developed for controlled-fusion or weapons research. Instead, the RCM is a one-group effort at a very specialized code. Our work has its roots in earlier qualitative and analytic theory, which began with Axford and Hines (1961), Dungey(1961), and Cole (1961). It was developed and made more quantitative by Fejer (1964), Taylor and Perkins (1971) and particularly by Vasyliunas (1970, 1972). In the seventies, emphasis gradually shifted to computer models (Wolf, 1970; Swift, 1971; Jaggi and Wolf, 1973), because it appeared that analytic calculations were too limited to solve the system of equations for reasonably realistic conditions.

Another distinguishing feature of the RCM is the extensive boundary condition requirements, especially as compared to the global MHD models. Aside from the initial-condition information, the MHD models require only very simple physical input data, namely ρ, T, $\underset{\sim}{B}$, $\underset{\sim}{v}$ of the solar wind. In addition to similar initial-condition information, the RCM also needs the following: potential distribution on the polar cap boundary; plasma sheet n_e, T_e, and T_i at the model's outer boundary; a global ionospheric conductivity model; and a three-dimensional magnetic field model. One result of the heavy reliance on boundary conditions is to tie the model strongly to data. We frequently model specific well-observed events. We use some data from the event as input; other data are compared with a wide variety of model predictions. These frequent direct theory-vs.-observation comparisons sometimes yield important hints about the physics, and generally help to keep us in touch with reality.

A second result of the heavy use of boundary conditions is that we do a lot of computer experiments. The model has many knobs to turn in order to probe the physics of the system -to determine what causes what. Thus the heavy reliance on boundary conditions, which initially seems like a disadvantage of our approach compared to the global MHD models, turns out actually to be an advantage, in many cases.

Section 2 of this paper states and discusses the limiting assumptions of the model, and gives the basic equations. Secton 3 describes the inputs to the model, and how we specify them. The overall logic of the program and the numerical methods used are described in Section 4. Section 5 lists what we regard as the major physical conclusions and insights gained from the modeling effort, and gives references to published discussions of them.

2. BASIC EQUATIONS AND PHYSICAL ASSUMPTIONS

Basic Equations

First consider the equation for bounce-averaged adiabatic drift of particles bouncing on closed field lines. We assume that the kinetic energy associated with particle drift is small compared to the particle's thermal motion (slow-flow approximation). Consequently, we can express the particle's kinetic energy in the form

$$E_K = E_K(\underset{\sim}{x}_e, \mu, J) \tag{1}$$

where $\underset{\sim}{x}_e$=equatorial crossing point and μ and J are the first two adiabatic invariants. Then the equation for bounce-averaged drift velocity becomes

$$\underset{\sim}{v}_e = \frac{\underset{\sim}{B}_e(\underset{\sim}{x}_e) \times \nabla_e E_K(\underset{\sim}{x}_e,\mu,J)}{qB_e^2} + \frac{\underset{\sim}{E}(\underset{\sim}{x}_e) \times \underset{\sim}{B}_e(\underset{\sim}{x}_e)}{B_e(\underset{\sim}{x}_e)^2} \tag{2}$$

where the first term represents gradient/curvature drift and the second is $\underset{\sim}{E}\times\underset{\sim}{B}$ drift. Subscript "e" refers to the equatorial plane. (For a derivation of the first term, see Northrop [1963] or Wolf [1983].) Instead of considering particles of each pitch angle separately, we usually take the pitch-angle distribution to be isotropic, for the sake of simplicity. We effectively assume that the particles' pitch-angles are scattered in a time short compared to the convection time. Particle energies are assumed not to change in these scattering processes. The particle's kinetic energy is then given by

$$E_K = \lambda S \tag{3a}$$

where

$$S = [\int ds/B]^{-2/3} \tag{3b}$$

and λ is called the energy invariant. The integral $\int ds/B$ extends along a field line from the southern ionosphere to the northern and represents the volume of a tube of unit magnetic flux. The corresponding total drift velocity is given by

$$\underset{\sim}{v}_e = \frac{\lambda \underset{\sim}{B}_e \times \nabla_e S}{qB_e^2} + \frac{\underset{\sim}{E}_e \times \underset{\sim}{B}_e}{B_e^2} \tag{4}$$

(Harel *et al.*, 1981a; Wolf, 1983).

Such a collection of adiabatically drifting particles satisfies a modified frozen-in flux theorem. Namely, consider particles drifting according to a law of the form

$$\underset{\sim}{v}_e(\underset{\sim}{x}_e) = \frac{\underset{\sim}{E}_e \times \underset{\sim}{B}_e}{B_e^2} + \frac{\nabla_e Y \times \underset{\sim}{B}_e}{B_e^2} \tag{5}$$

Let subscript "s" refer to a given particle species, specifically a particle of a given charge and given values of the relevant invariant(s). Let η_s= number of particles per unit magnetic flux for particles of species s. Then, neglecting loss, the continuity equation in the equatorial plane can be written

$$\frac{\partial(\eta_s B_e)}{\partial t} + \nabla_e \cdot (\eta_s B_e \underset{\sim}{v}_e) = 0 \tag{6}$$

with equation (5) and Faraday's law, equation (6) can be simplified to the form

$$\left(\frac{\partial}{\partial t} + \underset{\sim}{v}_e \cdot \nabla_e \right) \eta_s(\underset{\sim}{x}_e, t) = 0 \tag{7}$$

In other words, neglecting loss, the number of particles of a given species, per unit magnetic flux, is constant along a drift path.

Currents play a central role in our formulation of magnetospheric convection. The essential physics coupling the inner magnetosphere and ionosphere comes from equating the divergence of magnetospheric drift current, appropriately scaled, to minus the divergence of ionospheric conduction current. Consider the current carried by particles drifting in the magnetosphere. It is convenient to map these drift currents to the equatorial plane, and to discuss the current flowing in that plane per unit length perpendicular to the current. In this formalism the current is the sum of the magnetization current and the gradient/curvature-drift current. Using (4), we obtain

$$\underset{\sim}{j}_e(\underset{\sim}{x}_e) = \text{magnetization current} + \sum_s \eta_s(\underset{\sim}{x}_e)\, \lambda_s \hat{b}_e \times \nabla_e S \tag{8}$$

where $\hat{b}_e$ is a northward unit vector parallel to $\underset{\sim}{B}_e$. Taking the divergence (which makes the magnetization-current term vanish), we obtain

$$\nabla_e \cdot \underset{\sim}{j}_e(\underset{\sim}{x}_e) = - \hat{b}_e \cdot \nabla_e \left[\Sigma \eta_s \lambda_s \right] \times \nabla_e S \tag{9}$$

Since the pressure is given by

$$p = \frac{2}{3} \sum_s \eta_s \lambda_s S^{5/2} \tag{10}$$

Equation (9) becomes

$$\nabla_e \cdot \underset{\sim}{j}_e(\underset{\sim}{x}_e) = \hat{b}_e \cdot \nabla_e p(\underset{\sim}{x}_e) \times \nabla_e \left[\int ds/B \right] \tag{11a}$$

$$= \frac{2}{3} \sum_s \lambda_s S^{5/2}\, \hat{b}_e \cdot \nabla_e \eta_s \times \nabla_e \left[\int ds/B \right] \tag{11b}$$

$$= -2\, J_{\parallel e} \tag{11c}$$

where $J_{\parallel e}$ is the density of Birkeland current up from the ionosphere, mapped to the equatorial plane. The factor of 2 accounts for there being two ionospheric ends of the field line, one in the northern hemisphere, one in the southern (assumed to have equal conductivities, for simplicity). Equation (10) can actually be applied anywhere on the field line, including at the ionospheric ends. One can calculate the density of Birkeland current down into the ionosphere by taking gradients in p and $\int ds/B$ just above the ionosphere.

The same equation can be derived from the momentum equation of MHD, namely

$$\rho \frac{D\underset{\sim}{v}}{Dt} = -\nabla p + \underset{\sim}{J} \times \underset{\sim}{B} \tag{12}$$

Crossing with $\underset{\sim}{B}$ and solving for the component of $\underset{\sim}{J}$ perpendicular to $\underset{\sim}{B}$ gives

$$\underset{\sim}{J}_{\perp} = \frac{\underset{\sim}{B} \times \nabla p}{B^2} + \frac{\underset{\sim}{B} \times \left(\rho \frac{D\underset{\sim}{v}}{Dt} \right)}{B^2} \tag{13}$$

Within the slow-flow approximation, we can neglect the last term on the right side of (13). The condition that $\nabla \cdot \underset{\sim}{J}=0$ can be written

$$B \frac{d}{ds} \left(\frac{J_{\parallel}}{B} \right) + \nabla_{\perp} \cdot \underset{\sim}{J}_{\perp} = 0 \tag{14}$$

Integrating along the field line, and performing several vector-calculus manipulations, we obtain

$$\frac{J_{\parallel i}}{B_i} = \frac{J_{\parallel e}}{B_e} = - \frac{1}{2B_e} \hat{b} \cdot \nabla p \times \nabla \left[\int ds/B \right] \tag{15}$$

This equation, which is the same as we obtained above from drift theory (eqn. 11), was derived by Vasyliunas (1970).

For the case where there is no neutral wind, the equation for height-integrated horizontal ionospheric current can be written

$$\underset{\sim}{j} = \underset{\approx}{\Sigma} \cdot (-\nabla_h V) \tag{16}$$

where ∇_h represents the horizontal gradient operator in the ionosphere, and

$$\begin{aligned} \Sigma_{xx} &= \int \sigma_1 dh/\sin^2 I \\ \Sigma_{yx} &= - \Sigma_{xy} = \int \sigma_2 dh/\sin I \\ \Sigma_{yy} &= \int \sigma_1 dh \end{aligned} \tag{17}$$

Here x and y are in the magnetic northward and eastward directions, respectively, I is the magnetic dip angle, and σ_1 and σ_2 are the local Pedersen and Hall conductivities, respectively. These simplified forms for the height-integrated conductivities are valid for all of the ionosphere, except very near the dip equator. The general equations were derived by, for example, Fejer (1953). Equating the divergence of ionospheric current in (16) to the field-aligned current from (15) yields the fundamental equation for large-scale magnetosphere-ionosphere coupling:

$$J_{\parallel i} \sin I = - \frac{B_i \sin I}{2B_e} \cdot \hat{z} \; \nabla_e p \times \nabla_e \, [\, \int ds/B \,] = \nabla_h \cdot [\underset{\approx}{\Sigma} \cdot (-\nabla_h V)] \qquad (18)$$

where $\hat{z}$ is a northward unit vector perpendicular to the equatorial plane. (This equation was originally derived by Vasyliunas [1970].)

Table 1 compares our convection-model approach to standard ideal-fluid MHD, as follows:

1. The continuity equation exists in both formalisms. However, in our approach, it is expressed in terms of η_s, the number of particles of species s per unit magnetic flux. Also, in the inner magnetosphere, gradient/curvature drifts are typically comparable to $\underset{\sim}{E} \times \underset{\sim}{B}$ drifts, so that different particle species drift at different velocities. Consequently, a multifluid continuity equation is needed.
2. If we neglect the inertial term in the MHD momentum equation, its three components reduce to the condition that p is constant along a field line and equation (13) for the component of $\underset{\sim}{J}$ perpendicular to $\underset{\sim}{B}$.
3. We apply the adiabatic-expansion form of the MHD energy equation, with two modifications: (i) we apply it to each fluid species individually, and (ii) we apply it to an entire flux tube at once, using the fact that pressure and density are constant along a magnetic field line.
4. The input magnetic field models have $\nabla \cdot \underset{\sim}{B} \approx 0$.
5. The Ampere's law Maxwell equation is not satisfied in our model, in the sense that $\underset{\sim}{B}$ is not forced to be consistent with our computed $\underset{\sim}{J}$.
6. The Faraday's law Maxwell equation is included implicitly in our simulations. Because the input magnetic field model changes with time, the equatorial mapping point of a given point in the ionosphere changes with time. This motion is interpreted as $\underset{\sim}{E} \times \underset{\sim}{B}$ drift in an induction electric field.
7. The usual ideal-MHD perfect-conductor assumption is adopted, with regard to the component parallel to $\underset{\sim}{B}$. The components of $\underset{\sim}{E} + \underset{\sim}{v} \times \underset{\sim}{B} = 0$ perpendicular to $\underset{\sim}{B}$ are modified to allow the different plasma species to gradient/curvature drift perpendicular to $\underset{\sim}{B}$, in addition to the $\underset{\sim}{E} \times \underset{\sim}{B}$ drift.

TABLE 1. Comparison of Ideal-Fluid MHD with Conveciton Model

Ideal-Fluid MHD	Convection Model
1. $\frac{\partial\rho}{\partial t}+\nabla\cdot(\rho\underset{\sim}{v})=0$	$\frac{D\eta_s}{Dt}=0$
2. $\rho\frac{D\underset{\sim}{v}}{Dt}=-\nabla p+\underset{\sim}{J}\times\underset{\sim}{B}$	$\underset{\sim}{J}_\perp=\frac{\underset{\sim}{B}\times\nabla\sum_s P_s}{B^2}$, P_s=constant on field line
3. $\frac{D}{Dt}(p\rho^{-5/3})=0$	$P_s[\int ds/B]^{5/3}=2\lambda_s\eta_s/3$
4. $\nabla\cdot\underset{\sim}{B}=0$	$\nabla\cdot\underset{\sim}{B}=0$ in input $\underset{\sim}{B}$ model
5. $\nabla\times\underset{\sim}{B}=\mu_o\underset{\sim}{J}+\mu_o\varepsilon_o\frac{\partial\underset{\sim}{E}}{\partial T}$	Not included self-consistently
6. $\nabla\times\underset{\sim}{E}=-\frac{\partial\underset{\sim}{B}}{\partial t}$	Included implicitly in time-dependent mapping
7. $\underset{\sim}{E}+\underset{\sim}{v}\times\underset{\sim}{B}=0$	$\underset{\sim}{E}\cdot\underset{\sim}{B}=0$
	$\underset{\sim}{v}_s=\underset{\sim}{v}_\parallel+$ $\underset{\sim}{E}\times\underset{\sim}{B}$ drift + (grad/curv. drift)$_s$

Discussion of Assumptions

We now return to discuss the physical implications of our principal assumptions.

Slow-Flow (Quasi-Static) Approximation. Our neglect of inertial drifts compared to gradient and curvature drifts is equivalent to neglecting the inertial term in the MHD momentum equation (13). This approximation is valid if the plasma flow velocity is small compared to the sound speed, and the time scale for variations is long compared to typical wave travel times. It is one of our principal limitations compared to MHD. This approximation limits our ability to treat substorms in detail, because some substorm phenomena occur on very short time scales (less than a minute). Also, flow velocities close to the sound speed have sometimes been observed even in the innermost parts of the plasma sheet during substorms (Moore et al., 1981). We do not treat MHD waves — fast, intermediate, or slow — in our formalism. The neglect of the inertial terms is a non-trivial limitation, even for steady-flow situations. If we combine standard empirical magnetic field information (e.g., Behannon, 1970) with standard estimates of the cross-tail potential drop, we would estimate that the velocity of earthward convection exceeds the sound speed somewhere around lunar orbit (60 R_E geocentric distance). Thus our model could not, even for steady conditions, be applied out as far as lunar orbit. The model also cannot be applied close to an X-type neutral line, since plasma normally flows away from such a line at a speed that exceeds the Alfven speed in the external plasma, and the Alfven speed is of the same order as the local sound speed. Nevertheless, the model nearly always can be safely applied to phenomena that occur in the inner magnetosphere on time scales greater than a few minutes.

Magnetic Field Model Taken as Input. Historically, the theory of inner-magnetospheric convection was formulated (e.g., Vasyliunas, 1970; Wolf, 1970) assuming that the magnetic field was known, because, at the time, the magnetic-field configuration of the magnetosphere was much better known than the electric field, which was the quantity to be calculated. There were several useful and well-established semi-empirical magnetic-field models (e.g., Mead and Beard, 1964) available in card-deck form for scientific use. In contrast, only a few measurements of auroral-zone electric fields had been made (e.g., Mozer and Serlin, 1969), and there was still substantial debate over whether magnetospheric convection actually existed. This situation has now changed somewhat, due mainly to extensive electric field measurements made by polar-orbiting spacecraft and incoherent-backscatter radars. The RCM regards the magnetic field as known input for "practical" reasons. To calculate the magnetic field self-consistently would enormously increase the size and complexity of our calculation. It would change the present system of two coupled two-dimensional problems (one in the ionosphere, one in the magnetospheric equatorial plane) to a fully three-dimensional calculation. We avoid doing a full three-dimensional self-consistent calculation by using the magnetic field model to link the ionospheric and magnetospheric calculations, and to map magnetospheric pressure, etc. along field lines. Along with the slow-flow approximation, the use of an input magnetic-field model limits our model's region of applicability to the inner magnetosphere and near tail. Further out in the tail, the beta of the plasma is very large, and the relevant characteristics of the magnetic field structure are too highly variable to be reliably deduced from available magnetic-field models. G.-H. Voigt and Gary Erickson in our group are currently working on the development of magnetic field models that are consistent with convection. It is a difficult problem, in terms of both numerical analysis and physics.

The global MHD models do not share our difficulties with self-consistently modeling $\tilde{B}$ and convection. Their powerful computational machinery automatically computes magnetic-field configurations that satisfy both Ampere's law and the momentum equation. However, at present these models cannot effectively address the interesting physics of the inner magnetosphere and near tail regions (x=-4 R_E to -20 R_E). The MHD models are limited in this respect because (i) they necessarily employ large grid spacing, (ii) economics prevents them from modeling for particle-drift time scales (typically several hours), (iii) particle transport by gradient and curvature drift (which they neglect) plays a major role in the physics of the region, and (iv) the ionosphere, which also plays an important role, is very difficult to include in a global-scale MHD model.

Thus we continue to pursue the quasi-static approach, despite the difficulties involved. It continues to be the best way to model the inner magnetosphere and large-scale magnetosphere-ionosphere coupling. The quasi-static approach (slow-flow approximation) is, of course, only applicable near the Earth. Unlike the global-MHD models, it cannot be used legitimately to model the far tail or solar-wind/magnetosphere coupling. However, it is presently the best way to treat most large-distance-scale, long-time-scale phenomena in the inner and middle magnetosphere.

3. MODEL INPUTS

We now discuss the various input models needed for the main simulations. It is here that one can sense the strains and sacrifices involved in trying to realistically model actual events. Much of the detailed disagreement between model results and observations is undoubtedly due to inaccurate input.

Potential Distribution on the Poleward Boundary of the Model

The potential distribution on our poleward boundary is needed as a boundary condition, so that the elliptic equation (18) can be solved for V.

First consider the total potential drop across the boundary. This parameter, which measures the total strength of convective flow through the model system, is approximately equal to what observers call the "polar cap potential drop." We have considered three methods for estimating this crucial time-dependent parameter from observations. The most obvious and direct source is the electric field measured by a polar-orbiting spacecraft in dawn-dusk orbit. We used such data for our simulations of the substorm of September 19, 1976, but, unfortunately, such direct measurements are not usually available with sufficient frequency to be helpful for event simulations. A second approach, which is applicable to many more events, is to estimate the polar-cap potential drop from the observed interplanetary magnetic field and an empirical formula (see, e.g., Reiff et al., 1981). This basic approach was used for our simulations of the magnetic storms of July 29, 1977 (Wolf et al., 1982) and March 22, 1979 (Spiro and Wolf, 1983). (See top panel of Figure 2.) Another possible approach involves the use of a magnetogram-inversion scheme (Kamide et al., 1981). Given a conductivity model and a large number of ground magnetograms as input for a specific event, this scheme computes global electric current and electric-field patterns, and consequently, the polar-cap potential drop. Reiff et al. (1983) report remarkably good agreement between polar-cap potential drops inferred by the magnetogram-inversion procedure for the March 22, 1979 magnetic storm and values predicted by using IMF data and an empirical formula for the same event.

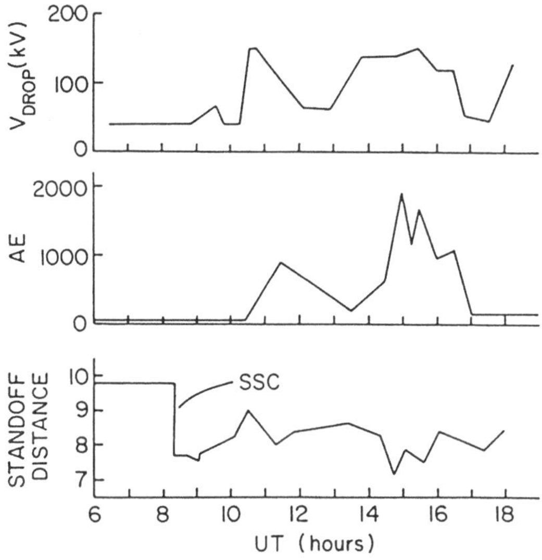

Figuer 2. Time-dependent input parameters for simulation of the March 22, 1979 magnetic storm. The top panel shows the polar-cap potential drop estimated from solar-wind parameters (P. H. Reiff, private communication, 1982). The middle panel shows the auroral-electrojet index, which we need to choose the conductivity model. The bottom panel shows the magnetopause standoff distance, as estimated from the solar-wind ram pressure (G.-H. Voigt, private communication, 1982).

With regard to the local-time variation of the potential, the classic pattern of magnetospheric convection corresponds to a maximum potential on the dawn side, a minimum on the dusk side. This established local-time variation is expressed very roughly by the simple equation

$$V = - V_0 \sin \phi \qquad (19)$$

where V_0 is a constant, and ϕ is the local-time angle (=0 at local noon, $\pi/2$ at dusk, etc.). Although there is some observational information on deviations from this simple law (see, e.g., Foster, 1983), there is very little systematic information on how the local-time variation depends on magnetic activity. However, substantial evidence of complex, non-classical convection patterns during periods when the interplanetary magnetic field is northward for a prolonged period have been presented (see, e.g., Burke et al., 1979). Of course, V_O is relatively small under these conditions anyway, and flows tend to be weak.

Ionospheric Conductances

The daytime ionospheric conductance (height-integrated conductivity) depends on the strength of ionizing solar UV radiation, which, in turn, depends on sunspot cycle. Sufficient data exist from the various incoherent backscatter radars around the world to provide adequate observational information on the sunspot-cycle dependence of

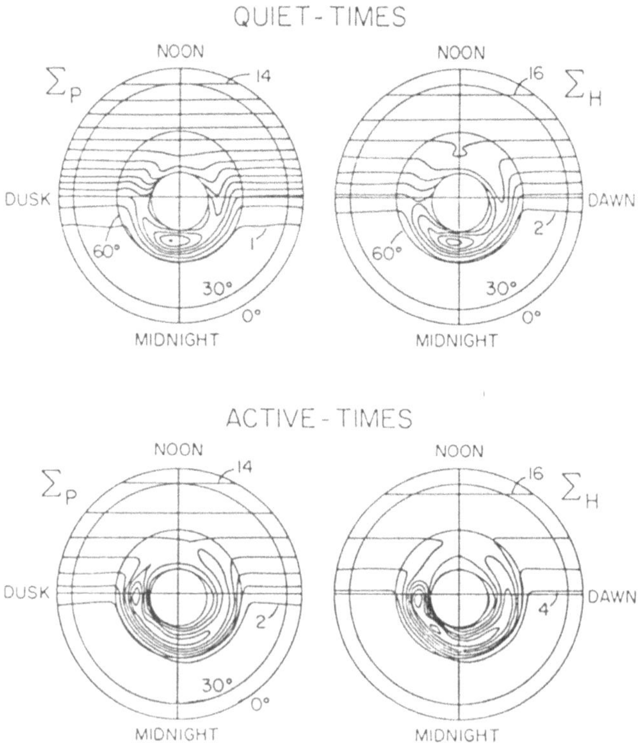

Figure 3. Ionospheric Pedersen and Hall conductances for quiet and active times, as used in our simulation of the March 22, 1979 event. The distributions are based on the empirical electron-flux model of Spiro et al. (1982), but the distributions are adjusted in latitude according to the computed electron inner edges. The "active time" diagrams pertain to an AE index near 1000 γ.

ionospheric conductance. Unfortunately, no one has systematically collected and compiled the information, and consequently, there are possible errors of nearly a factor of two in our best conductance estimates in the sunlight-dominated region.

Auroral electron precipitation has a major effect on ionospheric conductance at high latitudes. To estimate the effect of auroral precipitation, we have used parameterized results of ionospheric models to estimate the conductance corresponding to a given auroral electron flux (see Spiro et al. [1982] for detailed discussion and references). To estimate the global distribution of auroral electron fluxes as a function of time during the event, we have used two approaches. One approach is to use electron data from polar-orbiting spacecraft for the event in question. The second approach is to utilize a statistical study of auroral electron fluxes, sorted according to a magnetic index (the AE and AL indices seem most useful). Sample conductances based on such a statistical study are shown in Figure 3. We then combine the AE or AL index measured as function of time through the event with the statistical model to arrive at a global, time-dependent conductance model. The second panel of Figure 2 shows the idealized version of the preliminary AE index that was used to generate the conductance models for the March 22, 1979 event.

Both of these approaches to obtaining global conductance models have problems. The real-time spacecraft measurements should give accurate conductivities along the orbit track, but major interpolations and extrapolations are required to construct global models for all times during an event. Statistical models generally give smooth electron-flux distributions that do not accurately portray the physical situation at a given time. An AE index of 100 may correspond to the peak of a small, isolated substorm occuring during a long quiet period, or it may correspond to a lull between successive substorms in a large magnetic storm. The electron flux distributions will be quite different in these two situations, but the statistical analysis averages them together. For a comparison of statistical-average conductances with those derived from spacecraft overflight, see the paper by Reiff (1983) or Simons et al. (1983).

Magnetic Field Models

For our recent simulations, we have used time sequences of Voigt (1981) magnetic field models. For each model in the sequence, the magnetopause standoff distance is adjusted to correspond to the solar-wind ρv^2 observed at that time (see middle panel of Fig.2), and the ring-current strength is adjusted according to the observed Dst index. The location of the ring current, and the strength and location of the "neutral sheet current" are sometimes adjusted to agree with magnetic fields or particles observed by spacecraft at the time. Analogous sequences of models were previously constructed by Olson and Pfitzer (1982) for the July 29, 1977 magnetic storm.

Hot-Particle Distribution at our Outer (High-L) Boundary

In most cases, we have somewhat arbitrarily set our plasma-sheet boundary condition to correspond to nominal values of plasma-sheet parameters. In the case of the March 22, 1979 event, the ISEE-2 spacecraft was in the plasma sheet near our tailward boundary for most of the event, so that we could adopt a more realistic plasma boundary condition . Of course, single-satellite measurements give no information on the variation of plasma-sheet parameters along the boundary.

Initial Hot-Particle Distribution

As an initial condition, we specify the particle distribution function throughout the magnetosphere. This initial distribution is not very important for the plasma sheet, because the particles that initially populated the plasma sheet are soon replaced by others. However, the distribution of particles trapped near the Earth is important, because many of the particles that were initially on trapped orbits remain on trapped orbits throughout the event. In specifying the initial plasma condition, we have generally made reasonable guesses representing average conditions plus whatever fragmentary information was available for the specific day in question.

Other Possible Inputs

Non-zero neutral-wind velocity in the current-carrying layers of the ionosphere affects the ionospheric current-voltage relation (16). To specify the neutral-wind correction term for (16), we actually need a complete wind model — three-dimensional and time — for the E and F layers of the ionosphere. Observational summaries are not nearly that comprehensive, but reasonable three-dimensional theoretical neutral-wind models are now available (e.g., Roble et al., 1982). We expect to examine neutral-wind effects in the next few years. In models constructed so far, however, we have arbitrarily set the neutral-wind velocity equal to zero.

Given a correct algorithm for computing the magnetic-field-aligned potential drop on each field line, we could relatively easily incorporate these potential drops in our model. One simple, reasonable approach would be to use a statistical study of particle data (e.g., Yeh and Hill, 1981) to infer the distribution of field-aligned potential drops, and we hope to do that soon. In the meantime, we assume that there is no electric field parallel to the magnetic field.

Particle loss is another physical process that we have been neglecting, for simplicity. We do not know of any simple algorithm that would allow us to predict either ion or electron precipitation correctly. The assumption of strong pitch-angle scattering, with a full loss cone, is an optimally simple assumption (see, e.g., Kennel

[1969] for a discussion). It would tend to give an upper estimate of the actual precipitation rate. Electron precipitation can result in substantial loss of electrons from a flux tube in less than a typical convection time. However, the effect on the model is not enormous, because electrons contribute a small fraction of the total pressure in the plasma sheet and ring current. It should be noted that the most important effect of electron precipitation on the system, namely the effect on the conductivity of the ionosphere, is already taken into account, in a rough, empirical way, in the model of ionospheric conductances. Ions are lost by both precipitation and charge exchange. The latter process is a bit more predictable since it depends directly on atomic reaction rates and a neutral-atmosphere model, less directly on subtle plasma physics. We anticipate, in the next few years, including charge-exchange loss in the model.

Intense fluxes of kilovolt ions have been observed streaming up out of the ionosphere (Shelley et al., 1976). These fluxes seem to be associated with upward Birkeland currents. Correspondingly, the ionospheric ion species O^+ seems to contribute about half of the ring current energy, at least at energies below about 20 keV (see, e.g., Balsiger, 1983). Observational information is rapidly approaching the point where a reasonable observation-based algorithm could be formulated for estimating the upward ion fluxes. We anticipate incorporating such an empirical model of ionospheric injection in our model.

The present version of our model defines an idealized inner magnetosphere, with no neutral winds, no field-aligned potential drops, no particle loss from the plasma sheet and ring current, and no ion injection directly from the ionosphere to the magnetosphere. However, even with all these interesting and uncertain points of physics out, there is a lot of physics to investigate. We are still doing computer experiments to determine how all of the various input parameters in our simplified, idealized magnetosphere — the potential on the polar-cap boundary, the distribution of ionospheric conductance, the magnetic-field model, the plasma distribution at our outer boundary, our initial plasma distribution — affect the various calculated observable quantities (e.g., ionospheric currents, Birkeland currents, ionospheric electric fields, ring current strength, ring current location). These are very complicated functional relationships. After several years of work on the problem, we feel that we are only now nearing the point where we have sorted out most of the dominant causal connections. When we get to that point, we will begin adding new physical processes (field-aligned potential drops, etc.) and investigating their impact on this complicated system. We will add them one at a time, doing enough computer experiments on each to try to sort out the dominant cause-and-effect relationships.

4. PROGRAM LOGIC AND NUMERICAL METHODS

Program Logic

Figure 4 shows the essential logic of our program. We start at the top of the diagram and proceed counterclockwise. The initial magnetospheric plasma configuration is specified, as is the magnetic-field model for the initial time t_0. The magnetospheric current density can then be calculated from equation (8), and the density of Birkeland current is computed by taking the appropriate divergence (equation 11). Assuming no neutral wind, and with specified models for the ionospheric conductance and polar-boundary potential at time t_0, equation (18) is solved for the ionospheric potential

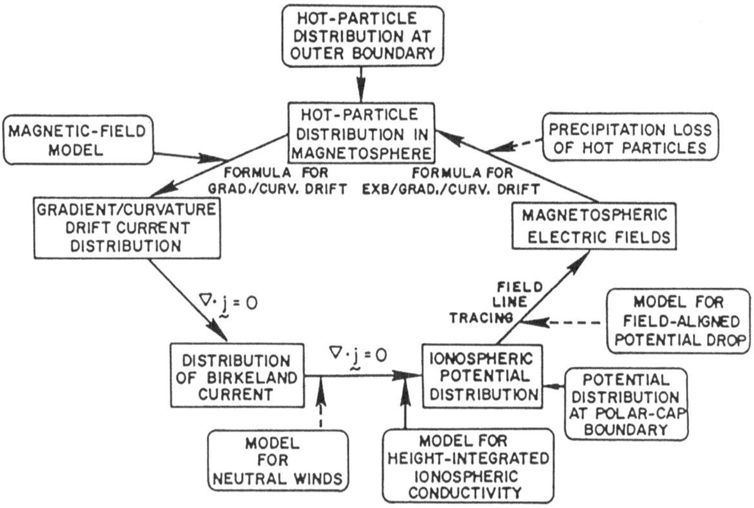

Figure 4. Basic logic diagram of our main program, an elaboration of a diagram published by Vasyliunas (1970). The program goes once around the diagram each time step. Rectangles represent primary computed quantities. Rounded boxes signify input information. Arrows with dashed lines indicate physics that could straightfowardly be included in the model, but has not been yet.

distribution. The second boundary condition on (18) is that there be no current across the equator; in practice, we actually require zero current across a low-latitude boundary at about 20^0 latitude. Taking the field lines to be equipotentials, we transform the potential to an "inertial" frame (not rotating with the Earth) using the equation

$$V_m = V - (92{,}400 \text{ volts}) \sin^2\theta \tag{20}$$

where θ=ionospheric colatitude. We then map this potential distribution out to the magnetospheric equatorial plane. The positions of magnetospheric particles are then advanced according to the law

$$\underset{\sim}{x}_e(t_o+\Delta t) = \underset{\sim}{x}_e(t_o) + \left[\frac{\underset{\sim}{B}_e \times \nabla_e V_m}{B_e^2} + \underset{\sim}{v}_{GC} + \underset{\sim}{v}_g \right] \Delta t \tag{21}$$

where Δt is the time step, V_m is the potential mapped up from the ionosphere, $\underset{\sim}{v}_{GC}$ is the bounce-averaged gradient/curvature drift velocity, and $\underset{\sim}{v}_g\Delta t$ represents the local motion of our equatorial grid between times t_o and $t_o+\Delta t$. (Our grid is fixed in the ionosphere. Our equatorial grid represents the equatorial map of the ionospheric grid, but it changes in time if the magnetic field model changes.) The $\underset{\sim}{V}_g\Delta t$ term in equation (21) can thus be regarded as $\underset{\sim}{E}\times\underset{\sim}{B}$ drift in an induction electric field. Neglecting particle loss or addition, we compute the new particle distribution for time $t_o+\Delta t$ using equation (7) with the total effective velocity from equation (21). The whole procedure is then repeated each time step.

Numerical Methods

At the time that our basic numerical method was developed (1975-1976), computing charge rates were sufficiently high that we could only afford to carry out the calculations for a coarse grid. Thus, the numerical procedure was selected for use on a grid with a relatively small number of grid points (~600). Since computing costs have dropped drastically since 1975, and core sizes have increased dramatically, it is now economically feasible to run the program with a much denser grid, for increased detail and accuracy. We are in the process of writing a more sophisticated program, using more elaborate numerical methods for efficient handling of the system with a denser grid. However, that work is not complete yet, and we will describe here the old numerical methods.

Representation of the Particle Distribution. We consider an isotropic plasma and represent the energy distribution in terms of a finite number of "energy channels." Each channel actually corresponds to a given value of the energy invariant λ (equation 3) and given charge q. In most of the simulations done so far, we have used 21 of these channels. Representing the plasma simply by densities at grid points seemed to offer inadequate spatial resolution, given the number of grid points that we could afford. Therefore, we decided to represent the particle distribution in terms of computed inner edges, not in terms of densities at grid points. For a given energy channel, a sharp density jump tends to form physically at the inner edge of the plasma sheet at the natural boundary between the region filled with fresh plasma that has flowed in from the tail, and trapped particles that have been circling the Earth for an extended period. The plasma pressure changes drastically over one grid space at this inner edge, which is just the region we wish to model carefully. In our most recent simulations, we have been assigning each invariant-energy level several different inner edges, each one corresponding to a different level of invariant density. In other words, we represent the distribution of particles with a given λ in terms of directly computed contours of constant η. This allows accurate treatment of very sharp jumps in pressure and density.

Distribution of Birkeland Current. According to equation (11), magnetic-field-aligned currents occur only where there is a gradient in some η_s, which, in our numerical scheme, occurs only at the inner edge for species s. Thus there is a sheet of Birkeland current flowing to or from the ionosphere along each inner edge, or each computed contour of constant η (see equation [A27] of Harel et al. [1981a]).

Current Conservation of the Ionosphere. We convert the differential equation (18) to the following difference equation:

$$V^{ij} = c_1{}^{ij}V^{i+1,j}+c_2{}^{ij}V^{i-1,j}+c_3{}^{ij}V^{i,j+1}+c_4{}^{ij}V^{i,j-1}+c_5{}^{ij} \qquad (22)$$

where the coefficients c_1-c_4 are given by equation (6) of Jaggi and Wolf (1973), and c_5, which represents the effect of Birkeland currents, is specified by equations (A29) and (A30) of Harel et al. (1981a). Equation (22) is solved by a successive over-relaxation method using

$$\delta V^{ij} = W[c_1{}^{ij}V^{i+1,j}+c_2{}^{ij}V^{i-1,j}+c_3{}^{ij}V^{i,j+1}+c_4{}^{ij}V^{i,j-1}+c_5{}^{ij}-V^{ij}] \qquad (23a)$$

The improved estimate of V^{ij} is given by

$$V'^{ij} = V^{ij}+ \delta V^{ij} \qquad (23b)$$

The procedure is repeated until the calculated potential changes are sufficiently small that

$$\sum_{i,j} | \delta V^{ij} |^2 < \varepsilon \qquad (24)$$

Actually, the method should perhaps be called successive under-relaxation, because we find that the weighting factor W has to be less than 1 for convergence, usually about 0.5. This simple procedure is adequately efficient for our situation, because we only have about 600 grid points, and we usually have an excellent initial guess for V, based on the potential distributions previously calculated for the last two time steps. Usually, variations from one time step to the next are relatively small, and the system converges in ~20 iterations.

Mapping from the Ionosphere to the Equatorial Plane. For our recent simulations of the March 22, 1979 magnetic storm, G.-H. Voigt supplied us with magnetic-field models for 16 key times during the event. For each of these models, field lines were traced from each of our ionospheric grid points to the equatorial plane. Results were entered into our program in the form of matrices specifying the position of the equatorial crossing point, the equatorial magnetic field strength, and the flux-tube volume for each grid point. To evaluate these parameters for arbitrary times during the event, we linearly interpolated between the key-time values stored in the matrices.

Computation of Boundary Motions. In the present version of the program, each inner edge is represented by a series of "test particles" (see appendix of Wolf *et al.* [1982]). The gradient/curvature-drift velocity of each test particle is computed by simple central differences and interpolations, using the value of flux-tube volume and equatorial magnetic field at the grid points. The $\tilde{E}\times\tilde{B}$ drift is treated more carefully because of the fine structure in the computed electric fields in the inner-edge region. The first step is to compute the electric field by straightforward central differences and interpolations applied to the potentials. Then we compute a fine-structure correction by calculating the electric field due to each nearby inner edge. This fine-structure correction compensates for the coarse grid, at least with respect to the electric-field effects of one species' inner edge on another (for a more detailed discussion, see Appendix 2 of Harel *et al.* [1981a]).

The particles move in time according to equation (21). The time step Δt is limited mainly by a numerical instability that occurs when the steplength exceeds approximately the physical decay time of the shortest-wavelength inner-edge ripple that our grid spacing allows us to represent. The very simple step procedure (21) is used because we have not found an alternative that deals more efficiently with the numerical instability and is also simple to adapt to our overall scheme.

Accuracy Tests. As in most computer simulations in physics, there are never enough precise numerical-accuracy tests to make one completely sure what the numerical accuracy is for an given physical situation. However, the following tests have provided some indications of accuracy levels:

1. For the case of a single inner edge, uniform ionospheric conductance, dipole magnetic field, and a boundary potential of the simple form (19), time-dependent analytic solutions can be constructed. (Such analytic solutions are helpful for physical discussions [*e.g.*, Siscoe, 1982; Senior and Blanc, 1983].) In the initial testing of our code, we ran a series of comparisons with such analytic solutions, and found good agreement.
2. We have repeated runs several times for different values of the convergence parameter ε (see equation [24]). Based on those results, we have generally chosen ε=1000 (volts)^2 as the best compromise between speed and accuracy.
3. We have repeated runs with decreased values of the time step Δt, and found very small differences. As a practical matter, if the time step is short enough to avoid the numerical instability

discussed earlier, it seems to give adequate accuracy.

4. The accuracy of the particle-moving algorithm can be tested by running the model with steady input parameters for a time long compared to a drift time, and comparing the computed inner edges with computed contours of constant effective potential

$$V_{eff} = V_m + \lambda S \tag{25}$$

Such a test can be performed for a fast-drifting species (i.e., high λ) in a few hours magnetosphere time. Agreement in such tests has been satisfactory, although problems can occur near local noon, where inner-edge shapes are particularly complex.

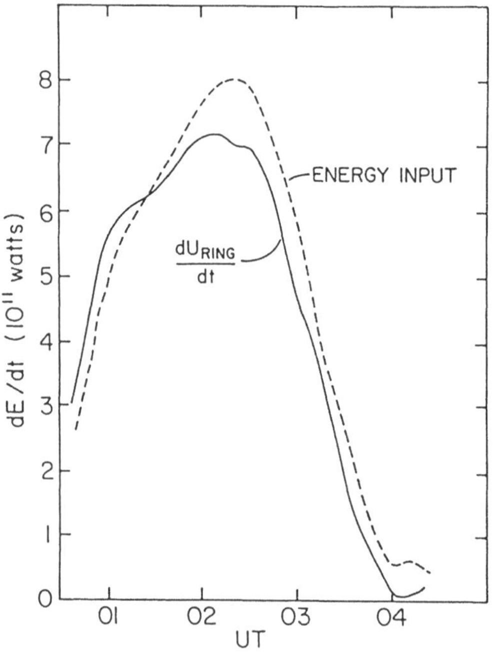

Figure 5. Conservation-of-energy test, applied to our simulation of the July 29, 1977 storm. The solid curve gives the rate at which energy is fed into the ring current, by means of currents from the ionosphere and currents through the boundary. The dashed curve gives the rate of change of ring-current energy in the model.

5. For the case where the input magnetic field model remains constant in time, there is a conservation-of-energy check on the system. With no particle loss, the rate of change of particle energy in the modeling region equals the rate at which energy is fed into the particles by Birkeland currents from the ionosphere and by currents and particles flowing through the boundary of the

calculation. Figure 5 shows this energy comparison for the case of the main phase of the July 29, 1977 storm. We feel that this level of error is about as low as we can get it with the present coarse-grid version of the program. Much higher accuracy should be possible with a fine-grid version.

Cost. The advantage of using a coarse grid is, of course, that it allows us to do a lot of runs at reasonable cost. The present version of our main program requires only about 500 kbytes of core, and the CPU cost of running it at night amounts to only about $10 per hour magnetosphere time.

5. REVIEW OF RESULTS

The purpose of this paper has been to discuss the physical basis of the Rice Convection Model, the relationship between the model and standard MHD, and the numerical methods used. We have not discussed the Earth's magnetosphere, only how we model it. In other papers, we have used the model to explain and clarify aspects of the physics of the Earth's magnetosphere. Rather than select one or two such aspects for detailed discussion, we instead present a catalogue to illustrate the variety of topics that the model addresses. Specifically, we list points of magnetospheric physics that have been illuminated through use of the RCM, and indicate where these points are discussed in more detail.

Magnetospheric and Ionospheric Electric Fields

1. Shielding Efficiency. Using idealized analytic models and order-of-magnitude arguments, Vasyliunas (1972) and others have shown that the inner edge of the plasma sheet tends to shield the region earthward of it from the magnetospheric-convection electric field. The RCM has strengthened the theoretical argument for the shielding process, by showing it to be a prominent and persistent feature of much more realistic model calculations (see, *e.g.*, Jaggi and Wolf, 1973; Harel *et al.*, 1981b; Spiro *et al.*, 1981).

2. Shielding Times. The RCM has provided estimates of the time scales for the shielding process, for realistic conditions where ionospheric conductance varies substantially along the shielding layer (Jaggi and Wolf, 1973).

3. Electric-Field Pattern at Subauroral Latitudes During Substorms. With its complicated conductivity model and detailed simulation of ring-current dynamics, the RCM makes predictions as to how the shielding process is disrupted during substorms, allowing magnetospheric convection electric fields to penetrate to lower ionospheric latitudes. The predictions agree well with the observations in most respects (see Spiro *et al.* [1981] or Maynard *et*

al. [1983] for comparisons). (An interesting semi-analytical model of electric-field penetration has recently been developed by Senior and Blanc [1984].)

4. Dawn-Dusk Convection Asymmetry. Observations indicate that sunward flow tends to be stronger in the duskside auroral zone that on the dawn side (Kelley, 1976; Foster, 1983). (This asymmetry is a standard feature of our model calculations, where it results mainly from the change in Hall conductance at the terminators [Harel et al., 1981b].)

5. Rapid Subauroral Flow. Very large poleward electric fields are frequently observed over a narrow latitude band just equatorward of the evening auroral zone (Spiro et al., 1974; Heelis et al., 1976; Smiddy et al., 1977; Maynard, 1978; Spiro et al., 1978). Using results from the RCM and related analytic arguments, these events are interpreted as being due to the inner edge of the ion plasma sheet being slightly equatorward of the electron inner edge on the dusk side (Southwood and Wolf, 1978; Harel et al., 1981b).

6. Effects of Sudden Commencement on Shielding. From results of our simulation of the magnetic storm of July 29, 1977, which involved a massive compression of the magnetosphere, we predict that the compression of the magnetosphere by increased solar wind dynamic pressure temporarily disrupts shielding and causes substantial sunward flow at subauroral latitudes. To our knowledge, this prediction has not been checked directly in mid-and low-latitude electric-field data.

Electric Currents

1. Overall Pattern of Ionospheric and Birkeland Current. The overall magnetosphere-driven current system, with region-1 and region-2 Birkeland currents, eastward and westward electrojets, and large meridional Pedersen currents, are intrinsic in our simulations (Wolf, 1974; Harel et al., 1981b; Karty et al., 1982). The same basic pattern emerges from detailed model analysis of ground magnetograms (e.g., Hughes and Rostoker, 1977; Kamide et al., 1981), and other models (e.g., Kamide and Matsushita, 1979).

2. Nightside-Overlap Region in Birkeland-Current Pattern. The characteristic down-up-down Birkeland-current pattern observed near midnight (Iijima and Poterma, 1978) usually occurs in our models, where it results from high-energy plasma-sheet ions having inner edges that are shaped differently from low-energy ions and electrons (Wolf et al., 1982; Spiro and Wolf, 1983).

3. Dawn-Dusk Asymmetry in Low-Latitude ΔH. It has long been known that the horizontal component of the low-latitude magnetic field decreases more on the dusk side than the dawn side during a substorm or early in the main phase of a storm. This observation was usually interpreted in terms of a partial ring current in the inner

magnetosphere, connected through Birkeland currents to the eastward electrojet (e.g., Kamide and Fukushima, 1972; Crooker and McPherron, 1972). When the large-scale Birkeland-current patterns were first discovered (Zmuda and Armstrong, 1974), it was not clear how those patterns related to the idea of a dusk-side partial ring current. Since the complicated current patterns computed in the RCM are consistent with both the observed large-scale Birkeland currents and with the classic dawn-dusk asymmetry observed in low-latitude magnetograms, we have been able to interrogate the model to determine which currents are responsible for which observational features. This exercise resulted in a modification of the dusk side partial-ring - current picture of the substorm current system (Harel et al., 1981b; Chen et al., 1982). Using hints from the computer results, Crooker and Siscoe (1981) have constructed an elegant analytic theory of this same asymmetry.

4. Birkeland Current Patterns Just After Strong Magnetospheric Compressions. Our simulations of the July 29, 1977 storm indicate that the general pattern of region-1/region-2 Birkeland current may be temporarily and dramatically disrupted during a strong magnetospheric compression. As far as we know, this theoretical prediction has not been tested observationally.

5. Generation of Region-1 Currents on Plasma-Sheet Flux Tubes. Region-1 currents have frequently been observed on field lines that are drifting slowly toward the Sun (e.g., Smiddy et al., 1980; Mozer et al., 1980). Model results and related analytic calculations indicate that such region-1 currents are a natural result of gradients in particle content among different plasma-sheet flux tubes (Karty et al., 1983).

Ring-Current Injection

1. Convection and Ring-Current Injection. It was suggested many years ago that enhanced convection during a magnetic storm injected some plasma-sheet plasma deep into the magnetosphere to form the storm-time ring current. This idea was supported by particle-trace calculations done with simple models of the magnetospheric electric field (see review by Kivelson et al. [1979]). Our simulations of the magnetic storms of July 29, 1977 and March 22, 1979 have shown that enhanced convection could inject plasma-sheet plasma to form a storm-time ring current, even considering that the injection is inhibited by the shielding effect, which we attempt to calculate realistically with the RCM (Wolf et al., 1982; Spiro and Wolf, 1983).

2. Relationship Between Joule Heating and Convection. The time-integrated Joule heating of the upper atmosphere through a substorm or storm main phase is, in our simulations, typically of the same order of magnitude as the change in ring-current energy. In events where the injection is entirely due to enhanced convection, the integrated Joule heating is larger. In events where the injection is

due in large part to magnetospheric compression, integrated Joule heating tends to be smaller than the ring current energy (Harel et al., 1981b; Wolf et al., 1982). The theoretical necessity for integrated Joule heating to be of the same order of magnitude as the ring-current increase in an injection event has now been derived by two analytic arguments (Siscoe, 1982; Harel et al., 1981b).

3. Factors Affecting Ring-Current Injection. A recent long series of computer experiments, done partly in collaboration with George Siscoe, has clarified a number of cause-and-effect relationships governing ring-current injection (Spiro and Wolf, 1983). The following results hold, within the physics presently included in our model:

(i) Increasing ionospheric conductance increases the energy of the injected ring current, though not dramatically. (Increasing conductance decreases shielding efficiency.)
(ii) Compression of the magnetosphere plays an important role in injection of the ring current.
(iii) Decreasing the pressure in the plasma sheet but keeping temperature constant causes ring-current plasma to be injected deeper into the magnetosphere (weaker shielding), but decreases the total strength of the ring current (less plasma available to inject).
(iv) Increasing the temperature of the plasma-sheet ions slightly decreases total ring-current energy, but allows deeper injection.

Convection and Magnetic Field Configuration

A research effort motivated by the need to obtain reliable plasma boundary values for the RCM showed that there is an inconsistency between the idea of steady, subsonic, sunward convection in the plasma sheet, and the requirement that the magnetic-field configuration be closed and in stress balance (Erickson and Wolf, 1980). This theoretical difficulty, which may, we think, contain the key to substorm occurrence, has been verified and further investigated by Schindler and Birn (1982) and Tsyganenko (1982).

6. CONCLUDING COMMENTS

The Rice Convection Model has provided substantial physical insights into how convection works in the Earth's inner magnetosphere. We are now close to the point where the major cause-and-effect relations are established, within the physics presently included in our model. The following things remain to be done:

1. Develop a version of the program that will deal efficiently with a denser grid, for more spatial detail and higher overall accuracy.

2. Add additional physics to the model, doing many computer experiments to determine the key cause-and-effect relationships involved. The most obvious additional physical effects are the following: particle loss by charge exchange and precipitation; magnetic-field-aligned electric fields; ion injection from the ionosphere; equatorial electrojet; Van Allen-belt particles; and neutral winds.
3. Combine the convection model with a time-dependent magnetic-field model that is consistent with the currents computed in the convection calculation.

The Rice Convection Model has proven to be an effective theoretical tool for investigating the physics of the inner magnetosphere. The continued development of the model to include additional physical processes promises to be just as fruitful.

Acknowledgements

Much of the research reviewed here was carried out in collaboration with M. Harel, P. H. Reiff, G.-H. Voigt, C.-K. Chen, J. L. Karty, and G. M. Erickson. We are grateful to G. L. Siscoe for motivating the computer experiments on the ring current and for many helpful discussions. Preparation of this paper was supported by the National Science Foundation under grant ATM82-06026 and by NASA grant NGR-44-006-137.

REFERENCES

Axford, W. I., and Hines, C. O., Can. J. Phys. 39, 1433, 1961.

Balsiger, H., in B. Hultqvist and T. Hafgors (eds.), High-Latitude Space Plasma Physics, Plenum Publ. Co., London, 1983.

Behannon, K. W., J. Geophys. Res., 75, 743, 1970.

Brecht, S. H. Lyon, J. G., Fedder, J. A., and Hain, K., NRL Memorandum Report 4690, 1981.

Burke, W. J.,Kelly, M. C., Sagalyn,R. C.,Smiddy, M., and Lai, S. T., Geophys. Res. Lett., 6, 21, 1979.

Chen, C.-K., Wolf, R. A., Harel, M., and Karty, J. L., J. Geophys. Res., 87, 6137, 1982.

Cole, K. D., Geophys. J. 6, 103, 1961.

Crooker, N. U.,and McPherron, R. L., J. Geophys. Res., 77, 6886, 1972.

Crooker, N. U., and Siscoe, G. L., J. Geophys. Res., 86, 11201, 1981.

Dungey, J. W., Phys. Rev. Lett, 6, 47, 1961.

Erickson, G. M., and Wolf, R. A., Geophys. Res. Lett., 7, 897, 1980.

Fejer, J. A., J. Atmos. Terr. Phys., 4, 184, 1953.

Fejer, J. A., J. Geophys. Res., 69, 123, 1964.

Foster, J. C., J. Geophys. Res., 88, 981, 1983.

Harel, M., Wolf, R. A., Reiff, P. H., Spiro, R. W., Burke, W. J., Rich, F. J., and Smiddy, M., J. Geophys. Res., 86, 2117, 1981a.

Harel, M., Wolf, R. A., Spiro, R. W., Reiff, P. H., Chen, C.-K., Burke, W. J., Rich, F. J., and Smiddy, M., J. Geophys. Res., 86, 2242, 1981b.

Heelis, R. A., Spiro, R. W., Hanson, W. B., and Burch, J. L., (abstract) EOS Trans. AGU 57, 990, 1976.

Hughes, T. J., and Rostoker, G., J. Geophys. Res., 82, 2271, 1977.

Iijima, T., and Potemra, T. A., J. Geophys. Res., 83, 599, 1978.

Jaggi, R. K., and Wolf, R. A., J. Geophys. Res., 78, 2842, 1973.

Kamide, Y., and Fukushima, N., Rep. Ions. Space Res. Jpn. 26, 79, 1972.

Kamide, Y., and Matsushita, S., J. Geophys. Res., 84, 4099, 1979.

Kamide, Y., Richmond, A. D., and Matsushita, S., J. Geophys. Res., 86, 801, 1981.

Karty, J. L., Chen, C.-K., Wolf, R. A., Harel, M., and Spiro, R. W., J. Geophys. Res., 87, 777, 1982.

Karty, J. L., Wolf, R. A., and Spiro, R. W., in T. A. Potemra(ed.) Magnetospheric Currents, AGU, Washington, D.C., 1984.

Kelley, M. C., Planet. Space Sci., 24, 355, 1976.

Kennel, C. F., Rev. Geophys. Space Phys. 7, 379, 1969.

Kivelson, M. G., Kaye, S. M., and Southwood, D. J., in S.-I. Akasofu (ed.) Dynamics of the Magnetosphere, D. Reidel, Hingham, MA, 1979.

Leboeuf, J. N., Tajima, T., Kennel, C. F., and Dawson, J. M., Geophys. Res. Lett., 5, 609, 1978.

Lyon, J., this volume, 1984.

Lyon, J., Brecht, S. H., Fedder, J. A., and Palmadesso, P., Geophys. Res. Lett., 7, 721, 1980.

Maynard, N. C., Geophys. Res. Lett., 5, 617, 1978.

Maynard, N. C., Aggson, T. L., and Heppner, J. P., J. Geophys. Res.,

88, 3991, 1983.
Mead, G. D., and Beard, D. B., J. Geophys. Res., 69, 1169, 1964.
Moore, T. E., Arnoldy, R. L., Feynman, J., and Hardy, D. A., J. Geophys. Res., 86, 6713, 1981.
Mozer, F. S., and Serlin, R., J. Geophys. Res., 74, 4739, 1969.
Mozer, F. S., Cattell, C. A., Hudson, M. K., Lipak, R. L., Temerin, M., and Torbert, R. B., Space Sci. Rev., 27, 155, 1980.
Northrop, T. G., The Adiabatic Motion of Charged Particles, Interscience Publ., New York, 1963.
Olson, W. P., and Pfitzer, K. A., J. Geophys. Res., 87, 5943, 1982.
Reiff, P. H., in T. A. Potemra(ed.) Magnetospheric Currents, AGU, Wshington, D.C., 1984
Reiff, P. H., Spiro, R. W., and Hill, T. W., J. Geophys. Res., 86, 7639, 1981.
Reiff, P. H., Spiro, R. W., and Kamide, Y., (abstract), EOS Trans. AGU 64, 293, 1983.
Roble, R. G., Dickinson, R. E., and Ridley, E. C., J. Geophys. Res., 87, 1599, 1982.
Schindler, K., and Birn, J., J. Geophys. Res., 87, 2263, 1982.
Senior, C. L., and Blanc, M., J. Geophys. Res., 89, 261, 1984.
Shelley, E. G., Sharp, R. D., and Johnson, R. G., Geophys. Res. Lett., 3, 654, 1976.
Simons, S. L., Jr., Reiff, P. H., Spiro, R. W., Hardy, D. A., and Kroehl, H. W.: submitted for publication in J. Geophys. Res, 1983.
Siscoe, G. L., J. Geophys. Res., 87, 5124, 1982.
Smiddy, M., Kelley, M. C., Burke, W., Rich, F., Sagalyn, R., Shuman, B., Hays, R., and Lai, S., Geophys. Res. Lett., 4, 543, 1977.
Smiddy, M., Burke, W. J., Kelley, M. C., Saflekos, N. A., Gussenhoven, M. S., Hardy, D. A., and Rich, F. J., J. Geophys. Res., 85, 6811, 1980.
Southwood, D. J., and Wolf, R. A., J. Geophys. Res., 83, 5227, 1978.
Spiro, R. W., and Wolf, R. A., in T. A. Potemra(ed.) Magnetospheric Currents, AGU, Washington D.C., 1984.
Spiro, R. W., Hanson, W. B., Sterling, D. L., and Hoffman, R. A., (abstract), EOS Trans. AGU 58, 1159, 1974.
Spiro, R. W., Heelis, R. A., and Hanson, W. B., J. Geophys. Res., 83, 4255, 1978.
Spiro, R. W., Harel, M., Wolf, R. A., and Reiff, P. H., J. Geophys. Res., 86, 2261, 1981.
Spiro, R. W., Reiff, P. H., and Maher, L. J., Jr., J. Geophys. Res., 87, 8215, 1982.
Swift, D. W., J. Geophys. Res., 76, 2276, 1971.
Taylor, H. E., and Perkins, F. W., J. Geophys. Res., 76, 272, 1971.
Tsyganenko, N. A., Planet. Space Sci., 30, 1007, 1982.
Vasyliunas, V. M., in B. McCormac (ed.) Particles and Fields in the Magnetosphere, D. Reidel, Hingham, MA, 1970.
Vasyliunas, V. M., in B. McCormac (ed.) Earth's Magnetospheric Processes, D. Reidel, Hingham, MA, 1972.
Voigt, G.-H., Planet. Space Sci., 29, 1, 1981.
Wolf, R. A., J. Geophys. Res., 75, 4677, 1970.
Wolf, R. A., in B. M. McCormac (ed.) Magnetospheric Physics, D.

Reidel, Hingham, MA, 1974.
Wolf, R. A., in R. L. Carovillano and J. M. Forbers(eds.) Solar-Terrestrial Physics, D. Reidel, Hingham, MA, 1983.
Wolf, R. A., Harel, M., Spiro, R. W., Voigt, G.-H., Reiff, P. H., and Chen, C.-K., J. Geophys. Res., 87, 5949, 1982.
Wu, C. C., this volume, 1984.
Wu, C. C., Walker, R. J., and Dawson, J. M., Geophys. Res. Lett., 8, 523, 1981.
Yeh, H.-C., and Hill, T. W., J. Geophys. Res., 86, 6706, 1981.
Zmuda, A. J., and Armstrong, J. C., J. Geophys. Res., 79, 4611, 1974.

HYBRID SIMULATION TECHNIQUES APPLIED TO THE EARTH'S BOW SHOCK

D. Winske

Department of Physics and Astronomy
University of Maryland
College Park, MD 20742, U.S.A.

and

M. M. Leroy

DESPA, Observatoire de Paris
92195 Meudon, FRANCE

April 1983

ABSTRACT

The application of a hybrid simulation model, in which the ions are treated as discrete particles and the electrons as a massless charge-neutralizing fluid, to the study of the earths bow shock is discussed. The essentials of the numerical methods are described in detail: movement of the ions, solution of the electromagnetic fields and electron fluid equations, and imposition of appropriate boundary and initial conditions. Examples of results of calculations for perpendicular shocks are presented which demonstrate the need for a kinetic treatment of the ions to reproduce the correct ion dynamics and the corresponding shock structure. Results for oblique shocks are also presented to show how the magnetic field and ion motion differ from the perpendicular case.

H. Matsumoto and T. Sato, Computer Simulation of Space Plasmas,

1. INTRODUCTION

When an obstacle is placed in a supersonic flow, a shock wave is created in front of it. The earth and its surrounding magnetic field and plasma configuration (the magnetosphere) is such an obstacle to the supersonic stream of plasma flowing out from the sun (the solar wind); the resulting shock is called the "bow shock". Other planets with magnetospheres, such as Jupiter, also have bow shocks. Excellent discussions of the earth's bow shock and its relation to the magnetosphere are given in the review articles by Greenstadt and Fredricks (1979) and Schindler and Birn (1978), respectively. The bow shock has been and is still actively studied both for its intrinsic interest as part of the solar-terrestrial system and also because of the variety of plasma physics behavior it exhibits. Processes which occur in the bow shock, such as wave generation, plasma heating and particle acceleration are generic to all of plasma physics and have wide application to both laboratory and astrophysical phenomena. However, the study of bow shock phenomena is often complicated by an overabundance of data. Unlike laboratory plasmas, the diagnostic probes (satellites) are so small and temporal response is so fast that phenomena on all time and distance scales (i.e. electrons as well as ions) are resolvable. It is often difficult to determine what are the most important effects which are occurring and need to be included in the physical model and which are the data that can be ignored.

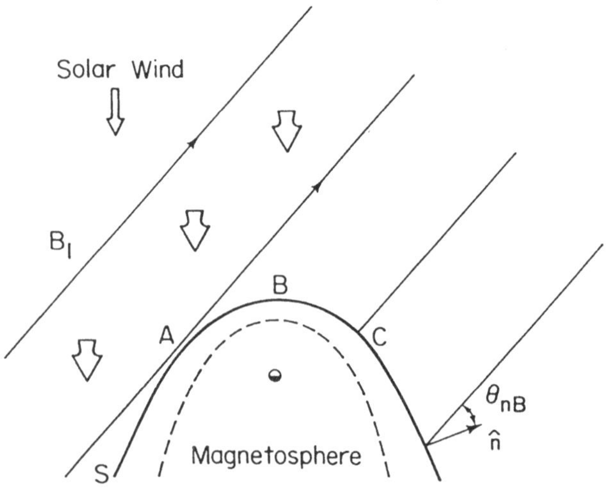

Figure 1. Basic geometry of the earth's bow shock showing the solar wing (arrows), interplanetary magnetic field (straight lines) and the bow shock surface (S). θ_{nB} is the angle between the magnetic field and the shock normal ($\hat{n}$).

Furthermore, the bow shock itself is not a homogeneous structure. A pictorial view of the bow shock is given in Figure 1. The magnetic field (straight lines) is carried along on the average at roughly 45^0 with respect to the solar wind (arrows). The bow shock is the curved solid line (S). The angle that the magnetic field makes with the shock normal ($\hat{n}$) is denoted θ_{nB}. This angle varies from 90^0 at point A (called the perpendicular shock) to about 45^0 at point B and 0^0 at C (the parallel shock). The portion of the shock where θ_{nB} lies in the range $50^0 < \theta_{nB} < 90^0$ is called 'quasiperpendicular' and for $\theta_{nB} < 45^0$ it is referred to as 'quasiparallel' (Greenstadt and Fredricks, 1979). The structure of quasiperpendicular shocks is characterized by a fairly narrow, well-defined transition from upstream (solar wind side) to downstream (magnetospheric side) and has been studied in much detail (e.g., Greenstadt et al., 1980). As θ_{nB} decreases, the structure becomes more complicated as large scale magnetic fluctuations appear and the transition from upstream to downstream becomes longer and less well defined. Systematic observations of quasiparallel shocks are much less extensive (e.g. Greenstadt et al., 1976).

Our ultimate goal is to understand the detailed physics of strong shock waves in magnetized plasmas (such as the bow shock). However, the problem appears formidable: there are a large number of degrees of freedom in the system, disparate time and length scales associated with the two species involved, nonlocal and nonlinear behavior occurring through turbulence arising as a result of numerous instabilities taking place over a wide range of frequencies, etc. Even when we realize (on observational grounds) that the phenomena associated with different scales (namely,large ion scales governed by macroscopic ion dynamics, small scales govered by microturbulence and dissipation) separate naturally, the theorist is faced with an inextricable analytical problem. Thus, we turn to simulation in the hope that it will not only duplicate and put into readable form the observations, but more importantly will improve our detailed understanding of the physics.

If one is going to study the bow shock (or any other space or laboratory configuration, for that matter) by simulation, one has to decide what scale of phenomena one wants to focus on and what processes one wishes to investigate in order to construct an appropriate model. For example, a macroscopic MHD code would be useful to look at the whole bow shock and its relation to the solar wind, but would not be appropriate to study microturbulence at the transition layer. In this chapter we thus briefly describe some of the observations which motivated this study and discuss a physical model appropriate to the problem. We then describe at length the numerical tools used to implement the physical model, give examples to show how they have been successfully applied, and illustrate the types of results and information which are produced. Obviously, there is not enough space here to discuss many of the results; hopefully,with this chapter as a background, the interested reader can the pursue the details, published elsewhere.

The simulation study described here was originally motivated by the high resolution data of the quasiperpendicular bow shock acquired with the ISEE spacecraft experiments. An example of the magnetic field profiles obtained by the UCLA high resolution magnetometers at one shock crossing (Nov. 7, 1977) is shown in Figure 2 (Greenstadt et al., 1981). The top panels show the magnetic field observed by each satellite as a function of time. (ISEE-2 follows ISEE-1 through the

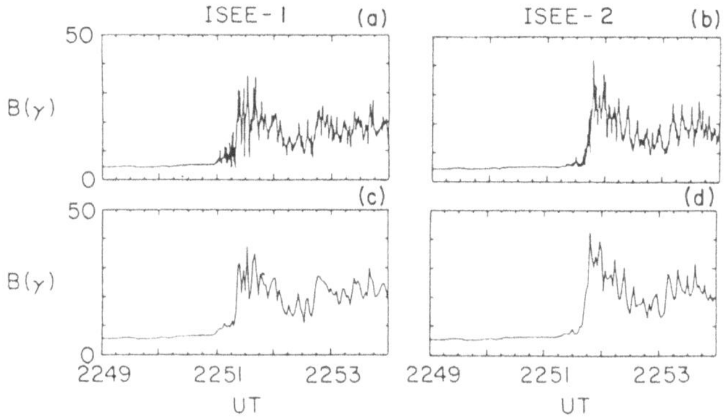

Figure 2. Magnetic field measurements of the November 7, 1977 bow shock crossing: (a) ISEE-1 and (b) ISEE-2; (c-d) two second averages of the above. (Figure courtesy of C. T. Russell, UCLA).

shock about half a minute later.) The bottom panels show the same data with some of the high frequency noise filtered out. These averaged profiles are obviously much simpler and better illustrate the overall structure of the shock profile. The most prominent feature of the magnetic field is the sharp transition where most of the jump between the upstream (left side of picture) and downstream (right side of picture) magnetic field occurs. We will sometimes refer to this at the 'ramp'. Just to the front of the ramp (to the left in the figures) the magnetic field rises slightly above its upstream value; this region is commonly called the 'foot'. The magnetic field just behind (to the right) of the shock front is somewhat larger than its downstream value. This region is referred to as the 'overshoot'; it is followed by an 'undershoot' where the magnetic field falls below its downstream value. Note that this overall foot-ramp-overshoot-undershoot configuration is observed both by ISEE-1 and 2, although the fine structure varies somewhat. Examples of the magnetic field structure from other shock crossings are given by Russell and Greenstadt (1979).

Furthermore, the use of two different spacecraft in the ISEE experiments has allowed reliable determinations of the length scales that characterize the shock transition layer. The overall structure has been found to extend over hundreds of kilometers, which is equivalent to many ion inertia lengths (c/ω_{pi}), or more relevantly, to a few ion gyroradii. The electron inertia length c/ω_{pe} is only about 2km, while the thickness of the ramp is somewhat bigger (14 km for the case shown in Figure 2). Hence the scale size for the overall shock structure is very much larger than c/ω_{pe}.

In addition to magnetic field data, there are a variety of other types of measurements. For example, the ion distribution function data show that some (10-20%) of the solar wind ions are reflected at the shock and eventually reach downstream without thermalizing rapidly (Paschmann et al., 1982). Measurements also show that the electrons are heated primarily isotropically at the shock, although a heat flux, which is directed upstream along the magnetic field, can be produced (Feldman et al., 1983).

This information suggests that the simulation model for studying the (quasiperpendicular) bow shock structure should contain the following ingredients. The observed ion reflection implies that kinetic effects will be important; hence, the ions should be treated as discrete particles, instead of as a fluid. The large scale lengths associated with the overall shock structure imply that the simulation should extend over many ion gyroradii. One is then faced with two choices: either include electron effects in the same way or in some approximate fashion. In the first case, to treat the electrons in the same manner as the ions implies compressing the widely differing electron and ion time and distance scales by the use of artifical parameters (e.g. mass ratio $m_i/m_e \sim 100$, instead of 1836, etc.). This can sometimes obscure the scales on which the physical processes occur. This strategy, however, has been successfully used in previous simulations of shocks (Papadopoulos et al., 1971; Forslund and Freidberg, 1971; Mason, 1972; Auer et al., 1971; Biskamp and Welter, 1972) and more recently in two-dimensional simulations by Forslund et al.(1984).

The second choice, which is the one used here, is to ignore electron kinetics and treat the electrons as a massless fluid. Electron scales are thus neglected, which then allows a rather easy numerical investigation of the large scales associated with the ions. Plasma heating is included through the use of a macroscopic resistivity. The numerical methods used to implement this model are described in detail in Section 2. Section 3 presents examples of this technique applied to the bow shock, first for a perpendicular ($\theta_{nB}=90^0$) shock and then for an oblique case ($\theta_{nB}=45^0$). A short summary follows in Section 4.

2. NUMERICAL METHODS

In this section the numerical methods are described in detail. Some of the techniques (e.g., particle-in-cell method) are standard and well known, while others (e.g., solution of the field equations) have been applied primarily to fusion-related problems and hence are generally not familiar to the space science community. The discussion is divided into five parts: (A) an overall description of the simulation methods, (B) treatment of the ions, (C) solutions of the field and electron fluid equations, (D) refinements to the basic model and (E) discussion of the initial and boundary conditions. The methods pertaining to the first four parts are rather general and could thus be applied to any problem where a 'hybrid' (kinetic ion, fluid electron) description is appropriate. Specific adaptation of the method to the bow shock enters primarily through the last topic, the initial and boundary conditions.

2.A. Overall Description

The simulation method is particle-in-cell (Morse, 1970) with the electrons treated as a fluid (i.e. cell) quantity rather than as discrete particles. The assumption of a massless, charge neutral electron fluid thus eliminates the restriction of time and spatial lengths to the inverse electron plasma frequency and electron Debye length, usually associated with full particle codes. The hybrid models discussed here were originally developed for and extensively applied to dynamical studies of high density pinch experiments by Chodura (1975), Sgro and Nielson (1976) and Hamasaki et al.(1977). These models apply strictly to the case of only one spatial dimension; extension to two dimensions is possible, although nontrivial, as discussed later. Even with only one spatial dimension (which is along the x-axis and thus implies $\partial/\partial y=\partial/\partial z=0$), all three velocity and magnetic and electric field components are included. The simulation region has length L, divided into N cells, each of length $\Delta x= L/N$. There is an additional ('ghost') cell at each end of the system, which helps in keeping track of the particles entering and exiting the system and in setting up the boundary conditions. The cell quantities (i.e., fields, electron fluid properties, ion velocity moments) are specified at the cell centers. Because the restriction to electron spatial and temporal scales has been eliminated, much larger time steps can be used. Typically, the time step (Δt) is limited by the condition that ion gyromotion is well defined and the cell size is constrained by the condition that the fastest ions do not traverse one cell in one time step. The cell size and system length are also chosen to resolve length scales of interest to the problem.

2.B. Ion Dynamics

The ion component is modeled by a discrete set of particles. The ion distribution is advanced in time by stepping forward each particle in time under the influence of the local, self-consistent Lorentz force. The motion of the ions in a four-dimensional phase space (vx, vy, vz, x) is solved by the particle-in-cell technique, using a second-order-accurate but non-reversible scheme (Nielson and Lewis, 1976). The epuations for the ion advance (with superscripts denoting the time level) are:

$$\bar{\underset{\sim}{v}}^0 = \underset{\sim}{v}^{-1/2} + (h/2)\, \underset{\sim}{E}^0$$

$$\underset{\sim}{v}^{1/2} = f\underset{\sim}{v}^{-1/2} + h\left(\underset{\sim}{E}^0 + g\underset{\sim}{B}^0 + \frac{\bar{\underset{\sim}{v}}^0 \times \underset{\sim}{B}^0}{c} + \underset{\sim}{P}^0 \right) \tag{1}$$

$$\underset{\sim}{x}^1 = \underset{\sim}{x}^0 + \Delta t\underset{\sim}{v}^{1/2}$$

where $h=\Delta t\ e/m_i$, $f=1-(h^2/2)\ \underset{\sim}{B}^0\cdot\underset{\sim}{B}^0$, $g=(h/2)(\underset{\sim}{v}^{-1/2}\cdot\underset{\sim}{B}^0)$, e is the charge and m_i is the mass of the ions. The electric ($\underset{\sim}{E}$) and magnetic fields ($\underset{\sim}{B}$) are evaluated at the particle position $\underset{\sim}{x}^0$. $\underset{\sim}{P}$ is a mean friction force, $\underset{\sim}{P}=-e\underset{\sim}{\eta}\cdot\underset{\sim}{J}$, exerted by the electrons as macroscopic force only, $\underset{\sim}{J}$ is the current and $\underset{\sim}{\eta}$ represents a phenomenological anomalous resistivity which gives rise to Ohmic heating, as will be explained later. After all of the ions have been advanced, the ion density (n_i) is updated, by averaging over the positions of all the particles in each cell. The ion mean velocities ($\underset{\sim}{V}_i$) are also needed at time level 1. The velocities are pushed ahead one half time step,

$$\underset{\sim}{v}^1 = \underset{\sim}{v}^{1/2} + (h/2)\left(\underset{\sim}{E}^1 + \frac{\underset{\sim}{v}^{1/2} \times \underset{\sim}{B}^1}{c} + \underset{\sim}{P}^1 \right) \tag{2}$$

using the fields at the new particle positions $\underset{\sim}{x}^1$, in order to accumulate the velocity moments $\underset{\sim}{V}_i$.

The assumption of charge neutrality then implies

$$n_e = n_i = n \tag{3}$$

and the continuity equation gives

$$\frac{\partial}{\partial t}(n_e + n_i) = 0 = -\frac{\partial}{\partial x} J_x = -\frac{\partial}{\partial x} en(V_{ix} - V_{ex}) \tag{4}$$

or

$$V_{ex} = V_{ix} \tag{5}$$

Thus, the electron density and one component of the electron fluid velocity are determined.

2.C. Field-Fluid Equations

With the assumption of zero electron mass the electron momentum equation reduces to

$$n_e m_e \frac{d\underset{\sim}{V}_e}{dt} = 0 = -en_e\left(\underset{\sim}{E} + \frac{\underset{\sim}{V}_e \times \underset{\sim}{B}}{c} \right) - \frac{\partial}{\partial x} P_e\hat{x} - n_e\underset{\sim}{P} \tag{6}$$

where p_e is the scalar pressure and $\underset{\sim}{V}_e$ is the velocity of the electron fluid. The transverse (y and z) components of (6) can be written as

Ohm's law

$$\underset{\sim}{J} = \underset{\approx}{\eta}^{-1} \cdot \left(\underset{\sim}{E} + \frac{\underset{\sim}{V}_e \times \underset{\sim}{B}}{c} \right) \tag{7}$$

In general, the magnetic field is inclined at some angle with respect to the x axis: $\underset{\sim}{B}=|B|(b_x,b_y,b_z)$ with $b_x^2+b_y^2+b_z^2=1$, and if $\sigma_{\parallel}(\sigma_{\perp})$ represents the conductivity parallel (perpendicular) to the magnetic field, the conductivity tensor is (Krall and Trivelpiece,1973):

$$\underset{\approx}{\sigma} \equiv \underset{\approx}{\eta}^{-1} = \begin{vmatrix} \sigma_1 & \sigma_y & \sigma_z \\ \sigma_y & \sigma_2 & \sigma_x \\ \sigma_z & \sigma_x & \sigma_3 \end{vmatrix} \tag{8}$$

where

$$\begin{aligned}
\sigma_1 &= \sigma_{\perp} + (\sigma_{\parallel} - \sigma_{\perp}) \, b_x b_x \\
\sigma_2 &= \sigma_{\perp} + (\sigma_{\parallel} - \sigma_{\perp}) \, b_y b_y \\
\sigma_3 &= \sigma_{\perp} + (\sigma_{\parallel} - \sigma_{\perp}) \, b_z b_z \\
\sigma_x &= (\sigma_{\parallel} - \sigma_{\perp}) \, b_y b_z \\
\sigma_y &= (\sigma_{\parallel} - \sigma_{\perp}) \, b_x b_z \\
\sigma_1 &= (\sigma_{\parallel} - \sigma_{\perp}) \, b_x b_y
\end{aligned}$$

The transverse fields can be expressed in terms of the vector potentials in the usual manner [$\underset{\sim}{B}=\underset{\sim}{\nabla}\times\underset{\sim}{A}$; $\underset{\sim}{E}=-(1/c)(\partial\underset{\sim}{A}/\partial t)$]:

$$\begin{aligned}
B_y &= -\frac{\partial A_z}{\partial x} \\
B_z &= \frac{\partial A_y}{\partial x} \\
E_y &= -\frac{1}{c}\frac{\partial A_y}{\partial t} \\
E_z &= -\frac{1}{c}\frac{\partial A_z}{\partial t}
\end{aligned} \tag{9}$$

and $\underset{\sim}{\nabla}\cdot\underset{\sim}{B}=0$ implies

$$B_x = \text{constant}. \tag{10}$$

(The last field component, E_x, is derived from the x-component of Eq.(6) as will be shown later.)

Neglecting the displacement current, Ampere's law can be expressed

$$\begin{aligned}
\frac{\partial^2 A_y}{\partial x^2} &= -\frac{4\pi}{c} J_y \\
\frac{\partial^2 A_z}{\partial x^2} &= -\frac{4\pi}{c} J_z
\end{aligned} \tag{11}$$

Substituting in Ohm's law (7) and the definitions of the field components, Eq. (11) becomes a coupled system of equations:

$$\frac{\partial^2 A_y}{\partial x^2} = f_1 \left(\frac{\partial A_y}{\partial t}, \frac{\partial A_z}{\partial t}, \frac{\partial A_y}{\partial x}, \frac{\partial A_z}{\partial x} \right)$$
$$\frac{\partial^2 A_z}{\partial x^2} = f_2 \left(\frac{\partial A_y}{\partial t}, \frac{\partial A_z}{\partial t}, \frac{\partial A_y}{\partial x}, \frac{\partial A_z}{\partial x} \right) \qquad (12)$$

In order to solve this system a fully implicit, space centered difference scheme is used (Richtmyer and Morton, 1973). If we let $u_j^n = A_y$ at the jth cell position and n-th time level (and $w_j^n = A_z$ in the same way), the left hand side of Eq. (12) is differenced as follows:

$$\frac{\partial^2 A_y}{\partial x^2} = \frac{u_{j+1}^{n+1} + u_{j-1}^{n+1} - 2u_j^{n+1}}{\Delta x^2} \qquad (13)$$

while on the right hand side

$$\frac{\partial A_y}{\partial t} = \frac{u_j^{n+1} - u_j^n}{\Delta t} \qquad (14)$$

and

$$\frac{\partial A_y}{\partial x} = \frac{u_{j+1}^{n+1} - u_{j-1}^{n+1}}{2\Delta x}$$

When this is substituted into Eqs. (12), the resulting system of equations can be written symbolically as

$$\underset{\approx}{A}_j \cdot \underset{\sim}{X}_{j+1} + \underset{\approx}{B}_j \cdot \underset{\sim}{X}_j + \underset{\approx}{C}_j \cdot \underset{\sim}{X}_{j-1} = \underset{\sim}{D}_j \qquad (15)$$

where

$$\underset{\sim}{X}_j = \begin{vmatrix} u_j^{n+1} \\ w_j^{n+1} \end{vmatrix} \qquad (16)$$

and $\underset{\approx}{A}_j, \underset{\approx}{B}_j$, and $\underset{\approx}{C}_j$ are 2×2 matrices that depend only on components of $\underset{\approx}{\sigma}$ and $\underset{\sim}{V}_e$ (evaluated at cell j at time level n) and constants, while $\underset{\sim}{D}$ also depends on $u^n{}_j$, $w^n{}_j$. Thus, all the coefficients $\underset{\approx}{A}, \underset{\approx}{B}, \underset{\approx}{C}$ and $\underset{\sim}{D}$ are explicitly known. Note that there are N cells in the computation mesh j=1,2,... N with ghost cells (j=0 and j=N+1) at each end.

The set of equations (15) is easily solved. Assuming a solution of the form

$$\underset{\sim}{X}_j = \underset{\approx}{E}_j \cdot \underset{\sim}{X}_{j+1} + \underset{\sim}{F}_j \qquad j = 0, 1, \cdots N \qquad (17)$$

and substituting it into Eqs. (15), it then follows that

$$\underset{\approx}{E}_j = - [\underset{\approx}{B}_j + \underset{\approx}{C} \cdot \underset{\approx}{E}_{j-1}]^{-1} \underset{\approx}{A}_j$$
$$\underset{\sim}{F}_j = [\underset{\approx}{B}_j + \underset{\approx}{C}_j \cdot \underset{\approx}{E}_{j-1}]^{-1} \cdot [\underset{\sim}{D}_j - \underset{\approx}{C}_j \cdot \underset{\sim}{F}_{j-1}] \qquad (18)$$

Because all the $\underset{\approx}{A}_j, \underset{\approx}{B}_j, \underset{\approx}{C}_j$ and $\underset{\sim}{D}_j$ are known, if $\underset{\approx}{E}_0$ and $\underset{\sim}{F}_0$ are known (from boundary conditions, as explained later), Eqs. (18) can be solved in ascending order to obtain $\underset{\approx}{E}_1, \underset{\sim}{F}_1$; $\underset{\approx}{E}_2, \underset{\sim}{F}_2, \cdots, \underset{\approx}{E}_N, \underset{\sim}{F}_N$. Then if $\underset{\sim}{X}_{N+1}$ can be determined (again from boundary conditions), Eqs. (17) can be solved in descending order to obtain $\underset{\sim}{X}_N, \underset{\sim}{X}_{N-1}, \cdots, \underset{\sim}{X}_0$.

Once the components of the vector potential are known, the new values of the electric and magnetic field follow from Eqs. (9) and the components of the current from Eqs. (11), again using the differencing in Eqs. (13) and (14). The transverse components of $\underset{\sim}{V}_e$ are then easily found from the current and the ion velocity moments:

$$V_{ej} = V_{ij} - J_j / en \qquad j = y \text{ or } z \qquad (19)$$

The last electron fluid quantity to be calculated is the temperature. The electron energy equation can be written as

$$\frac{3}{2} \frac{\partial}{\partial t} (n_e T_e) + \frac{\partial}{\partial x} (\frac{3}{2} n_e T_e V_{ex}) + n_e T_e \frac{\partial V_{ex}}{\partial x} = Q \qquad (20)$$

The source term Q includes resistive heating(ηJ^2) and loss mechanisms, such as thermal conduction or radiation, depending on the application. Letting

$$p_e = n_e T_e \qquad (21)$$

Eq. (20) can be rewritten in the form (with γ=5/3)

$$(\frac{\partial}{\partial t} + V_{ex} \frac{\partial}{\partial t}) p_e = - \gamma p_e \frac{\partial V_{ex}}{\partial x} + (\gamma - 1) Q \qquad (22)$$

The differencing for $\partial p_e/\partial t$ and $\partial V_{ex}/\partial x$ is identical to Eq. (14); for stability reasons "donor cell" differencing is used for the convective term (Richtmyer and Morton, 1967):

$$V_{ex} \frac{\partial p_e}{\partial x} = \frac{V^n_{exj}}{\Delta x} \times \begin{cases} (p^{n+1}_{ej} - p^{n+1}_{ej-1}) & V_{exj} > 0 \\ (p^{n+1}_{ej+1} - p^{n+1}_{ej}) & V_{exj} < 0 \end{cases} \qquad (23)$$

Again, a tridiagonal system is obtained,

$$A_j X_{j+1} + B_j X_j + C_j X_{j-1} = D_j \qquad (24)$$

except that now all quantities are simple scalars. As before, an assumed solution of the form

$$X_j = E_j X_{j+1} + F_j \qquad (25)$$

leads to

$$E_j = - [B_j + C_j E_{j-1}]^{-1} A_j$$
$$F_j = [B_j + C_j E_{j-1}]^{-1} [D_j - C_j F_{j-1}] \quad (26)$$

Appropriate boundary conditions lead ot a determination of E_0, F_0 (and then to all E_j, F_j in ascending order, using Eqs. (26)) and X_{N+1} (and thus to all X_j in descending order through Eq. (25)). Since $n_e (=n_i)$ is known, T_e is then obtained from Eq. (21). An alternative to Eq. (22) which is sometimes useful (Sgro, 1978) is to write the energy equation (20) in terms of the entropy $[S=\ln(p_e n_e^{-\gamma})]$ instead of the pressure.

Finally, the x component of the electron momentum equation (6) can be solved for Ex:

$$E_x = - \frac{1}{c} (\underset{\sim}{V}_e \times \underset{\sim}{B})_x - \frac{1}{ne} \frac{\partial p_e}{\partial x} \quad (27)$$

This electric field is needed to maintain charge neutrality.

The computational loop through one time step can thus be summarized as follows.

1. Advance the ions one time step and then calculate the ion velocity moments $(n_i = n_e, V_{ix} = V_{ex}, V_{iy}, V_{iz})$.
2. Compute the plasma conductivity $\underset{\approx}{\sigma}$, according to some prescription. (The resistivity $\underset{\approx}{\eta}$ is $\underset{\approx}{\sigma}^{-1}$).
3. Solve the coupled equations for A_y, A_z; we can then easily calculate $E_y, E_z, B_y, B_z (B_x = \text{constant})$. From the currents (Ampere's law) and the ion moments calculate the other two electron velocity components, V_{ey} and V_{ez}.
4. The electron temperature is obtained next from the solution of the differenced energy equation.
5. E_x is then calculated from the x-component of the electron momentum equation.
6. Thus, all field and electron fluid components are known and we are ready to move the ions again.

2.D. Refinements

The simulation scheme described thus far is rather general, the only assumptions being quasineutrality, zero electron mass and one spatial dimension. All of these conditions can be replaced by more appropriate ones, as the physical situation dictates. For example, for the study of low frequency ion waves an adiabatic electron model is more appropriate; such a model has been successfully used by Okuda <u>et al.</u> (1978). In some situations phenomena in the lower hybrid frequency range are of interest. In this case electron inertia effects are non-negligible; they have been included in the one-dimensional model of Liewer (1976) and the two-dimensional model of Hewett and Nielson (1978). Two-dimensional simulation models with me = 0 have been successfully employed by Byers <u>et al.</u> (1978), Hewett (1980) and Harned (1982).

The bow shock simulations described in this chapter take the conductivity to be constant. A further level of sophistication is to include anomalous processes due to the microphysics, which is occurring on time and distance scales shorter than those resolved in the hybrid model, by means of more complicated transport coefficients. This is done by expressing the conductivity (or resistivity) as a sum of two terms, one representing classical (Spitzer) effects and the other due to anomalous effects arising from microinstabilities due to cross-field currents. Two types of anomalous transport coefficients have been used successfully to model the behavior of laboratory plasmas: one type uses a semi-empritical expression (Chodura, 1975; Sgro and Nielson, 1976), the other is based on a quasilinear analysis of known instabilities (Davidson and Krall, 1977). The transport processes can be further refined to include electron thermal conductivity (Sgro, 1980) and anomalous ion heating (Hamasaki et al., 1977). Multispecies ions can also be included in a straightforward manner (Sgro, 1980; Sgro and Winske, 1981).

2.E. Initial and Boundary Conditions

In order to simulate the bow shock environment the hybrid model described previously must be implemented by appropriate initial and boundary conditions. There are several ways to produce the type of shock structure shown in Figure 2. One method, which has been used in many simulations (Papadopoulos et al., 1971; Forslund and Freidberg, 1971; Mason, 1972; Auer et al., 1971; Biskamp and Welter, 1972; Forslund et al., 1984), is to drive a magnetic piston from one end of the system by means of a rapidly rising electric field. As the piston moves inward, a shock is formed in front of it. A preferable method, which is employed here, is to initialize the simulation with two uniform regions representing the upstream and downstream conditions separated by a thin intermediate region. This allows eliminating any eventual effect of the driving piston on the shock formation, and also to produce a shock at rest in the simulation frame, simplifying the diagnostics. In the upstream state (the left region in the simulation), the particle ions are uniformly distributed in x and given random velocities to approximate a Maxwellian distribution convecting towards the intermediate region with appropriate density (n_1), temperature (T_{i1}) and flow speed (V_1) in the x direction. The upstream magnetic field $\underset{\sim}{B}_1=(B_x,0,B_{z1})$ and temperature of the fluid electrons (T_{e1}) are also assumed to be uniform. The downstream region is prepared similarly with density (n_2), ion temperature (T_{i2}), flow speed (V_2), magnetic field $\underset{\sim}{B}_2=(B_x,0,B_{z2})$, and electron temperature (T_{e2}). The thin intermediate region is prepared such that the density is a linear function of x which matches n_1 at the left end and n_2 at the right end, and satisfies $B_z/B_{z1}=n/n_1=V_1/V_x$, some temperatue profile being assumed. The downstream quantities are computed from the upstream values using the Rankine-Hugoniot relations with $\gamma=5/3$ and an assumed initial value of T_{e2}/T_{i2} (Tidman and Krall, 1971). The aim of this method is to prepare at time t=0 a shock transition "sufficiently" close to the final state so that the system can eventually reach this final state by allowing the shock transition to relax in time.

The system is allowed to evolve in space and time subject to the following boundary conditions. The magnetic field is taken to be constant and equal to its upstream (downstream) value at the left (right) boundary. Recalling the definitions (9) and the fact that cell quantities are defined in the center of the cells,

$$\begin{aligned} B_y(x=0) &= 0 = -\frac{A_z(1) - A_z(0)}{\Delta x} \\ B_z(x=0) &= B_{z1} = \frac{A_y(1) - A_y(0)}{\Delta x} \end{aligned} \qquad (28)$$

where (1) implies cell number 1 and (0) refers to the ghost cell. We can thus write (28) in the notation of Eq. (16) as

$$\underset{\sim}{X}_0 = \underset{\sim}{X}_1 + \underset{\sim}{H}_L \qquad (29)$$

with

$$\underset{\sim}{H}_L = \begin{vmatrix} -\Delta x\ B_{z1} \\ 0 \end{vmatrix} \qquad (30)$$

Thus, by comparing with Eq. (17) we see

$$\underset{\approx}{E}_0 = \underset{\approx}{1} \quad \text{and} \quad \underset{\sim}{F}_0 = \begin{vmatrix} -\Delta x\ B_{z1} \\ 0 \end{vmatrix} \qquad (31)$$

This is the initial condition we need to solve Eq. (18) for all j.

In similar fashion at the right end,

$$\begin{aligned} B_y(x=L) &= 0 = -\frac{A_z(N+1) - A_z(N)}{\Delta x} \\ B_z(x=L) &= B_{z2} = \frac{A_y(N+1) - A_y(N)}{\Delta x} \end{aligned} \qquad (32)$$

thus, with

$$\underset{\sim}{X}_N = \underset{\sim}{X}_{N+1} + \underset{\sim}{H}_R \qquad (33)$$

we find

$$\underset{\sim}{H}_R = \begin{vmatrix} -\Delta x\ B_{z2} \\ 0 \end{vmatrix} \qquad (34)$$

This can be combined with Eq.(17) at j=N

$$\underset{\sim}{X}_N = \underset{\approx}{E}_N \cdot \underset{\sim}{X}_{N+1} + \underset{\sim}{F}_N \qquad (35)$$

to eliminate $\underset{\sim}{X}_N$:

$$\underset{\sim}{X}_{N+1} = [\, \underset{\approx}{1} - \underset{\approx}{E}_N \,]^{-1} \cdot [\, \underset{\sim}{F}_N - \underset{\sim}{H}_R \,] \qquad (36)$$

Since $\underset{\sim}{F}_N, \underset{\approx}{E}_N$ are known, $\underset{\sim}{X}_{N+1}$ can be explicitly evaluated and then

Eq.(17) used to find all the $\underset{\sim}{X}_j$. Other boundary conditions (e.g. A_y and A_z, and thus E_y and E_z, =constant) can also be used. For other applications, periodic boundary conditions ($\underset{\sim}{X}_0=\underset{\sim}{X}_N$, $\underset{\sim}{X}_1=\underset{\sim}{X}_{N+1}$) are desirable. The same technique can be used, although it is somewhat more complicated, because the expressions at the two ends are coupled.

In similar fashion a boundary condition for the energy equation (22) is needed. For the present calculations p_e=constant or $\partial p_e/\partial x$=constant at x=0 and L has been used. Similarly, we use $\partial n_i/\partial x=\partial \underset{\sim}{V}_i/\partial x=0$ at the boundaries to set n_i and $\underset{\sim}{V}_i$ in the ghost cells equal to their values in the first real cell.

A final boundary condition is that a constant flux n_1V_1 of upstream plasma is maintained at the left end. While for all cases of interest, the upstream ion velocity is much larger than the thermal speed so that no upstream ion has a negative x-velocity, the parameter $T_{i2}/T_{e2}(t=0)$ is often large enough so that some of the downstream ions have negative velocities and can thus drift upstream. Thus, the downstream ion distribution has a negative x-velocity wing. The flux F_1 leaving the system at the right end is easily computed as

$$F_1=\int_0^\infty v_x f_i(v_x)dv_x=n_2V_2[1+\frac{1}{2}(\frac{\exp(-m_iV^2/2T_{i2})}{\sqrt{\pi}V_2(m_i/2T_{i2})^{1/2}}-\mathrm{erfc}(\frac{V_2}{(2T_{i2}/m_i)^{1/2}}))] \quad (37)$$

assuming a Maxwellian for f_i. Particles which exit downstream at the right end are then reinserted into the system such that the upstream flux at the left end is maintained, as is the negative x-velocity wing at the right end. Then, at each time step a fraction $(1-n_2V_2/F_1)$ of the particles escaping the system are readmitted into the right end, with negative random x-velocities approximating the negative wing of f_i, and the remaining fraction n_2V_2/F_1 of ions are injected into the left end. The net flux leaving the system at the right end is thus exactly n_2V_2 and balances the incoming upstream flux at the left end. This implies that the total number of particles is conserved, and also that there is no source or sink of mass, momentum or energy in the system, since downstream and upstream states are related by Rankine-Hugoniot relations. This procedure is accurate as long as the downstream distribution remains Maxwellian, i.e., as long as the perturbations coming from the transition region (shock region) and convected downstream with approximately a velocity V_2 do not reach the right end of the system. In practice, the size of the system is chosen to be large enough for magnetic effects on the ions to be investigated during the time needed for an initial perturbation in the transition region to reach the right end.

3. APPLICATIONS

In this section applications of hybrid simulation methods to bow shock phenomena are presented. The purpose of the discussion is to illustrate the kind of information that such simulations provide and how such information can be used to draw meaningful conclusions. The details of the calculations and in depth discussions of the background, relevance and physics of the examples can be found in the cited references. The examples presented show the structure of (A) the perpendicular and (B) the oblique portion of the bow shock.

3.A. Perpendicular Bow Shock

As remarked in Sec.1, our primary motivation for doing hybrid simulations of the earth's bow shock was to interpret the ISEE observations of the shock structure, such as shown in Figure 2. We first describe results of a simulation with parameters approximating those of the shock crossing of Figure 2 (Nov.7,1977): $\theta_{nB}=90^0$, $n_1=8cm^{-3}$, $B_1=5\gamma$, $T_{e1}=T_{i1}=4.7eV$, $V_{x1}=304km/sec$, corresponding to beta values $\beta_e=8\pi n_1 T_{e1}/B_1^2=0.6=\beta_i$ and an Alfven Mach number $M_A=V_{x1}/v_A=8$ ($v_A^2=B_{z1}^2/4\pi n_1 m_i$). Further details of this simulation are found in Leroy et al. (1981).

In the case of a perpendicular shock, where the upstream magnetic field is $\underset{\sim}{B}_1=(0,0,B_1)$, the solution of the field equations simplify because the equation (9) for A_y and A_z do not couple and $A_z(x,t)\equiv 0$. Recent calculations (Leroy and Winske, 1983) show that the overall structure of the shock is not changed significantly for smaller angles, down to $\theta_{nB}=\tan^{-1}(B_z/B_x)\simeq 50^0$, even though a sizeable B_y is present.

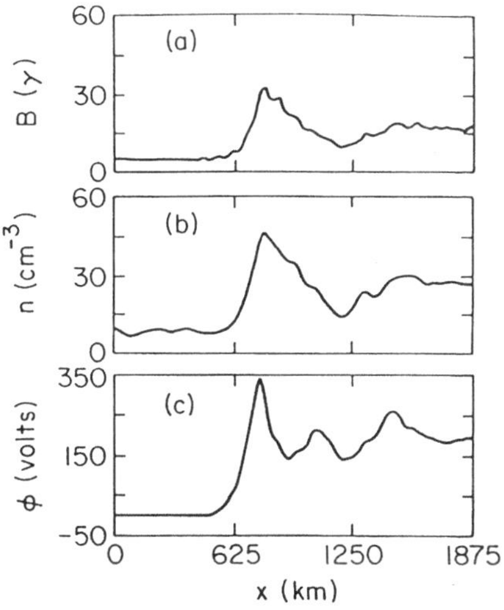

Figure 3. Profiles of perpendicular shock simulation at t=8 sec: (a) magnetic field, (b) density, and (c) potential as a function of x.

Results of the simulation, showing the spatial profiles of magnetic field, density, and electric potential ϕ are presented in Figure 3 at one instant of time, t=8 sec $\simeq 4\Omega_i^{-1}$ (Ω_i is the upstream ion gyrofrequency), long after the initial transistory period. At this time the shock structure is self-sustaining and these profiles are largely independent of time and the initial value of T_{e2}/T_{i2}. We have used realistic rather than dimensionless units in Figure 3 (B in γ, n in cm^{-3}, ϕ in volts and x in km) to facilitate the comparison of the results with the ISEE observations. The magnetic field (Figure 3a) exhibits the overall structure of the observations (Figure 2), including a foot region (540 km$\leqslant$x$\leqslant$700 kms), a magnetic ramp (700 km$\leqslant$x$\leqslant$800 km) and a large, well-defined overshoot (700 km$\leqslant$x$\leqslant$1100 km). The value of the magnetic field at the peak of the overshoot is about twice B_2 and its length is about 400 km. The magnetic ramp, here about 100 km long, is much larger than in the observations; its width is determined by the resistivity. While the density profile (Figure 3b) closely follows that of the magnetic field (since the resisitivity is small), the potential profile does not (Figure 3c). Although an overshoot also exists in the potential, it is much narrower.

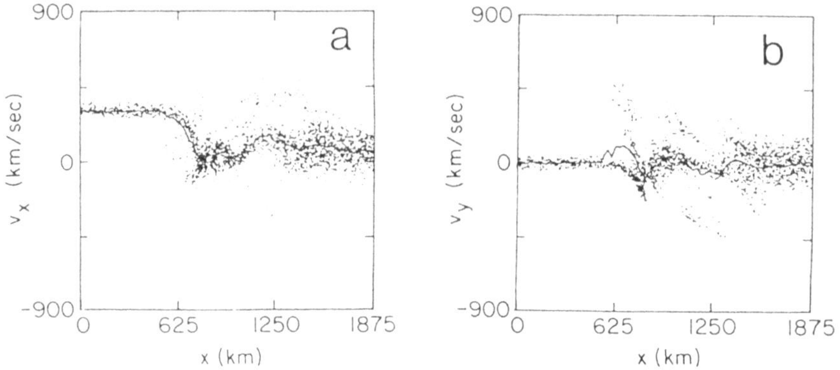

Figure 4. Phase space at t=8 sec.:
(a) v_x versus x, solid curve is V_x;
(b) v_y versus x, solid curve is V_y.

The necessity of a kinetic ion description can be seen from ion phase space. The results are presented in Figure 4 at the same instant of time as Figure 3. The v_x versus x phase space (Figure 4a) shows the upstream ions incident from the left slowed as they enter the shock region (x= 540 km). The small number of particles with smaller x velocities in this region are the ones which are reflected from the shock. These same particles have large values of v_y (v_y versus x phase space, Figure 4b). These reflected particles constitute part of a gyrating stream, which can be seen in Figure 4a by following the small group of particles starting at x=540 km, below the main incident beam, upward to large v_x and then across larger x (~900 km) at large v_x (~500 km/sec). The stream continues at x≃1100 km, starting at large v_x, continuing to negative v_x (at the same x), then upstream. Similarly, a second gyrating stream can be seen at larger x (starting at x=1100 km). That these particles are gyrating is evident when comparing with Figure 4b. In regions where the gyrating particles have small v_x, they have large v_y and vice versa. The average gyroradius of these particles is about 300 km. This gyrating stream can be correlated with the overall structure of the shock by comparing Figure 4a with Figure 3a. The ramp region of the magnetic field corresponds to the region of reflected particles in front of the shock, while the length of the overshoot of the magnetic field is roughly the gyroradius of the gyrating particles. The solid curves in Figure 4 are the mean velocities, V_x and V_y, obtained by averaging over all of the ions in each cell. There are large deviations, especially in the foot region, between these quantities and most of the ions, indicating the importance of kinetic effects.

The presence of these "reflected-gyrating" ions results in highly nonthermal ion distributions with large effective thermal spreads, which are consistent with observations described earlier (Paschmann et al., 1982). Most of the incident ions, however, penetrate into the ramp region and are heated by compression. Note that the thermal spread of the ions in both the v_x and v_y phase space plots (Figure 4) is much broader for x> 1500 km. This does not represent thermalization; rather, it is a remnant of the initial downstream plasma.

In addition to identifying the overall shock structure, and describing the characteristics of the reflected ions, the hybrid simulations have led to an increased understanding of both the macrostructure and microstructure of the perpendicular shock. For example, an analytic theory of the ion reflection process, which provides a quantitative explanation of the magnetic field and potential overshoots, has been developed (Leroy, 1983). The theory is based on a cold beam approximation to the ions dynamics observed in the simulation.

Because of the relative simplicity of the simulations of perpendicular shocks, it has been possible to carry out an extensive survey of parameter space. This has resulted in a systematic study (Leroy et al., 1982) of how the shock structure varies with the parameters of the model, including the upstream conditions (electron and ion beta and the Mach number) and the magnitude of the resistivity. Such parameter surveys have been a valuable complement to similar observational studies (e.g. magnetic field overshoots by Livesey et al., 1982). It should be noted in this regard that the existence of a resistivity which is known exactly, is kept fixed, and is independent of other parameters (in contrast to being self consistently determined by some algorithm, time and space dependent, and explicitly or implicitly dependent on other parameters) was a decided advantage in the analysis.

A third aspect of the hybrid simulations which has had far reaching consequences has been the utilization of output from the simulations as a basis for analyzing the microstability of the shock structure. An understanding of what are the various plasma instabilities which can grow at the shock is the first step in determining self-consistently the anomalous transport processes responsible for plasma resistivity, heating, and acceleration. The output from the simulation (ion distributions, field profiles, electron properties) give the local plasma parameters as a function of time. Such snapshots of the plasma can then be used in the stability analysis. The snapshots are meaningful because the microprocesses occur on time scales fast compared to those in the simulation. For example, instead of plotting the ions in velocity-position phase space (Figure 4) it is possible to plot the ions in velocity space in various regions in the shock (e.g. the foot, ramp, overshoot) separately. The microscopic stability of these ion distributions in the presence of the cross-field current, temperature/density/magnetic field gradients, etc. can then be analyzed using local linear Vlasov theory. The results of an extensive investigation of this type have been summarized in a recent review article (Wu et al., 1984), which contains references to the detailed calculations. The results of this study have been used to compare with ISEE observations of wave signatures, electron acceleration, and electron and ion heating at the bow shock. They could also be used to develop a self-consistent transport model for the simulation, similar to that described in Sec.2.D.

3.B. Oblique Bow Shock

As a second example, the oblique portion of the earth's bow shock ($\theta_{nB}\simeq 45^0$) is examined using the same simulation techniques (Leroy and Winske, 1983). In this case there are several reasons why application of the basic hybrid model could be somewhat questionable. First, since whistlers are a predominant feature of the shock structure for oblique shocks, it is desirable that the whistler mode be present in the simulation in as accurate a form as possible. The linear dispersion curves for a uniform, Maxwellian plasma with either zero or finite electron mass are identical at low frequencies and begin to differ from each other at a frequency of the order of the lower hybrid frequency. However, the highest frequency of interest we can reasonably expect is the frequency of the standing whistler precursor generated by the shock front in order to balance the nonlinear steepening of the shock wave, i.e.,$\omega\simeq k\ V_1$. If the Mach number is not too high, the straight line $\omega=k\ V_1$ intersects the whistler dispersion curve in the (ω,k) dispersion diagram at a frequency smaller than the lower hybrid frequency, i.e., in a domain accurately described by the hybrid code. The amplitude of the whistler turbulence can be controlled somewhat by varying the resistivity. This is useful for testing the effect of the whistlers on ion heating and, as such, complements the two-dimensional implicit particle simulations of Quest et al. (1983).

A second criticism of the model for oblique shocks is that the simplified fluid treatment of the electrons is expected to be inaccurate. In oblique geometries the electrons can move along the field lines while convected through the shock. Thus, it has been observed (see, e.g., Feldman et al., 1983) that the electrons have anisotropic distribution functions and carry a significant heat flux along the field lines. Moreover, electron trapping can occur behind the shock front. Such a highly kinetic behavior is not taken into account in the present calculation, where a scalar electron pressure is assumed and no electron heat flux is included. In order to test the magnitude of the inaccuracies involved with the electron fluid treatment, and since the inclusion of a realistic electron heat flux is expected to inhibit electron heating, simulations have been run with and without the Ohmic heating term in Eq. (20). The fact that the results are changed only slightly gives confidence that the shock structure and ion dynamics are generally correct even though the electrons may not be described very accurately.

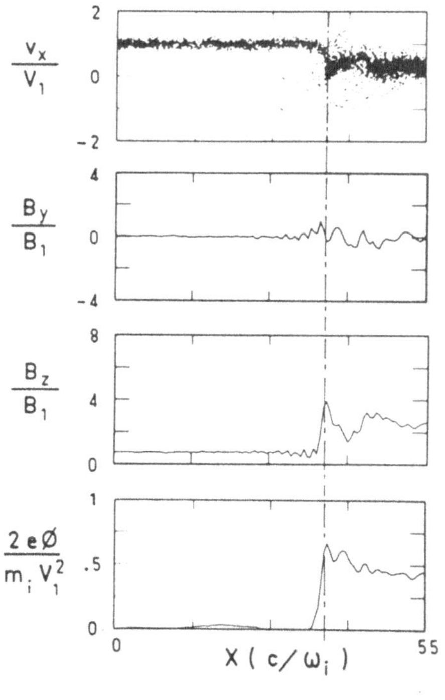

Figure 5. Simulation run $\theta_{nB}=45^0$ (a) v_x versus x phase space, (b) y component of magnetic field, (c) z component of magnetic field and (d) electric potential profile. All profiles are functions of x. The dashed vertical line indicates the position of the shock front.

Figure 5 shows the results in the oblique shock case. Plotted are the ion x-v_x phase space, the components B_y and B_z of the magnetic field profile and the electric potential profile at time $t=11\Omega_i^{-1}$. In this case quantities are plotted in dimensionless units in terms of upstream quantities. A time analysis of the simulation run shows that after a time $t \sim 7.5\Omega_i^{-1}$ some ions are emitted from the shock front and travel backstreaming upstream (Figure 5-a). Their density is estimated to be 5% of the solar wind ion density. However, Figure 5-a also shows the existence of a non-thermal ion distribution function behind the shock front, whose high energy tail corresponds to the presence of reflected-gyrating ions, similar to those seen in the perpendicular case (Figure 4). A right-hand polarized, small amplitude standing whistler precursor of wavelength 1.5 c/ω_i develops upstream of the shock front (Figure 5b-c). The amplitude of By increases from ~0.2 B_1 in the whistler precursor to ~B_1 in the vicinity of the shock front, where the behavior of the whistler wave is highly non-linear. At the shock front B_z makes a jump, while B_y reverses its sign. The electric potential jump at the shock front (Figure 5-d) remains of the same order of magnitude as in the perpendicular case, but the potential overshoot appears less clearly. An inspection in time of the simulation run shows that the potential overshoot is sometimes absent, hence the potential overshoot is no longer a characteristic of the electric field structure. However, the shock is still quasi-stationary, and the shock front, as before, does not move significantly during the simulation run. The magnetic field within the shock front is mildly turbulent, with temporal fluctuations of at most 0.5 B_1, which occur on a typical time scale of Ω_i^{-1}.

An important difference between the oblique ($\theta_{nB} \leqslant 45^0$) and the (quasi-) perpendicular ($\theta_{nB} > 50^0$) case is that some of the reflected ions go back upstream in the oblique case. The mechanism by which such ions are reflected is the subject of controversy at present (Sonnerup, 1969; Schwartz et al., 1983; Tanaka et al., 1983; Thomsen et al., 1983). One advantage of the hybrid simulations is that individual ion orbits can be followed and the various reflection mechanisms can be tested. A typical trajectory in x-v_x, x-v_y and x-v_z phase space of one of these backstreaming ions is shown in Figure 6. The incoming ion is once reflected off the shock front, then gyrates while crossing the shock front several times (2 or more; 4 in Figure 6), acquiring larger and larger v_z, and eventually leaks upstream. Such a trajectory may be referred to as "multiple shock traversal" trajectory. Note that the average ion gyroradius is much larger than the ramp thickness (Figure 5), which explains why the adiabatic theory result that the magnetic moment should be conserved is not verified. In addition to the backstreaming ion population, a large number of reflected ions behave as the "reflected-gyrating" ions described previously. It is interesting to point out that the multiple shock traversal trajectory can not be predicted by one commonly accepted reflection mechanism, the specular reflection model (Gosling et al., 1982). Varying the incoming ion pitch angles in the specular reflection model results in trajectories in which the reflected ion either bounces one more time off the shock front and escapes, or it transmitted through the front and eventually leaks upstream, after having spent only half a gyration in the downstream region (dashed line in Figure 6).

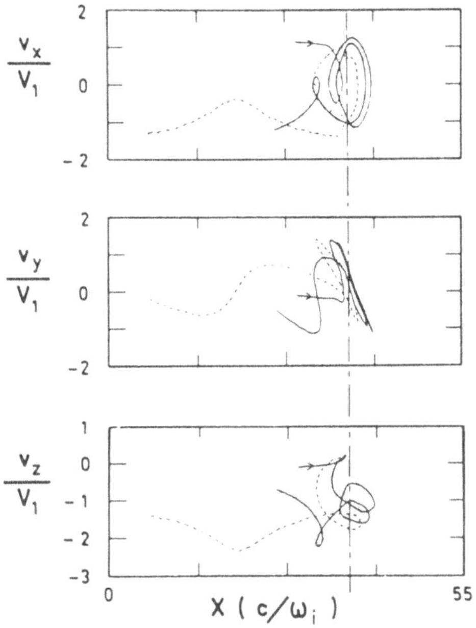

Figure 6. Simulation run $\theta_{nB}=45^0$. Solid lines: trajectory in (a) x-v_x, (b) x-v_y, (c) x-v_z phase space of one backstreaming ion. Dashed lines: similar trajectory as predicted by the specular reflection model. The vertical dashed line represents the position of the shock front.

Finally, simulations at even more oblique angles ($\theta_{nB}=30^0$) have been done (Leroy and Winske, 1983). In this case the magnetic field is highly turbulent and the shock structure is less stationary. Because electron dynamics increase in importance as θ_{nB} decreases, the present model becomes even more questionable. Work on a better representation of the electron physics is underway.

4. SUMMARY

In summary, a hybrid fluid-kinetic simulation model that has broad application to space plasma physics has been described. Examples demonstrating its use in explaining features of the earth's bow shock have been presented. In spite of its relative simplicity (by present day standards) significant conclusions have been drawn concerning the scale lengths associated with the bow shock structure and the importance of reflected ions (both the gyrating ions that eventually go downstream as well as the truly reflected ions which return upstream for $\theta_{nB}<45^0$). Furthermore, the simulations have produced a broad survey of the parameter space of perpendicular shocks (Leroy et al., 1982), inspired an analytical theory of the reflection process as well as a quantitative explanation of of the magnetic field overshoot (Leroy, 1983) and formed the basis of a microstability analysis of high Mach number perpendicular shocks (Wu et al., 1984).

These successes with the hybrid simulation model have led to other investigations, which are presently in progress. Studies for the quasiperpendicular shock include a detailed comparison of the observed ion distribution functions with those generated by means of a mechanically accurate 'ion detector' in the simulation and an analysis of alpha particle dynamics in the shock. For the quasiparallel shock a detailed study of ion trajectories is underway to determine under which conditions and by what means ions are reflected back upstream. In addition, progress is being made on improving the electron description so that the nearly parallel shock can be investigated. The simulations are also aiding in general studies to determine the microprocesses by which the directed solar wind energy is converted into electron and ion heating at the shock, which is the ultimate reason for studying the earth's bow shock in the first place.

ACKNOWLEDGMENTS

The authors thank their colleagues, Drs. C. Goodrich, A. Mangeney, K. Papadopoulos and C. S. Wu, for their advice and encouragement. This work was supported by NASA Solar Terrestrial Theory Program Grant NAGW-81.

REFERENCES

Auer, P.L., R. W. Kilb, and W. F. Crevier, J. Geophys. Res., 76, 2928, 1971.

Biskamp, D. and H. Welter, Nucl. Fusion, 12, 663, 1972.

Byers, J. A., B. I. Cohen, W. C. Condit, and J. D. Hanson, J. Comput. Phys., 27, 363, 1978.

Chodura, R., Nucl. Fusion, 15, 55, 1975.

Davidson, R. C., and N. A. Krall, Nucl. Fusion, 17, 1313, 1977.

Feldman, W. C., R. C. Anderson, S. J. Bame, S. P. Gary, J. T. Gosling, D. J. McComas, M. F. Thomsen, G. Paschmann and M. M. Hoppe, J. Geophys. Res., 88, 96, 1983.

Forslund, D. W., and J. P. Freidberg, Phys. Rev. Lett, 27, 1189, 1971.

Forslund, D. W., K. Quest, J. U. Brackbill, and K. Lee, J. Geophys. Res., 89, 2142, 1984.

Gosling, J. T., M. F. Thomsen, S. J. Bame, W. C. Feldman, G. Paschmann, and N. Sckopke, Geophys. Res. Lett., 9, 1333, 1982.

Greenstadt, E. W., C. T. Russell, V. Formisano, P. C. Hedgecock, F. L. Scarf, M. Neugebauer, and R. E. Holzer, J. Geophys. Res., 82, 651, 1976.

Greenstadt, E. W. and R. W. Fredricks, Shock systems in collisionless space plasmas, in Solar System Plasma Physics, Vol. III, edited by L. J. Lanzerotti, C. F. Kennel and E. N. Parker, p. 3, North-Holland, Amsterdam, 1979.

Greenstadt, E. W., C. T. Russell, J. T. Gosling, S. J. Bame, G. Paschmann, G.K. Parks, K. A. Anderson, F.L. Scarf, R. R. Anderson, D. A. Gurnett, R. P. Lin, C. S. Lin, and H. Reme, J. Geophys. Res., 85, 2124, 1980.

Greenstadt, E. W., R. W. Fredricks, C. T. Russell, F. L. Scarf, R. R. Anderson, and D. A. Gurnett, J. Geophys. Res., 86, 4511, 1981.

Hamasaki, S., N. A. Krall, C. E. Wagner, and R. N. Byrne, Phys. Fluids, 20, 65, 1977.

Harned, D. S., J. Comput. Phys., 47, 452, 1982.

Hewett, D. W., J. Comput. Phys., 38, 378, 1980.

Hewett, D. W. and C. W. Nielson, J. Comput. Phys., 29, 219, 1978.

Krall, N. A. and A. W. Trivelpiece, Principles of Plasma Physics, p. 119, McGraw-Hill, New York, 1973.

Leroy, M. M., C. C. Goodrich, D. Winske, C. S. Wu, and K. Papadopoulos, Geophys. Res. Lett., 8, 1269, 1981.

Leroy, M. M., D. Winske, C. C. Goodrich, C. S. Wu and K. Papadopoulos, J. Geophys. Res., 87, 5081, 1982.

Leroy, M. M., Phys. Fluids, 24, 2742 1983.

Leroy, M. M. and D. Winske, Ann. Geophys., 1, 527 1983.

Liewer, P. C., Nucl. Fusion, 16, 817, 1976.

Livesey, W. A., C. F. Kennel and, C. T. Russell, Geophys. Res. Lett., 9, 1037, 1982.

Mason, R. J., Phys. Fluids, 15, 1082, 1972.

Morse, R. L., Multidimensional plasma simulation by the particle-in-cell method, in Methods of Computational Physics, vol. 9, edited by B. Alder, S. Fernbach and M. Rotenberg, p. 213, Academic Press, New York, 1970.

Nielson, C. W. and H. R. Lewis, Particle-code models in the nonradiative limit, in Methods of Computational Physics, Vol. 16, edited by J. Killeen, B. Alder, S. Fernbach and M. Rotenberg, p. 367, Academic Press, New York, 1976.
Okuda, H., J. M. Dawson, A. T. Lin and C. C. Lin, Phys. Fluids, 21, 476, 1978.
Papadopoulos, K., C. E. Wagner, and I. Haber, Phys. Rev. Lett, 27, 982, 1971.
Paschmann, G., N. Sckopke, S. J. Bame, and J. T. Gosling, Geophys. Res. Lett., 9, 881, 1982.
Quest, K. B., D. W. Forslund, J. U. Brackbill, and K. Lee, Geophys. Res. Lett., 10, 471, 1983.
Richtmyer, R. D., and K. W. Morton, Difference Methods for Initial Value Problems, Ch. 8, Interscience, New York, 1967.
Russell, C. T. and E. W. Greenstadt, Space Sci. Rev., 23, 3, 1979.
Schindler, K. and J. Birn, Magnetospheric physics, Physics Reports (Phys. Lett.), 47, 109. 1978.
Schwartz, S. J., M. F. Thomsen, and J. T. Gosling, J. Geophys. Res., 88, 2039, 1983.
Sgro, A. G. and C. W. Nielson, Phys. Fluids, 19, 126, 1976.
Sgro, A. G., Phys. Fluids, 21, 1410, 1978.
Sgro, A. G., Phys. Fluids, 23, 1055, 1980.
Sgro, A. G. and D. Winske, Phys. Fluids, 24, 1156, 1981.
Sonnerup, B. U. O., J. Geophys. Res., 74, 1301, 1969.
Tanaka, M., C. C. Goodrich, D. Winske and K. Papadopoulos, J. Geophys. Res., 88, 3046, 1983.
Thomsen, M. F., S. J. Schwartz and J. T. Gosling, J. Geophys. Res., 88, 7843, 1983.
Tidman, D. A. And N Krall, Shock waves in collisionless plasmas, John-Wiley-Interscience, New York, 1971.
Wu, C. S., D. Winske, Y. M. Zhou, S. T. Tsai, P. Rodriguez, M. Tanaka, K.Papadopoulos, K. Akimoto, C. S. Lin, M. M. Leroy, and C. C. Goodrich, Space Sci. Rev., 37, 63, 1984.

VLASOV SIMULATIONS OF ION ACOUSTIC DOUBLE LAYERS

G. Chanteur

CNET/CRPE - 92131 Issy-les-Moulineaux (France)

ABSTRACT

We present a detailed description of a Vlasov code which has been used to study the formation of weak ion acoustic double layers and we discuss the physical process responsible for the formation of these structures.

1. INTRODUCTION

Recent spacecraft observations in the auroral regions (Temerin et al., 1982) have again focussed the attention of scientists on current driven double layers. Yet both the scarecity of the data and the experimental difficulties and limitations advocate the use of computer simulations to understand their formation in space. For a global presentation of auroral physics the reader is referred to the review article by Sato (1982). We will not discuss the double layers initiated by applying an external difference of potential to the plasma since we believe that such conditions are more interesting for laboratory than for space plasmas. Moreover we will not discuss the formation of double layers in the Buneman's regime (De Groot et al., 1977, Sigov, 1982; Singh and Schunk, 1982) but we will concentrate on the ion acoustic case.

Since the pioneer work of Sato and Okuda (1980), substantial progress has been made and a rather simple physical understanding of the problem has emerged. Recently Okuda and Ashour-Abdalla (1982) have proposed an interesting extension of the simulation studies in this field.

Section 2 is devoted to the statement of the problem and to a detailed description of the "Vlasov" code we have used to solve it numerically. Section 3 contains a presentation of our results, a comparison with similar results from particle simulations and a discussion of the formation of ion acoustic double layers.

H. Matsumoto and T. Sato, Computer Simulation of Space Plasmas,

2. DESCRIPTION OF THE "VLASOV" CODE

2.1 What is the problem?

We want to integrate numerically the following Cauchy problem (hereafter we use the rationalized MKS unit system):

$$\frac{\partial f_\alpha}{\partial t} + v\frac{\partial f_\alpha}{\partial x} + \frac{q_\alpha}{m_\alpha} E \frac{\partial f_\alpha}{\partial v} = 0$$

for $\alpha = (e, i)$ with $q_\alpha = \mathrm{sign}\, q_\alpha \cdot q$ $(q>0)$

$$\frac{\partial E}{\partial x} = \frac{1}{\varepsilon_0}\sum_\alpha q_\alpha \int_{-\infty}^{+\infty} f_\alpha(x,v,t)dv$$

$$f_\alpha(x,v,t_{=0}) = \frac{n(x)}{v_{th,\alpha}\sqrt{2\pi}} \exp\{-\frac{(v-v_{d,\alpha})^2}{2v_{th,\alpha}^2}\}$$

with the periodic boundary conditions

$$f_\alpha(x_+L,v,t) = f_\alpha(x,v,t)$$

The notations are obvious; let us just say that $v_{d,\alpha}$ and $v_{th,\alpha}$ are respectively the drift and the thermal velocities of the species α. The initial conditions need some comments. We use products of functions in x and v variables; this just reflects our inability to prepare an initial microscopic state with true modes of the system.

Next, the same density function n(x) is used in f_e and f_i to avoid the excitation of undesirable large amplitude Langmuir waves at the beginning of the computation. Section 3 will present simulations corresponding to different choices of $n(x)=n_0+\delta n(x)$.

Physical quantities of interest are, as usual : the temperatures $T_\alpha = m_\alpha v^2{}_{th,\alpha}$, the plasma frequency $\omega_p=(n_0 q^2/\varepsilon_0 m_e)^{1/2}$ and the Debye length $\lambda_D=v_{th,e}/\omega_p$. Let us define standard dimensionless variables:

$$\tilde{n}_\alpha = \frac{n_\alpha}{n_0}, \quad \tilde{v} = \frac{v}{v_{th,e}}, \quad \tilde{x} = \frac{x}{\lambda_D}, \quad \tilde{t} = \omega_p t, \quad \tilde{E} = \frac{\varepsilon_0 E}{n_0 q \lambda_D}$$

The above system of equations is now written in dimensionless form.

$$\frac{\partial f_\alpha}{\partial \tilde{t}} + \tilde{v}\frac{\partial f_\alpha}{\partial \tilde{x}} + \mathrm{sign}\, q_\alpha \cdot \frac{m_e}{m_\alpha} \tilde{E} \frac{\partial f_\alpha}{\partial \tilde{v}} = 0 \ , \ \alpha = (e,i) \tag{1a,b}$$

$$\frac{\partial \tilde{E}}{\partial \tilde{x}} = \sum_\alpha \mathrm{sign}\, q_\alpha \cdot \int_{-\infty}^{+\infty} f_\alpha(\tilde{x},\tilde{v},\tilde{t})d\tilde{v} \tag{1c}$$

$$f_\alpha(\tilde{x},\tilde{v},\tilde{t}=0) = \frac{\tilde{n}(\tilde{x})}{\sqrt{2\pi}}\left(\frac{m_\alpha T_e}{m_e T_\alpha}\right)^{1/2} \exp\{-\frac{m_\alpha T_e}{m_e T_\alpha}\frac{(\tilde{v}-\tilde{v}_{d,\alpha})^2}{2}\} \tag{1d,e}$$

$$f_\alpha(\tilde{x}_+\tilde{L},\tilde{v},\tilde{t}) = f_\alpha(\tilde{x},\tilde{v},\tilde{t}) \tag{1f,g}$$

We will now describe a finite difference scheme which approximates the above differential system (1). We will not make a mathematical study of the convergence and of the stability of this scheme, this still remains to be done and this is a formidable task. We will rather present this scheme step by step from a very empirical point of view. When the opportunity arises we refer the reader to recent related works in numerical analysis.

2.2 The discretization of the variables

First we introduce the velocity cut-offs and the phase space discretization. We define $v_{\alpha,min}$ and $v_{\alpha,max}$ for each species α usually by $v_{d,\alpha}\pm 5v_{th,\alpha}$, but an a priori knowledge of the gross features of the physical evolution may be a guide to choose the velocity intervals. For example for the ion acoustic instability it is well known that the space-averaged velocity distribution of the ions develops an high energy tail towards the mean electron velocity, so we can take the ion velocity cut-offs at e.g. $-3V_{th,i}$ and $7V_{th,i}$. The velocity cut-offs could be a serious bias because the phase space fluid of the species α is not allowed to escape out of the interval $(V_{\alpha,min};V_{\alpha,max})$ in order to have an exact conservation of the integral

$$\int_{x_o}^{x_o+L}\int_{v_{\alpha,min}}^{v_{\alpha,max}} f_\alpha(x,v,t)dvdx.$$

Hence, when the species α is accelerated locally towards a velocity cut-off the phase space fluid α is artificially accumulated just above $V_{\alpha,min}$ or just below $V_{\alpha,max}$: during the course of the computation we have to check that these accumulations near the velocity cut-offs only concern negligible fractions of the local distribution functions, when they occur. Each interval$(V_{\alpha,min};\ V_{\alpha,max})$is then divided into NV identical velocity cells (we choose the same number NV for the ions and for the electrons but this restriction may be released without difficulty if needed), the width ΔV_α of each cell being of the order of $0.1\ V_{th,\alpha}$. The length L of the system is divided into NX cells of width $\Delta x=\lambda_D$ for both species. This description of the phase spaces requires 2.NX.NV memory words, it may be huge and this is one of the main limitations of Vlasov codes.

Second, the time is discretized with a time step Δt which has to be small enough to ensure the numerical stability of the computation. As we will see later we integrate the system (1) along the characteristics, hence we have no Courant's condition but nevertheless the time-step Δt is limited, as usual for electrostatic codes by $\omega_p\cdot\Delta t < \pi$ which guarantees a correct sampling of the plasma oscillations. To reduce the phase errors to an acceptable level we choose $\omega_p\cdot\Delta t=0.5$. We have made two simulations of the ion acoustic instability with $V_{d,i}=0$, $V_{d,e}=0.8V_{th,e}$ 0.5 but with the same initial microscopic state. The ion and electron densities and the electric field for these two computations were exactly superposable after $\omega_p t=512$.

2.3 The numerical treatment of the Poisson's equation

From now on we will use only the dimensionless quantities defined above and we drop the superscript "~" for the sake of simplification. Let us now introduce the following standard notations: $t^n = n.\Delta t$, (x_l, v_m) coordinates of the center of the (l,m)-th phase space cell

$$E_l^n = \frac{1}{\Delta x} \int_{-\frac{\Delta x}{2}}^{+\frac{\Delta x}{2}} E(x + x_l, t^n)dx$$

$$f_{\alpha,l,m}^n = \frac{1}{\Delta x \cdot \Delta v_\alpha} \int_{-\frac{\Delta x}{2}}^{+\frac{\Delta x}{2}} \int_{-\frac{\Delta v_\alpha}{2}}^{+\frac{\Delta v_\alpha}{2}} f_\alpha(x+x_l, v+v_m, t^n)dvdx$$

The integration of Poisson's equation (1c) over the l-th space cell gives, at time t^n, taking into account the velocity cut-offs and discretization

$$E(x_l + \frac{\Delta x}{2}, t^n) - E(x_l - \frac{\Delta x}{2}, t^n) = \Delta x \sum_\alpha \mathrm{sign} q_\alpha \cdot \Delta v_\alpha \sum_{m=1}^{NV} f_{\alpha,l,m}^n$$

now we approximate

$$E(x_l \pm \frac{\Delta x}{2}, t^n) \text{ by } \frac{1}{2} (E^n_{l+\frac{1}{2}\pm\frac{1}{2}} + E^n_{l-\frac{1}{2}\pm\frac{1}{2}})$$

which leads to the discretized Poisson's equation

$$E_{l+1}^n - E_{l-1}^n = 2\Delta x \sum_\alpha \mathrm{sign} q_\alpha \cdot \Delta v_\alpha \sum_{m=1}^{NV} f_{\alpha,l,m}^n \tag{2}$$

The left member of the above equation recalls the second order approximation of $2\Delta x(\partial E/\partial x)^n{}_l$. For "particle codes" the effect of the spatial grid has already been discussed in great details (see e.g. Birdsall and Maron, 1980 and references therein), but for "Vlasov codes" the situation differs by the discretization of the whole phase space and not only of the space coordinate. In principle it should be possible to evaluate the approximation of Poisson's equation by equation (2) by considering an exact nonlinear solution of the Vlasov-Poisson system; Bernstein, Greene and Kruskal (1957) have shown how to construct such solutions which are now commonly designated as BGK solutions. This study is left for a future work.

2.4 The numerical treatment of the Vlasov equation

Let us come now to the main difficulty, that is the integration of the Vlasov equation itself. The scheme we describe pertains to a rather wide class proposed some years ago by Boris and Book (1976b). In such methods there are two basic ingredients which preserve separately the global conservation and positivity of f_e and f_i. First the Vlasov equation is splitted up into two advection equations (one over x, one over v) using a Strang splitting scheme (Strang, 1968) and second a "Flux Corrected Transport" algorithm (Boris and Book, 1976b) is used to integrate the advection equations.

2.4a The splitting scheme

We want to compute $f_\alpha(x,v,t_o+\tau)$ from $f\alpha(x,v,t_o)$ for $\alpha=(e,i)$ using (1a,b,c). In this case the Strang splitting consists of the following cycle of integrations (Boris, 1970, Boris and Book, 1976, Cheng and Knorr, 1976).

-integrate

$$\frac{\partial f_\alpha{}^*}{\partial t} + v\frac{\partial f_\alpha^*}{\partial x} = 0 \text{ from } t_o \text{ to } t_o + \frac{\tau}{2} \qquad (3a,b)$$

with the initial condition

$$f_\alpha^*(x, v, t_o) = f_\alpha(x, v, t_o) \qquad (4a,b)$$

for $\alpha = (e, i)$

-compute the electric field at time $t_o+\tau/2$ by substituting

$$f_\alpha^*(x, v, t_o+ \frac{\tau}{2}) \quad \text{in equation (1c).}$$

-integrate

$$\frac{\partial f_\alpha^{**}}{\partial t} + \operatorname{sign} q_\alpha\cdot\frac{m_e}{m_\alpha} E(x, t_o+ \frac{\tau}{2})\frac{\partial f_\alpha^{**}}{\partial v} = 0 \qquad (5a,b)$$

from t_o to $t_o+\tau$ with the initial condition

$$f_\alpha^{**}(x, v, t_o) = f_\alpha^*(x, v, t_o+ \frac{\tau}{2}) \qquad (6a,b)$$

for $\alpha = (e, i)$

-integrate

$$\frac{\partial f_\alpha^*}{\partial t} + v\frac{\partial f_\alpha^*}{\partial x} = 0 \text{ from } t_o+ \frac{\tau}{2} \text{ to } t_o+\tau \qquad (7a,b)$$

with the initial condition

$$f_\alpha^*(x, v, t_o+ \frac{\tau}{2}) = f_\alpha^{**}(x, v, t_o+ \tau) \qquad (8a,b)$$

for $\alpha = (e, i)$

-finally let an approximation of $f_\alpha(x, v, t_o+\tau)$ be $f^*{}_\alpha(x, v, t_o+\tau)$ for $\alpha = (e, i)$.

Integrating formally the whole system (3a, b to 8a, b) we obtain

$$\begin{aligned} f^*_\alpha(x,\ v,\ t_0+\tau) &= f^*_\alpha(x-v\tfrac{\tau}{2}\ ,\ v,\ t_0+\ \tfrac{\tau}{2}) \\ &= f^{**}_\alpha(x-v\tfrac{\tau}{2}\ ,\ v,\ t_0+\tau) \\ &= f^{**}_\alpha(x-v\tfrac{\tau}{2}\ ,\ v-E_\alpha\tau,\ t_0) \\ &= f^*(x-v\tfrac{\tau}{2}\ ,\ v-E_\alpha\tau,\ t_0+\ \tfrac{\tau}{2}\ ) \\ &= f^*_\alpha(x-v\tau+E_\alpha\tfrac{\tau^2}{2}\ ,\ v-E_\alpha\tau,\ t_0) \\ f^*_\alpha(x,\ v,\ t_0+\tau) &= f_\alpha(x-v\tau+E_\alpha\tfrac{\tau^2}{2}\ ,\ v-E_\alpha\tau,\ t_0) \end{aligned} \tag{9a,b}$$

with $E_\alpha = \text{sign}q_\alpha\cdot\frac{m_e}{m_\alpha}\ E(x-v\frac{\tau}{2}\ ,\ t_0+\ \frac{\tau}{2})$

To demonstrate that the proposed approximation of $f_\alpha(x,\ v,\ t_0+\tau)$ is second order accurate with respect to τ we can recognize with Cheng and Knorr (1976) that it is equivalent to take f_α=constant along the approximate characteristics of the Vlasov-Poisson system obtained by a leap-frog integration which is known to be second order accurate in time. Another way, more systematic, is to compute the Taylor development of $f_\alpha(x,v,t_0+\tau)$ up to the second order in τ using

$$\frac{\partial f_\alpha}{\partial t} = -\hat{L}_\alpha f_\alpha \quad \text{where} \quad \hat{L}_\alpha = +v\frac{\partial}{\partial x} + \text{sign}q_\alpha\cdot\frac{m_e}{m_\alpha}E\cdot\frac{\partial}{\partial v}$$

and $$\frac{\partial^2 f_\alpha}{\partial t^2} = \hat{L}^2_\alpha f_\alpha - \frac{\partial\hat{L}_\alpha}{\partial t}\cdot f_\alpha$$

and then to compare it with the Taylor development of the right member of (9a, b) up to the second order in τ; it appears that these two developments are identical. Of course during the implementation of the code the two adjacent integrations over x from the n-th and (n+1)-th cycles are combined together; hence we only need an half time step integration over x to start the computation as with the usual leap-frog algorithm. The above result concerning the second order accuracy of the splitting scheme is interesting, yet limited because it disregards totally the actual schemes used to integrate the advection equations; the above statement means only that the second order accuracy is the best we can expect with integration schemes of the advection equations which do not alter this accuracy. To evaluate the accuracy and convergence of a splitting scheme taking into account the full discretization of all variables (and not only of the time variable as above) and the particular numerical algorithms used for each fractional step is a difficult problem. Rigorous results have been demonstrated by Marchuk (1970) for the Crank-Nicholson schemes applied to various equations, and more recently by Crandall and Majda (1980) for conservation equations approximated by monotone schemes.

Lax-Wendroff schemes and the Glimm scheme combined with either Strang or Godounov splitting.

2.4b The numerical integration of the advection equation

Equations (3a,b) and (5a,b) have a common property : the coefficient of the "spatial" derivative $\partial f^*/\partial x$ or $\partial f^{**}/\partial v$ does not depend upon the "space" variable (respectively x or v); thus, in both cases, we have to integrate the advection equation with constant velocity

$$\frac{\partial \rho}{\partial t} + v\frac{\partial \rho}{\partial x} = 0 \tag{10}$$

with $\rho(x, t) \geqslant 0$, positivity constraint (11)

and $\int \rho(x, t)dx$ = constant, conservation constraint (12)

The very simple equation (10) is a special case of

$$\frac{\partial \rho}{\partial t} + \frac{\partial}{\partial x}f(\rho) = 0$$

which has been the subject of a huge amount of work in numerical analysis. What we want to do here is just to present in details one of the simplest FCT codes proposed by Boris and Book (1976b), that is an antidiffused donorcell algorithm. The reader interested in the numerous developments of antidiffusion is referred to the following classical papers : Boris and Book (1973); Book, Boris and Hain (1975), Boris and book (1976a, b). Until now the justification of these algorithms lay on rather intuitive ground, although rigorous results have been demonstrated, for example Ikeda and Nakagawa (1979) have studied the convergence and the stability of the SHASTA FCT algorithm (Boris and Book, 1973) (slightly modified). Starting from a very different point of view, Harten (1983) has proposed explicit second order finite difference schemes which, in some cases, recall the FCT algorithms.

The donor-cell algorithm is the simplest approximation to (10,11,12). Let $\rho^n{}_j$ be the value of the density in the j-th cell of width Δx at a time $t^n = n\Delta t$. Suppose that during the time step Δt the profile $(\rho^n{}_j)$ is shifted to the right ($v > 0$) or to the left ($v < 0$) by the amount $v\cdot\Delta t = (I+\varepsilon)\Delta x$ where I is the nearest integer to $v\,\Delta t/\Delta x$, so $-0.5 \leqslant \varepsilon \leqslant 0.5$. First the integral part $I.\Delta x$ of the displacement is done exactly by shifting the profile $(\rho^n{}_j)$ of I cells, and second the remaining fractional part is taken into account by

$$\rho^{n+1}_j = |\varepsilon|\rho^n_{j-\mathrm{sign}\varepsilon} + (1-|\varepsilon|)\rho^n_j \tag{13}$$

The boundary conditions have already been discussed. This scheme is obviously conservative (i.e. the discretized version of (11) holds) and preserves the positivity because $0 \leqslant |\varepsilon| \leqslant 0.5$, that is $\rho^{n+1}{}_j \geqslant 0$ for any j when $\rho^n{}_j \geqslant 0$ for any j : stability and monotonicity are achieved under the same condition. It is a first order approximation to (10) in Δx and Δt which is obtained when approximating $(\partial\rho/\partial t)^n{}_j$ by $1/\Delta t(\rho^{n+1}{}_j - \rho^n{}_j)$ and $(\partial\rho/\partial x)^n{}_j$ by $1/\Delta x(\rho^n{}_{j+1/2-1/2\text{sign}\varepsilon} - \rho^n{}_{j-1/2-1/2\text{sign}\varepsilon})$. The smearing of the transported profile results in a undesirable strong diffusion. To look into this point, let us suppose that $\rho^0{}_j=\delta_{j,o}$ (Kronecker symbol) and that I = 0, then

$$\rho^n_{j\text{sign}\varepsilon} = C^j_n|\varepsilon|^j(1 - |\varepsilon|)^{n-j} \qquad \text{for } 0 \leqslant j \leqslant n$$

and $\rho^n{}_j=0$ otherwise, where the $C^j{}_n$ are the binominal coefficients. Asymptotically for large n,

$$\rho^n_{j\text{sign}\varepsilon} \sim \frac{1}{\sqrt{2\pi n|\varepsilon|(1-|\varepsilon|)}} \exp\{-\frac{(j - n|\varepsilon|)^2}{2|\varepsilon|(1-|\varepsilon|)n}\}$$

it appears that the initial pulse spreads out during the propagation : $\rho^n{}_{j\text{sign}\varepsilon}$ is maximal for $j{\sim}n|\varepsilon|$, i.e. for x=vt, thus the velocity of the pulse is exact and the spreading of the pulse is described by the diffusion coefficient

$$D = \frac{1}{2}|\varepsilon|(1-|\varepsilon|). \tag{14}$$

To remedy this diffusion we insert an unphysical stage in the computation; we consider $\rho_j{}^{n+1}$ given by (13) as an intermediate value $\bar\rho^n_j$ and we apply an antidiffusion operator to (ρ^{-n}_j), that is the provisional values $\bar\rho^n_j$ can be seen as the result of a normal diffusion process (with a coefficient $\eta > 0$) over the profile $(\rho_j{}^{n+1})$. Approximating the diffusion equation

$$\frac{\partial\rho}{\partial t} = \eta\frac{(\Delta x)^2}{\Delta t}\frac{\partial^2\rho}{\partial x^2}$$

by the explicit first order method (see Potter), we obtain

$$\bar\rho^n_j = \rho^{n+1}_j + \eta(\rho^{n+1}_{j+1} - 2\rho^{n+1}_j + \rho^{n+1}_{j-1}) \tag{15}$$

this is known as implicit antidiffusion (for other types of antidiffusion and discussion of their respective advantages see the "classical" papers). From the above discussion of the diffusion associated with the donor cell transport algorithm we can guess that we have to choose (see 14) $\eta=|\varepsilon|/2(1-|\varepsilon|)$ this is confirmed by the analysis of the linear amplification factor. Let us consider an harmonic wave $\rho_j{}^n=e^{i(jk\Delta x-n\omega\Delta t)}$ where k is the wave number, $\omega=\omega_r+i\gamma$ the complex frequency and $i=\sqrt{-1}$. Substituting for $\rho_j{}^n$ in

$$|\varepsilon|\rho^n_{j-\text{sign}\varepsilon} + (1-|\varepsilon|)\rho^n_j = \rho^{n+1}_j + \eta(\rho^{n+1}_{j+1} - 2\rho^{n+1}_j + \rho^{n+1}_{j-1}) \tag{16}$$

we obtain the following dispersion relation for the linearly and implicitly antidiffused donor cell algorithm:

$$|\varepsilon|\ e^{-ik\Delta x\cdot\text{sign}\varepsilon} + (1-|\varepsilon|) = \{1 - 2\eta(1 - \cos k\Delta x)\}e^{-i\omega\Delta t}$$

The square of the modulus of the amplification factor is

$$e^{2\gamma\Delta t} = \frac{1 - 2|\varepsilon|(1-|\varepsilon|)(1-\cos k\Delta x)}{\{1 - 2\eta(1-\cos k\Delta x)\}^2}$$

For $k\Delta x \ll 1$

$$e^{2\gamma\Delta t} = 1 - [2\eta-|\varepsilon|(1-|\varepsilon|)](k\Delta x)^2 + O\{(k\Delta x)^4\}$$

It appears that our choice $\eta=(1/2)|\varepsilon|(1-|\varepsilon|)$ reduces the diffusion (or damping) to terms of the order $(k\Delta x)^4$ for the longer wavelengths and guarantees $\gamma \leqslant 0$ for any $k\Delta x$.

Another way of looking at this point is to check that (16) approximates (10) with second order accuracy when $\eta=(1/2)|\varepsilon|(1-|\varepsilon|)$. This is done with Taylor expansions like

$$\rho^{n+1}_{j+1} = \rho^n_j + \Delta t(\frac{\partial\rho}{\partial t})^n_j + \Delta x(\frac{\partial\rho}{\partial x})^n_j + \Delta t\cdot\Delta x(\frac{\partial^2\rho}{\partial t\partial x})^n_j \ldots$$

$$\ldots + \frac{\Delta t^2}{2}(\frac{\partial^2\rho}{\partial t^2})^n_j + \frac{\Delta x^2}{2}(\frac{\partial^2\rho}{\partial x^2})^n_j + O\{\Delta x^3, \Delta t^3, \Delta x^2\Delta t, \Delta x\Delta t^2\}$$

and with the help of $\partial/\partial t=-(v)\partial/\partial x$ The dispersion of the scheme is governed by (Boris and Book, 1973)

$$t_g(\omega_r\cdot\Delta t) = \frac{\varepsilon\sin(k\Delta x)}{1 - |\varepsilon|(1-\cos k\Delta x)}$$

For $k\Delta x \ll 1$

$$\omega_r = kv\{1 - \frac{1}{6}(|\varepsilon|-1)(2|\varepsilon|-1)(k\Delta x)^2 + O((k\Delta x)^4)\}$$

The relative phase error is of the order $(k\Delta x)^2$ for the longer wavelengths. It is possible to obtain a relative phase error of the order $(k\Delta x)^4$ by adding an appropriate diffusion term to the transport stage (13). Book, Boris and Hain (1975) have discussed the advantages of diffusive transport with regard to the phase properties of a scheme.

The constraint $0\leqslant|\varepsilon|\leqslant 0.5$ gives $\eta_{max}=1/8$, hence we may consider it as a small parameter and we will look for an approximation to (16), accurate to second order in ε and η, to avoid the solution of a tridiagonal system. Looking for a solution like

$$\rho^{n+1}_j = \sum_{k=-2}^{+2}(a_k+b_k\varepsilon+c_k\eta+d_k\varepsilon\eta+e_k\varepsilon^2+f_k\eta^2)\rho^n_{j+k}$$

we find that ρ_j^{n+1} is given by the following sequence

$$\bar\rho^n_j = |\varepsilon|\rho^n_{j-\mathrm{sign}\varepsilon} + (1-|\varepsilon|)\rho^n_j \quad (17a)$$

$$\tilde\rho^n_j = \bar\rho^n_j + \eta\Delta^n_{j-\frac{1}{2}} - \eta\Delta^n_{j+\frac{1}{2}} \quad (17b)$$

$$\rho^{n+1}_j = \bar\rho^n_j + \eta\tilde\Delta^n_{j-\frac{1}{2}} - \eta\tilde\Delta^n_{j+\frac{1}{2}} \quad (17c)$$

with
$$\Delta^n_{j+\frac{1}{2}} = \rho^n_{j+1} - \rho^n_j$$
$$\bar{\Delta}^n_{j+\frac{1}{2}} = \tilde{\rho}^n_{j+1} - \tilde{\rho}^n_j$$

We have reduced the diffusion but the positivity of ρ_j^{n+1} given by (17c) is no more guaranteed. To preserve this important property Boris and Book (1973) have introduced a limiter on the antidiffusion fluxes $f^n{}_{j+1/2} = \eta\bar{\Delta}^n{}_{j+1/2}$ (see section IV in Book, Boris and Hain, 1975) : (17c) is replaced by

$$\rho^{n+1}_j = \bar{\rho}^n_j + f^c_{j-\frac{1}{2}} - f^c_{j+\frac{1}{2}} \tag{18}$$

c stands for corrected. Whenever the transported profile $(\bar{\rho}^n{}_j)$ is not strictly increasing (or decreasing) over the four points j-1, j, j+1, j+2, the corrected flux $f^c{}_{j+1/2}$ is equal to zero; in all other cases $f^c{}_{j+1/2}$ is such that $\rho_j{}^{n+1}$ given by (18) is increasing (respectively decreasing). In fact this prescription does more than preserve the positivity, it guarantees that the antidiffusion stage neither creates new extrema nor accentuates already existing ones. The corrected flux is expressed by (19)

$$f^c_{j+\frac{1}{2}} = S_{j+\frac{1}{2}} \cdot \max\{0,\ \min(S_{j+\frac{1}{2}}\ \bar{\Delta}^n_{j-\frac{1}{2}}\ ,\ \eta|\bar{\Delta}^n_{j+\frac{1}{2}}|,\ S_{j+\frac{1}{2}}\ \bar{\Delta}^n_{j+\frac{3}{2}})\}$$

with $\quad S_{j+\frac{1}{2}} = \mathrm{sign}\bar{\Delta}^n_{j+\frac{1}{2}}\ ,\ \bar{\Delta}^n_{j+\frac{1}{2}} = \bar{\rho}^n_{j+1} - \bar{\rho}^n_j$

Finally equation (10) with constraints (11) and (12) is approximated by (17a,17b,18,19). The description of the code we have used is now complete. Let us say to close this section that this code could be made faster by implementing the electron subcycling algorithm proposed by Adam, Gourdin-Serveniere and Langdon (1982).

3. THE FORMATION OF ION ACOUSTIC DOUBLE LAYERS

Theoretical studies and computer simulations of the ion acoustic instability have been mainly devoted in the past either to cases with constant current or to cases where a constant electric field is imposed throughout the plasma. The effects of the dimensionality and of the magnetic field have been discussed in details (see the review by Dum, 1981, and references therein).

Ishihara and Hirose (1981), Appert and Vaclavik (1981) have studied the formation of the ion tail in the frame work of the quasilinear theory. This section is devoted to a study of the relaxation regime; the system is not influenced by any external factor, its evolution is governed by the initial condition only. We will distinguish between weak and strong initial perturbations. The evolution in the latter case will be compared with the evolution of an assembly of particles.

3.1 Weak initial perturbations

We will at first describe the simulation results we have obtained for noisy initial states, that is when the initial density perturbation $\delta n(x)$ is built as the superposition of NX/4 sinusoids having a common amplitude $\delta n_k=0.01$, wave numbers ranging from $2\pi/L$ to NX/4. $2\pi/L$ and random phases. We use it as an approximation to white noise. At the very beginning of the computation (say until $\omega_p t \sim 100$) the system builds up velocity perturbations consistent with the density perturbations; during this self-organization stage the electrostatic energy density decreases down to 10^{-5} - 10^{-6} T_e before increasing, due to the ion acoustic instability. This rearrangement in phase space is accompanied by electron diffusion around v=0; δn_k has to be chosen small enough to avoid the formation of a plateau around v=0, otherwise the instability could be quenched during this transient stage. Hence we are limited to weak initial perturbations. For comparison, the initial electrostatic energy density for a one-dimensional electrostatic particle code with 100 particles per λ_D is of the order of 10^{-3} T_e. We have made a series of runs with $T_e/T_i=20$, $m_i/m_e=100$, $L=512\lambda_D$ and $1024\lambda_D$ and with the drift speed of the electrons being equal to 0.6, 0.8 and 1.0 $V_{th,e}$. In all these cases the system evolves qualitatively in the same way, which is somewhat different from that exhibited by similar particle simulations. In the relaxation regime, that is when the electric current is not sustained by a current generator or by an external inductive electric field, the particle simulations show an asymptotic decrease of the spatially averaged current density equal to a few tens of per cent. Greater is the drift speed, greater and faster is the current interruption (see for example Nishihara et al., 1983). The same tendency appears in our computations but the current interruption is only equal to a few percent. The saturation levels of the electrostatic energy W also differ by an order of magnitude or more; e.g. for $V_d=0.6$, W reaches 4.10^{-5} instead of 2.7 10^{-3} T_e in the particle case (Sato and Okuda, 1980). Nevertheless in all cases we have simulated, we have observed localized negative potential structures similar to those observed in particle simulations (Chanteur et al., 1980, Sato and Okuda, 1981, Kindel et al., 1981, Hudson and Potter, 1981, Nishihara et al., 1983).

In both types of computations these potential spikes grow from the ion acoustic noise but their growth is much slower in the Vlasov case, and moreover they do not produce important differences of potential between their upstream and downstream sides ("upstream" and "downstream" refer to the mean electron flow which is going from the left to the right, i.e. increasing x, for all the relevant figures). They propagate in the direction of the mean electron flow with a velocity equal to the phase velocity of the ion acoustic waves in the beginning and later with a somewhat lower velocity. These potential wells reflect the electrons and accelerate the ions, thereby producing characteristic signatures in the phase spaces. Figure 1 shows the electron phase space at $\omega_p t=1024$ for $L=512\lambda_D$ and $V_d=0.8V_{th,e}$. A strong acceleration region is clearly visible near x=130 where the isodensity curves are squeezed. Especially the penetration towards v=0 of the curves related to the lower phase space density levels for v<0 indicates a pronounced dip in the electron density which is lowered down to 0.85 (unperturbed value $n_o=1$). Because of the drift

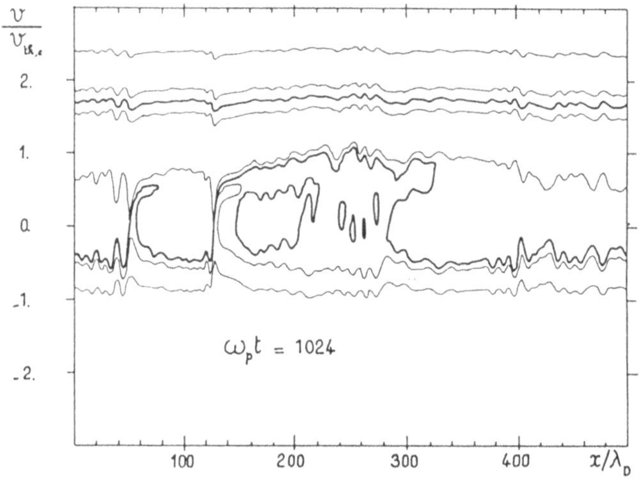

Figure 1. Electron phase space at $\omega_p t=1024$ for a Vlasov simulation with noisy initial condition: $L=512\lambda_D$, $V_d=0.8\ V_{th,e}$

velocity there are more electrons reflected on the upstream side than on the downstream side, and this results in a slight asymmetry of $n_e(x)$. The ion density looks like the electron density (a typical profile $n_i(x)$ is displayed by the upper part of Figure 4), but the space charge created by the reflection of the electrons is not exactly balanced by the massive ions, hence the upstream side having an excess of negative charge acquires a negative potential with respect to the downstream side where positive charges dominate. This rather weak difference of potential could be inferred from the electron phase space which shows that the domain of the reflected electrons is more inflated (in the velocity direction) on the downstream side, indicating that the potential barrier is higher seen from the down stream side than from the upstream one. Figure 2 represents the ion phase space at the same time but we have represented only the lower isodensity curves. The core of the ion distribution function represented by the white central band exhibits the sloshing motion of

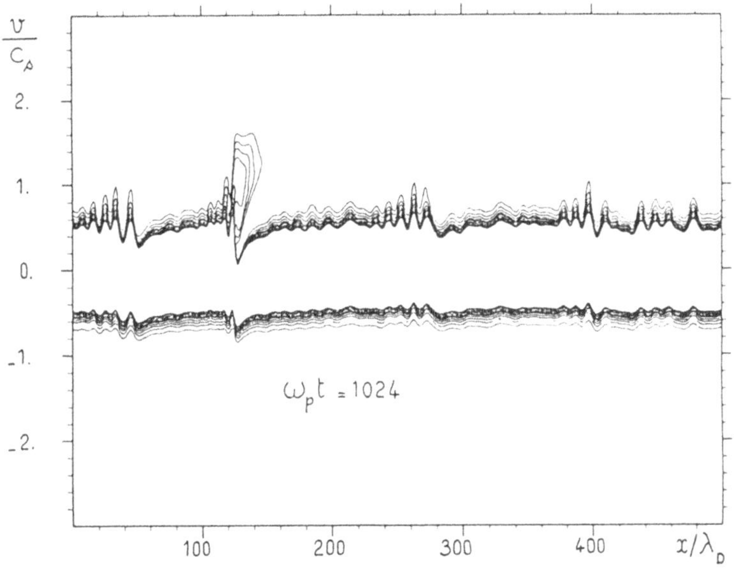

Figure 2. Ion phase space (same case as Figure 1). C_S is the ion acoustic velocity and only the lower isodensity levels are represented

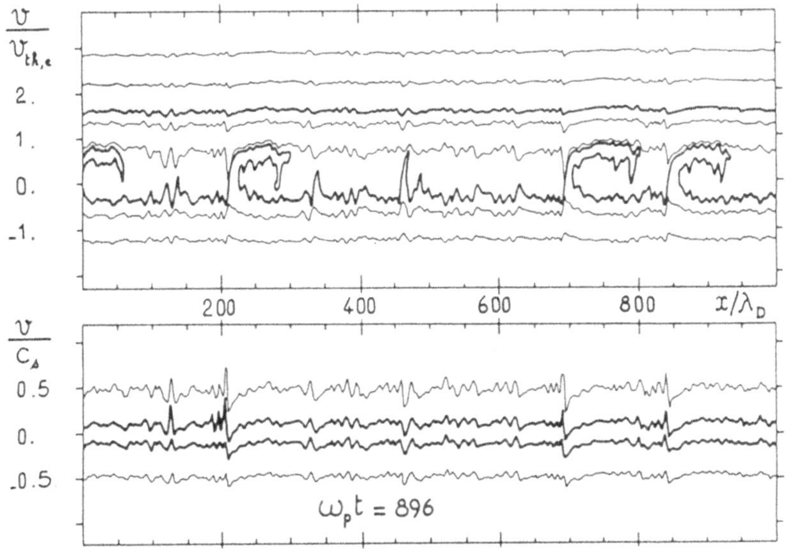

Figure 3. Upper : same as Figure 1 but for $L=1024\lambda_D$
lower : same as Figure 2 but for $L=1024\lambda_D$ (heavy lines delineate the core of the distribution function)

the ions in the ion acoustic waves and displays clearly the main features of the acceleration region near x=130, in particular the variation of the ion bulk velocity through the structure is consistent with the observations made for the electrons. Another interesting feature appears from the representation : as the potential well grows and propagates, it leaves an oscillating tail behind it. Two other acceleration regions are visible with their oscillating tails near $x=40\lambda_D$ and $x=270\lambda_D$ but they are less developed. The amplitude $-\phi_{min}$ of a potential spike is limited to $e\phi_{min} \sim m_e v_d^2$, the saturation occurs by electron trapping in the oscillating tail which grows to a level comparable to ϕ_{min}. Varying the length of the system does not change this picture, as shown by Figure 3 which represents the electron and ion phase spaces for $L=1024\lambda_D$, the other parameters being left unchanged. Increasing the drift velocity of the electrons tends to decrease the separation between the acceleration regions. The long time behavior in these relaxation experiments is characterized as in the particle case by an absence of resistivity and a very slow decay of the wave trains produced; Figure 4 shows the ion density and the space charge density at $\omega_p t=1280$ for the case $L=512\lambda_D$, $V_d=0.6\ V_{th,e}$. The situation does not evolve significantly until $\omega_p t=1792$, the time

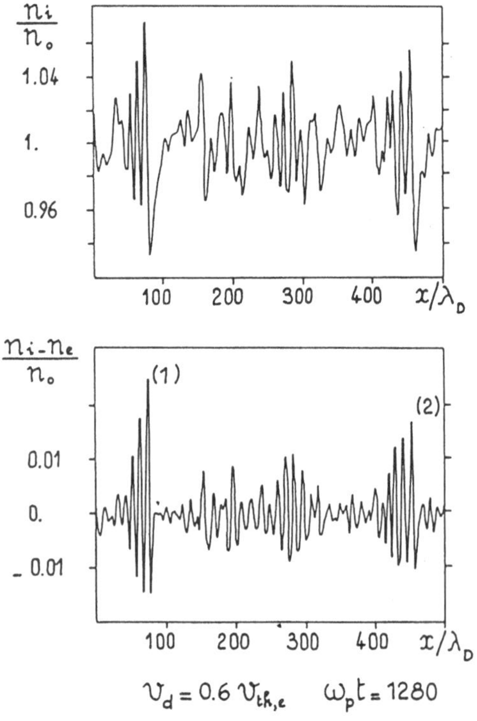

Figure 4. Ion density (upper) and space charge density (lower) for a Vlasov simulation with noisy initial condition : $L = 512\lambda_D$ and $V_d = 0.6\ V_{th,e}$

at which we have stopped the computation. The space charge density plot shows that the field has been broken in a few coherent nonlinear wave packets which have a sharply defined front and a decreasing tail (see packet marked (1) and (2)), moreover an examination of the phase spaces indicates that the electrons and the ions are trapped in these oscillations. It is likely that this behaviour is strongly related to the one dimensionality of the system.

3.2 Strong initial perturbations

The acceleration regions we have discussed above are so weak that they hardly deserve the name of double layers but basically they are driven by the same mechanism as we will see now. To get round the undesirable electron diffusion which occurs during the self-organization stage, we start the computation with a localized and smoothly shaped density depression having an amplitude of 0.1. The advantage of a localized density perturbation is the following : the velocity perturbations generated during the transient rearrangement of the system do not spoil the incoming plasma on the upstream side of the initial density dip. Due to the periodic boundary conditions, these ballistic perturbations reappear after some time on the upstream side of the density trough but the system is long enough to guarantee that the instability is well developed at that time. The length of the system is $L=512\ \lambda_D$ and the drift speed V_d of the electrons is

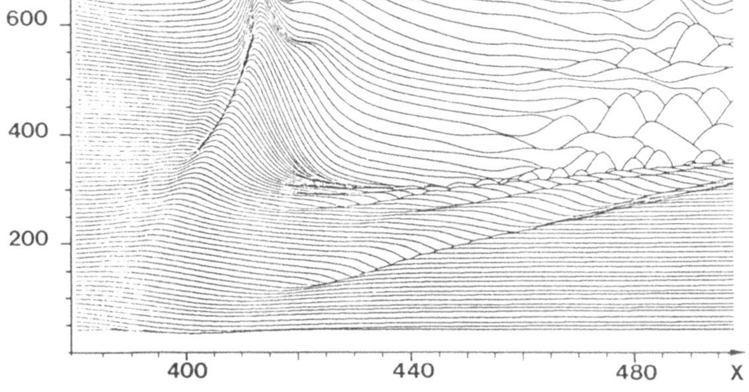

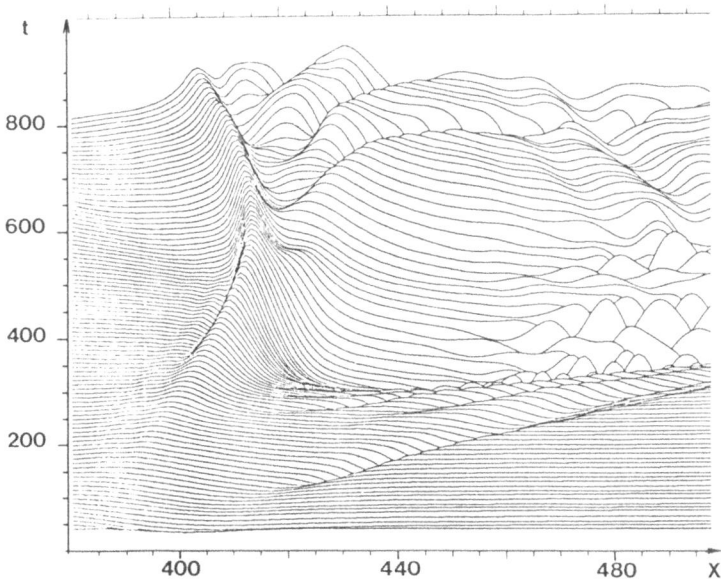

Figure 5. Potential energy of the electrons versus space and time for the Vlasov simulation with smooth initial condition

equal to 0.8 $V_{th,e}$, the other parameters having the values already mentioned. This simulation is discussed in detail by Chanteur et al. (1983). Let us recall the main features the evolution and the proposed interpretation. Figure 5 shows a three-dimensional plot of -ϕ, the potential energy of the electrons versus x (unit λ_D) and time (unit ω_p^{-1}). A potential energy barrier for the electrons grows in the depressed density region and moves in close association with it. At the beginning they propagate in the direction of the main electron flow with a velocity equal to 0.8 C_S (C_S is the ion acoustic speed which is equal to 0.1 $V_{th,e}$ in our case), a value corresponding to the linear group velocity of waves having the same spatial scale as the present one. The most interesting and puzzling points are the following : first the potential barrier suffers a very strong slowing down until it stops and starts propagating backwards with an increasing speed (a similar behavior has been noticed in particle simulations, see below); second, as the barrier grows and interrupts the electron flow, a positive difference of potential ϕ downstream -ϕ upstream is built up by the unbalanced space charge; third, it could be seen that the upstream flank of the barrier becomes steeper than the downstream flank before the reversal of the motion of the

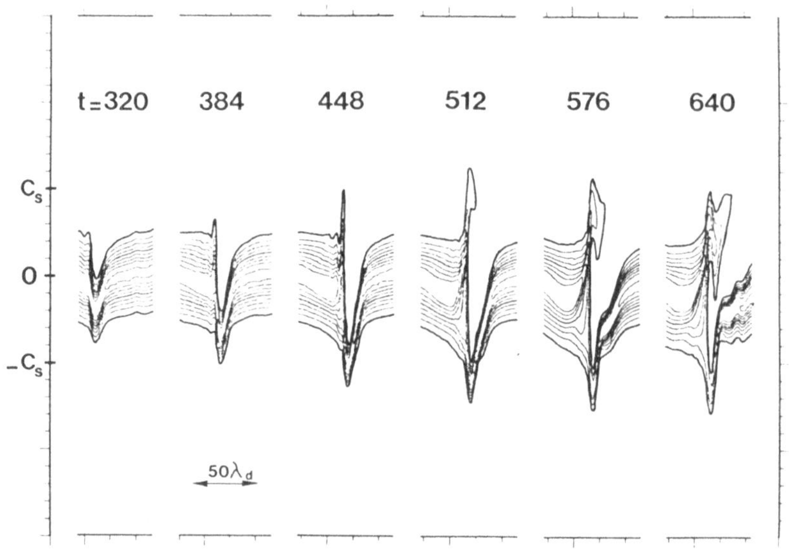

Figure 6. Same part of the ion phase space at different times for the same case as for Figure 5

structure. Figure 6 shows the same part of the ion phase space in the vicinity of the electric potential well at different times. Until $\omega_p t=400$ the velocity of the structure is almost constant and the main feature in the ion phase space is the localized modification of the mean velocity over the potential well. In the frame of reference moving with the localized wave the ions enter the potential region from the downstream side with a negative velocity, they are successively accelerated and decelerated as they fly above the well. This is the reason why the potential well produces a negative velocity perturbation in the laboratory frame. When the velocity of the wave is reversed we gain the opposite signature in the ion phase space (see below). Ion trapping occurs after $\omega_p t \sim 400$ when the velocity of the localized wave begins to decrease noticeably and an ion hole is forming after $\omega_p t \sim 550$ when the whole structure is almost stationary in the ion frame. The spatial variation of the ion bulk velocity through the structure also supports evidence for a localized difference of potential between the downstream and upstream sides.

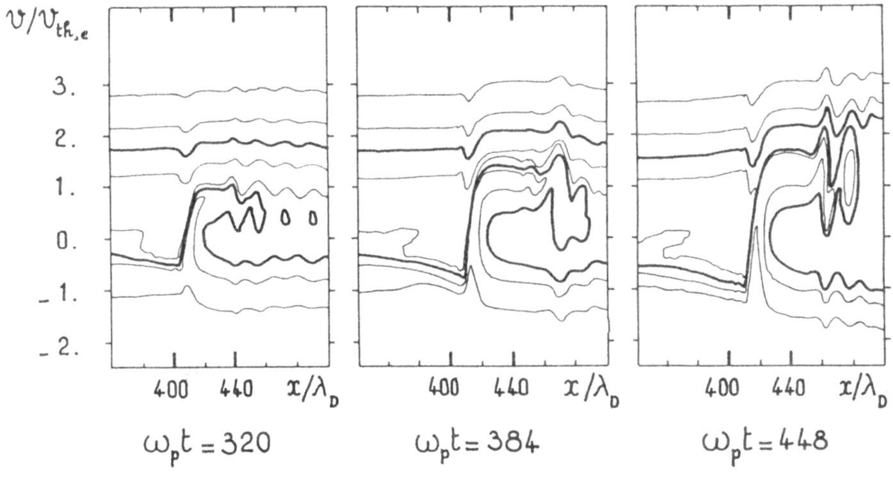

Figure 7. Same part of the electron phase space at different times for the same case as Figure 5

Figure 7 displays the same part of the electron phase space at three different times during the forwards propagation of the potential barrier. The reflections of the particles are clearly visible on both sides of the acceleration region, together with the resulting asymmetry between the upstream and downstream sides. The similarity with Figure 1 is obvious and the localized difference of potential is also responsible for a more extended velocity range of the reflected particles on the downstream side. Before discussing these results let us recall a particle simulation (Chanteur et al., 1980) done with the same physical parameters and the same length of the system. The main difference is that the particle computation is started from the

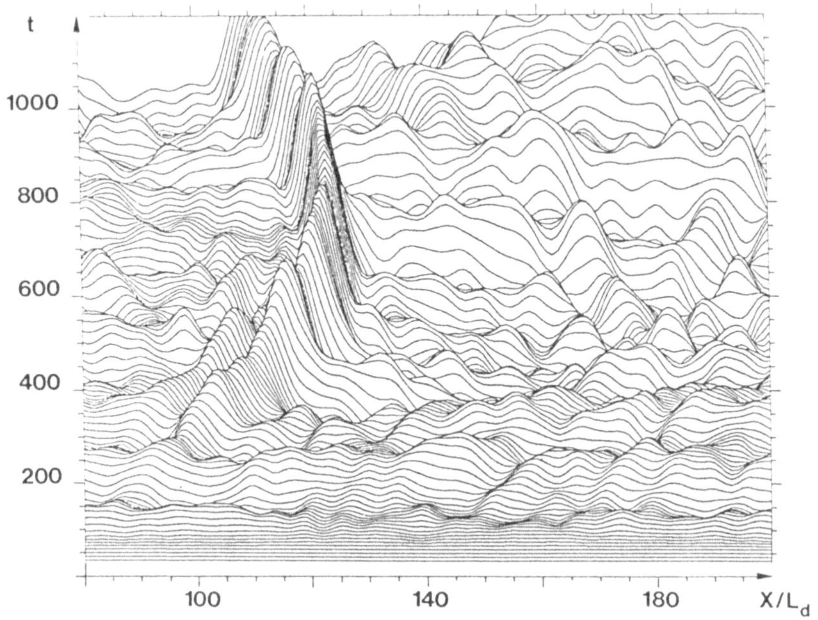

Figure 8. Potential energy of the electrons for the particle simulation with $L=512\lambda_D$ and $V_d=0.8\ V_{th,e}$

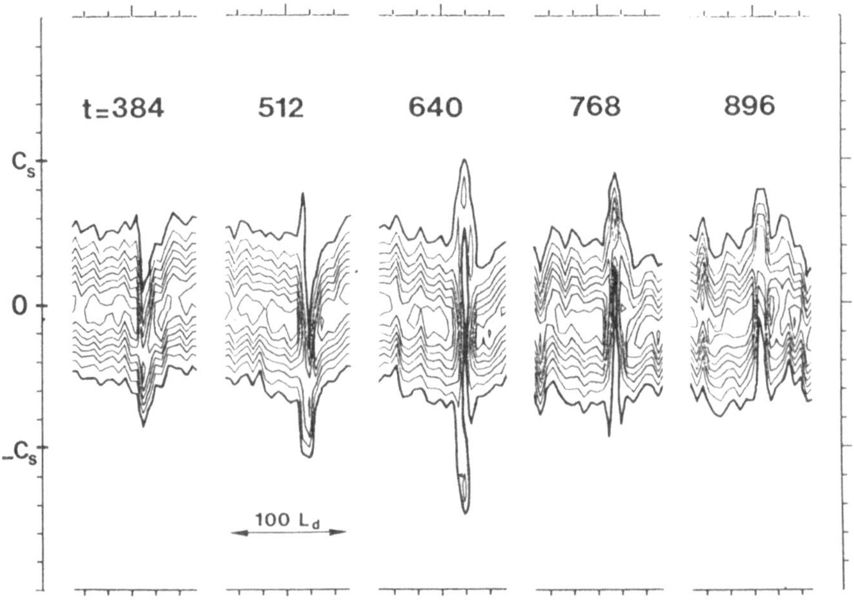

Figure 9. Same as Figure 6 for the particle simulation.

"thermal" density fluctuations (we have used 100 particles of each species per Debye length). The physical evolution of the system is dominated by the growth of a potential barrier for the electrons (see Figure 8 and compare with Figure 5), in particular the current interruption is mainly caused by this barrier. Figure 9 represents the same part of the ion phase space at different times and has to be compared with Figure 6. The first and second plots are taken when the localized wave propagates towards the right (notice the negative velocity dip). The third plot is for $\omega_p t = 640$ when the potential structure is still almost stationary in the ion frame and an ion hole could be seen forming. The two last plots are relative to the backwards propagation of the wave, and as was already discussed above, a positive velocity perturbation is apparent. This strange motion of the potential barrier was also noticed by Kindel et al. (1981), Nishihara et al. (1983), and in a different way by Sato and Okuda (1981). Figures 12 and 13 in Sato and Okuda (1981) are particularly interesting in this respect. The spatial variation of the frequency spectrum of the turbulence discussed by Sato and Okuda is now easy to

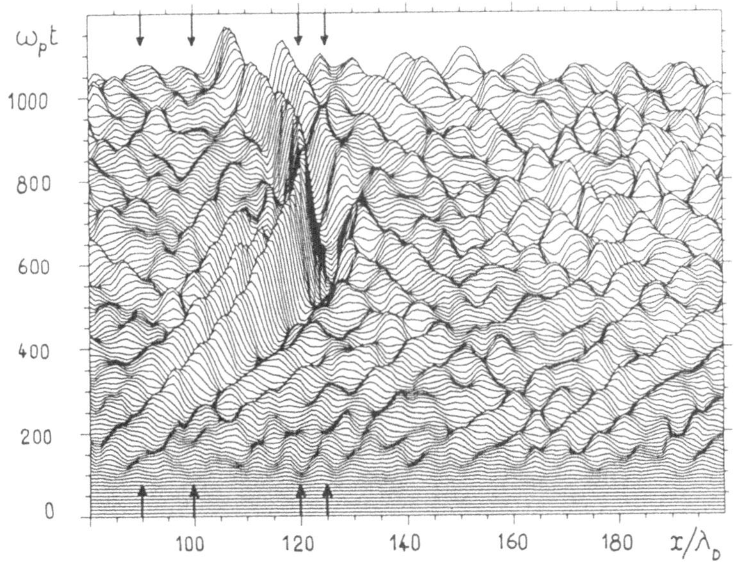

Figure 10. Electric field versus space and time for the particle simulation

understand. Figure 10 is a three-dimensional plot of the electric field versus space and time in our particle simulation. The huge electric field of the localized wave is clearly identified as well as the boomerang motion of the wave. The vertical black arrows indicate four positions, namely x = 90, 100, 120 and 125 λ_D; the corresponding temporal variations of the electric field at these particular places are shown on Figure 11 (which is very similar to Figure 13 of Sato and Okuda (1981). Thus we conclude that the spatial variation of the frequency spectrum of the electric field is mainly due to the slowing down of the coherent localized wave which has emerged from the turbulence.

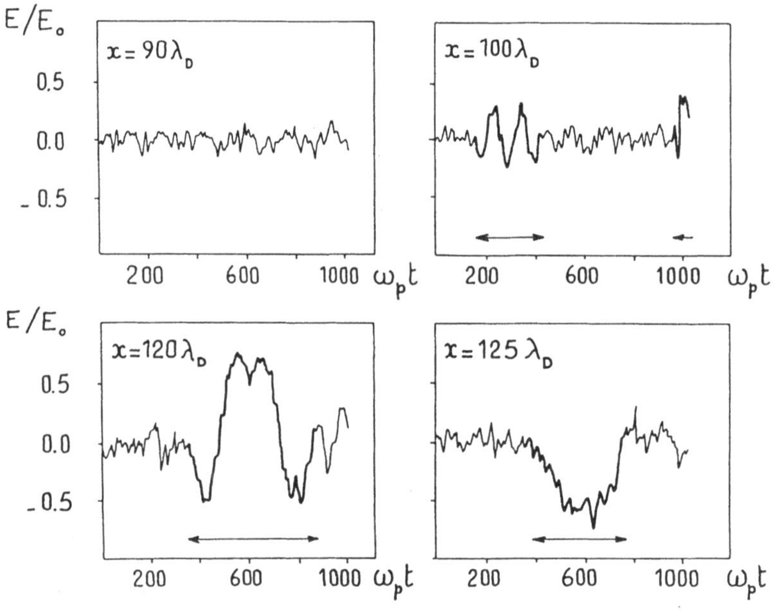

Figure 11. Temporal variations of the electric field at four different places for the particle simulation. Arrows under the plots indicate the coherent localized wave.

3.3 Discussion and conclusion

Nishirara et al. (1983) and Chanteur et al. (1983) have proposed the same interpretation of the physical process discussed above. They have shown that the propagation, the growth and the distorsion of the potential barrier could be explained by a modified Korteweg de Vries equation for the potential, taking into account the reflections of the electrons. A similar equation has been proposed by Karpman et al. (1979) and Lotko (1983) to interpret the damping of standard compressive ion acoustic solitons. Moreover, Nishihara et al. (1983) have integrated numerically both their modified KdV equation and the usual KdV equation for negative and localized initial conditions. In the pure KdV case which has also been studied numerically in detail by

Fornberg and Whitham (1978) and by Okutsu and Nakamura (1979) the negative perturbation is damped by the "radiation" of a dispersive tail while the localized wave is slowed down by the nonlinear convective term $\phi.\partial\phi/\partial x$. This last term also acts to steepen the trailing edge of the potential well and to splay its leading edge. Similar conclusions have been obtained experimentally for rarefactive ion acoustic pulses propagating in a quiescent plasma by Okutsu and Nakamura (1979) and by Saxena et al. (1981). When the amplification term caused by the asymmetry of the electronic distribution function is taken into account the work of Nishihara et al. (1983) indicates, as expected theoretically, the amplification of the potential well but also a stronger slowing down of the solution. This slowing down is a nonlinear hydrodynamic effect as in the pure KdV case but is in some way enhanced by the reflections of the electrons which amplify the potential well. Both interpretations of Nishihara et al. (1983) and Chanteur et al. (1983) are based on a fluid treatment of the ions. In conclusion, both particle and "Vlasov" simulation results concerning the formation of ion acoustic double layers could be accounted for by a fluid description of the ions until the wave has sufficiently decreased its velocity to have resonant interaction with the core of the ion distribution. It is likely that the ion trapping which occurs then accentuates the slowing down and is the key for understanding the reversal of the motion of the structure, but this last point has still to be checked. It appears that ion trapping is important in the late evolution of the system; in the beginning ion trapping is insignificant and does not explain the slowing down, as recently proposed by Hudson et al. (1983) on the basis of time-stationary models (see also Hudson and Potter, 1981). We are still far from a good theoretical understanding of the space observations related to double layers because the real situation in auroral regions is much more complex, due to the interpenetration of ionospheric and magnetospheric populations of particles. The work of Lotko and Kennel (1981) on rarefactive ion acoustic solitons supported by two electron populations is interesting in this respect. Care must be taken of stationary models invoking ion trapping because -as demonstrated by simulations -the growth of the structures occurs on a time scale comparable to the transit time of the ions through these structures.

Acknowledgements.

The author is greatly indebted to the Organizational Committee of the first ISSS and to URSI for financial support; he is also greatly grateful to Dr. R. Gendrin for many encouragements and discussions and to Drs. J. C. Adam and R. Pellat for their criticisms and constant help.

REFERENCE

Adam, J. C., A. Gourdin-Serveniere, and A. B. Langdon, J. Comput. Phys., 47, 229, 1982.
Appert, K., and J. Vaclavik, Plasma Phys. 23, 763, 1981.
Bernstein, I. B., J. M. Greene, and M. D. Kruskal, Phys. Rev., 108, 546, 1957.
Birdsall, C. K., and N. Maron, J. Comput. Phys., 36, 1, 1980.
Book, D. L., J. P. Boris, and K. Hain, J. Comput. Phys., 18, 248, 1975.
Boris, J. P., Proc. Conf. Numer. Simul. Plasmas, 4th, 3, 1970.
Boris, J. P., and D. L. Book, J. Comput. Phys., 11, 38, 1973.
Boris, J. P., and D. L. Book, J. Comput. Phys., 20, 397, 1976a.
Boris, J. P., and D. L. Book, Methods of Comp. Phys., 16, 85, 1976b.
Chanteur, G., J. C. Adam, and R. Pellat, Current Interruption by localized ion acoustic waves, Chapman Conference on the Formation of Auroral Arcs, Fairbanks, Alaska, 1980.
Chanteur, G., J. C. Adam, R. Pellat, and A. Volokhitin, On the formation of ion acoustic double layers, Phys. Fluids, 26, 1584, 1983.
Cheng, C. Z., and G. Knorr, J. Comput. Phys., 22, 330, 1976.
Crandall, M., and A. Majda, Numer. Math., 34, 285, 1980.
De Groot, J. S., C. Barnes, A. E. Walstead, and O. Buneman, Phys. Rev. Lett, 38, 1283, 1977.
Dum, C. T., in Phys. of Auroral Arc Formation, Geophys. Monogr. Ser., 25 S. I. Akasofu and J. R. Kan eds., 408, American Geophysical Union, Washington, D. C., 1981.
Fornberg, B., and G. B. Whitham, Phil. Trans. Roy. Soc. London A, 289, 373, 1978.
Harten, A., J. Comput. Phys., 49, 357, 1983.
Hasegawa, A., and T. Sato, Phys. Fluids., 25, no 4, 632, 1982.
Hudson, M. K., W. Lotko, I. Roth, and E. Witt, J. Geophys. Res., 88, no A2, 916, 1983.
Hudson, M. K., and D. W. Potter, in Physics of Auroral Arc Formation, Geophys. Monogr. Ser., 25, S. I. Akasofu and J. R. Kan eds., 260, American Geophysical Union, Washington, D. C., 1981.
Ikeda, T., and T. Nakagawa, Math. Comp., 33, 1157, 1979.
Ishihara, O., and A. Hirose, Phys. Rev. Lett., 46, no 12, 771, 1981.
Karpman, V. I., J. P. Lynov, P. Michelsen, H. L. Pecseli, J. J. Rasmussen and V. A. Turikov, Phys. Rev. Lett., 43, 210, 1979.
Kindel, J. M., C. Barnes, and D. W. Forslund, in Physics of Auroral Arc Formation, Geophys. Monogr. Ser., 25, S. I. Akasofu and J. R. Kan eds., 296, American Geophysical Union, Washington, D. C. 1981.
Lotko, W., Reflection dissipation of an ion acoustic soliton to appear in Phys. Fluids, 1983.
Lotko, W., and c. F. Kennel, in Physics of Auroral Arc Formation, Geophys. Monogr. Ser., 25, S. I. Akasofu and J. R. Kan eds., 437, American Geophysical Union, Washington, D. C. 1981.
Marchuk, G. I., in Numerical Solutions of Partial Differential Equations, vol.II, B. Hubbard ed., Synspade 1970, Academic Press, 469, 1971.

Nishihara, K., H. Sakagami, T. Taniuti, and A Hasegawa, Formation of weak double layers in ion acoustic turbulence, to appear in Phys. Fluids, 1983.

Okuda, H., and M. Ashour-Abdalla, Phys. Fluids, 25, no 9, 1564, 1982.

Okutsu, E., and Y. Nakamura, Plasma Phys., 21, 1053, 1979.

Potter, D., Computational Physics, John Willey & Sons Ltd., 1973.

Sato, T., Auroral Physics, Chap. 4, Magnetospheric Physics, A. Nishida ed., Center for Academic Publications Japan, Tokyo, 1982, and D. Reidel Pub. Co.

Sato, T., and H. Okuda, Phys. Rev. Lett., 44, 740, 1980.

Sato, T., and H. Okuda, J. Geophys. Res., 86, 3357, 1981.

Saxena, Y. C., S. K. Mattoo, A. N. Sekar, and V. Chandna, Phys. Lett., 84A, 1981.

Sigov, Yu.S., Physica Scripta, T2/2, 367, 1982.

Singh, N., and R. W. Schunk, Geophys. Res. Lett.,9, no 12, 1345, 1982.

Strang, G., Siam J. Numer. Anal., 5, no 3, 506, Sept. 1968.

Temerin, M., K. Cerny, W. Lotko, and F. S. Mozer, Phys. Rev. Lett., 48, 1175, 1982.

SIMULATION MODELS FOR SPACE PLASMAS AND BOUNDARY CONDITIONS AS A KEY TO THEIR DESIGN AND ANALYSIS

Christian T. Dum

Max-Planck-Institut für Physik und Astrophysik
Institut fur extraterrestrische Physik
D 8046 Garching, FRG

ABSTRACT

The basic types of simulation models as they derive from a variety of space plasma problems are examined. The design of these models and the evaluation of their potential in advancing physical theories, including also the caveats in their physical interpretation and generalization, are emphasized rather than a review of existing simulation results. Boundary condition play the key role in this analysis. Along with the choice of a particle, fluid or hybrid description for the plasma volume and of a direction or plane of computation in one dimensional or two dimensional models, respectively, boundary conditions largely determine the characteristics of a simulation model. Boundary conditions contain and control important physical processes and thus should be derived from the actual physical situation, rather than being chosen merely for numerical convenience. The analysis of physical problems and corresponding simulation models presented here puts some emphasis on double layers, anomalous resisitivity and reconnection, not only because these processes are fundamental and thus also actively studied in simulations, but also because they are outstanding examples for a strong dependence on boundary conditions. This strong dependence must also be considered in the physical analysis of simulation experiments and especially in their application to space plasmas. Space plasmas have few well defined boundaries and artificial boundaries must generally be introduced in order to limit the computational domain. It has been found very helpful to consider schematic laboratory experiments in defining and analyzing well posed boundary conditions, especially for artificial boundary surfaces. Thus, related laboratory experiments and corresponding simulations are also presented. Space plasmas, however, have no lumped electrical circuits. Global simulations are therefore considered as a means of providing boundary conditions which express electrodynamic coupling between the small subsystem to be studied in detail and the global system.

H. Matsumoto and T. Sato, Computer Simulation of Space Plasmas,

1. Introduction

The aim of a successful simulation is to elucidate selected physical processes deemed to be essential in accordance with an a priori theoretical understanding, rather than an attempt at reproducing physical systems in their entire complexity. In contrast to real experiments, simulations offer the possibility of turning on and off some of these processes at will. For example, the role of a magnetic field may be studied by switching off the magnetic field for the ions while retaining magnetized electrons. Space charge electric fields may also be switched on and off to see their role. As a result, a few dominant physical processes may emerge. These processes often are of surprising simplicity, thus making it possible a posteriori to develop a more usable theoretical model. It is this interaction between computation, physical theory, and observations that makes simulation on exciting activity of great potential.

The particular physical aspects that are to be illuminated by a simulation determine the minimum level of sophistication that has to be retained in the mathematical model. The actual physical system provides effective guidance in developing the mathematical model and in selecting input data. The mathematical model is, however , neccesarily strongly simplified. This is especially true of boundary conditions. In physical terms they mean that outside the chosen volume the relatively detailed description, say by partial differential equations for a continuous medium, is replaced by a much simpler lumped description. The computational effort alone dictates such simplifications.

Computational restraints may have the very benificial effect of forcing the computational physicist into a more careful analysis of experiments or observations, in order to find the truly essential physical processes that absolutely must be retained in the model. In particular, boundary and initial conditions often remain quite nebulous in purely theoretical studies. Many of these studies avoid this problem altogether by considering only stationary states in a infinite homogeneous system. It is usually expected that simplifications in the nature of the B.C. do not grossly affect the behavior of the mathematical model and thus the correspondence of the model with the physical system, at least at some distance from the boundaries. Asymptotic expansions are a well known device for the demonstration of such properties in mathematical physics (e.g. Carrier and Pearson, 1976, chapter 15)

Most simulations require a numerical solution. This implies not only that one is restricted to finite computational domains, but that due to limitations in computing capacity these domains are nearly always much smaller than one would desire. B.C. in computational physics thus play an even more important role than in mathematical physics or in the actual physical system where boundaries are more gradual and much further away. Often, a finite computational domain has to be created at the expense of introductional artificial boundaries. Especially for problems involving the flow of fluids or radiation, the B.C. at these surfaces have to be carefully designed in order to reduce spurious numerical reflections. At outflow (open) boundaries the computational domain may be extended by adding a region of wave damping (Sato and Hayashi, 1979; Hashimoto et al., 1983). Asymptotic radiation B.C. which describe the flow out of the computational domain and towards infinity can also reduce reflections from artificial boundaries. Computational boundaries of unknown shape (free boundaries) may be used in order to avoid numerical difficulties in resolving strong gradients associated e.g. with shocks or interfaces between different media. Grid points of the computational mesh that are on or near boundaries generally require special attention in order to make numerical boundary conditions consistent with the numerical scheme used inside the domain of computation. The algorithm may connect interior grid points to exterior grid points where by definition dependent variables are not computed. Careless extrapolation may cause a significant loss of overall numerical accuracy or even lead to a numerical instability. It may be necessary to introduce numerical B.C., e.g. at open boundaries, which are not specified for the original partial differential equations but must be directly formulated for the finite difference scheme. The implementation of fluid B.C. and their effect on overall numerical accuracy and stability was the topic of a recent symposium on "Numerical Boundary Condition Procedures and Multigrid Methods" (Abarbanel, Murman and others, 1982).

It is seen that computational physicists very soon become acutely aware of difficult problems related to B.C. Yet, at least in papers dealing with the physical results, one often finds almost no mention on how these boundary problems were resolved more or <u>less</u> satisfactorily. This article aims at filling this gap for the simulation of plasma dynamics by showing how simulation models for various classes of plasma physics problems are derived and in particular how B.C. are implemented. The choice of B.C. is nearly always restricted by numerical considerations. This restriction may severely affect the physics exhibited by the simulation. It is hoped therefore, that even readers who are solely interested in physical results make it a habit to critically examine B.C. in numerical simulations, before accepting conclusions for actual physical systems at face value. B.C. are a key feature in the design and analysis of simulation models.

In plasma physics the variety of problems is greatly increased by the inclusion of self-consistent electromagnetic fields and the various levels of collisionality for particle species. Rather complex B.C. for particles or their fluid equivalents have to be satisfied in addition to the more familiar B.C. for the electromagnetic field. The requirement of self-consistency between electromagnetic fields and charged particle motion leads to a coupled system of nonlinear equations which in general does not fit entirely into the standard classification schemes for partial differential equations and associated B.C. Numerical methods must be tailored to the specific problem at hand. No adequate summary of numerical algorithms or of existing simulation results can be attempted here. It is hoped, however, that the plasma physics problems and corresponding simulation models discussed in this article are representative of how, in the face of usually severe numerical constraints, simulation models in space plasma physics are designed, demonstrate the great potential of simulation models for advancing physical theories, but also indicate caveats to be taken into account in their physical interpretation and application to other physical situations. It is also hoped that the development of new simulation models is stimulated. The references given here are selected on a basis that should help in tapping sources for details on numerical algorithms, and also on physical details not directly related to the construction of simulation models. Simulation oriented papers or papers that most directly translate to simulation models were preferred. Still, to keep this list manageable it regrettable had to be restricted generally to the most recent papers that include a reasonable account of important earlier work on a particular point.

Section 2 outlines the physics of some boundary value problems as they derive from laboratory experiments. B.C. are more obvious and more easily explored in this case but as is demonstrated, great care must be exercised in translatating laboratory experiments to space plasmas with very different B.C. Simulation models and boundary problems specific of space plasmas are analyzed in Sec.3. A particular problem is the frequent necessity of introducing artificial boundaries in order to limit the computational domain. B.C. on these surfaces, if chosen for computational convenience, often correspond more directly to schematic laboratory experiments. It is necessary to carry out a careful analysis of the effects of introducing conducting plates, emitting or absorbing electrodes etc. into simulation models for space plasmas which do not possess such boundaries. The system size in the simulation is nearly always too small for these effects simply to be neglected. Section 4 considers the simplest way of constructing simulation models, i.e. by essentially ignoring boundaries with the help of a periodic continuation of the computation box. This method appears to be most appropriate for the analysis of waves and instabilities. The only difference to the infinite homogeneous plasma preferred in theoretical work arises then from the restriction of wave numbers to a discrete set, $k=2\pi n/L$ where L is the box size and n is an integer which in practice is also restricted, $|n|< N$, e.g. by the finite grid spacing. These wave numbers arise directly in Fast Fourier Transforms (FFT) which also form the basis of many efficient numerical algorithms, e.g. for the solution of Poisson's equation. The restriction to a finite set of wave numbers, however, may severely affect the outcome of a simulation. We demonstrate that this question must be examined with great care when a simulation experiment is to be designed or when more generally valid physical conclusions should be drawn from simulation runs. Moreover, with periodic B.C. only a few possibilities remain for controlling the outcome of simulations other than by initial conditions. Some B.C.

may be incorporated by adding analytic solutions e.g. for the homogeneous Laplace equation while still using FFT for the solution of Poisson's equation with a given charge distribution (Decyk and Dawson, 1979). Other problems involving inhomogeneous plasmas may be simplified by considering a slab that is bounded in one direction and periodic in the other. Use of FFT reduces then e.g. the finite difference approximation of Poisson's equation to a tridiagonal matrix equation which for fairly general B.C. in the inhomogeneity direction is solved by very efficient algorithms. Owing to their great numerical convenience, periodic B.C. and FFT are used also for a variety of problems other than waves including even global simulations of the magnetosphere (Leboeuf et al., 1979). The physical interpretation of periodic B.C. in these cases is that the image of the computational domain is effectively repeated throughout space and that phenomena within the computational domain may interact with the repeated images of the domain. Symmetry relations which are used with advantage in reducing the computational effort and certain other B.C. may also be interpretated in terms of a periodic continuation, even FFT may not actually be used. It should be noted that rapid elliptic solvers and efficient direct algorithms for field equations exist also for nonperiodic B.C., provided that boundary surfaces can be reduced by mapping or patching to a simple (rectangular) geometry that allows a separation of variables. Iterative methods may have to be used for other geometries (Hockney and Eastwood, 1981, Ch. 6).

A full discussion of some B.C. can only be carried out in conjunction with the specific mathematical model chosen for the plasma volume, such as a particle, fluid or hybrid description. The numerical algorithm for the solution of the resulting equations also influences the implementation of B.C. The subtile problem of B.C. for particle simulation of kinetic effects is discussed in Sec.5. Problems which are simplified by using a fluid description are considered in Sec.6. The computational effort restricts full simulations of kinetic effects to comparatively small domains of typically 1000 Debye lengths in one dimension and much smaller size in two or three dimensions. More efficient computing schemes and computers may, of course, help in easing these restrictions. Hybrid schemes in which electrons with their small characteristic time and length scales are treated as a fluid while retaining the full ion particle dynamics are also of enormous help. For other extended systems, however, the grid size in fluid calculations is at best comparable to the size of the entire simulation box of a particle simulation. Clearly, other methods must then be found for the coupling of macroscopic and microscopic dynamics (Dum, 1983a). This largely unsolved question is taken up in the concluding section.

2. SIMULATION MODELS AND BOUNDARY CONDITIONS DERIVED FROM LABORATORY PLASMAS

Development of a simulation model should include the careful analysis of B.C., as they derive from the physical system, primarily free of the additional simplifications that may be imposed by the numerics. Consideration of B.C. for laboratory plasmas guided the development of many simulation models that are now also applied to space plasmas. Many points on B.C. are more easily explored for laboratory plasmas, thus their consideration is helpful also for the development of B.C. in space plasmas. Self-consistent and well posed B.C. can be defined for schematic experimental setups, as described in this section. Simulation models which relate to space phenomena, yet for various reasons use no configurations (other than coordinate conventions) or B.C. specific of space plasmas (Sec.3) are mostly also included in this part.

2.1 Mainly Electrostatic Phenomena: Sheaths, Heat Conduction, Double Layers and Anomalous Resistivity in Collisionless Plasmas

For Laboratory plasmas B.C. are imposed in a seemingly obvious way by material walls. The structure of the plasma and the associated space charge potential in the plasma-wall boundary layer remains still an important problem. Wall reflectivity, primary or secondary emission, as well as particle distribution functions and the electric field in the presheath plasma determine the sheath structure. An oblique magnetic field splits the plasma sheath into a magnetic presheath where the electric field deflects only ions from their spiral motion along the magnetic field, and an electrostatic Debye sheath where the space charge electric field completely dominates particle motion (Chodura, 1982).

If an electric current is drawn, not only the sheath structure near the electrically biased electrodes changes, as is well known from Langmuir probe theory, but localized potential drops may suddently also appear in the plasma body. The Vlasov-Poisson equations admit stationary double layer (sheets of opposite charge) solutions, provided appropriate reflected electron and ion distributions exist in addition to the current carrying transiting particle distributions. Although the assumption of laminar, scatter free particle motion implies a perfect memory of boundary and initial conditions, these are usually not considered in sufficient detail when different experiments, space observations, simulation results, and theory are compared.

Experiments with triple plasma devices (Figure 1) where the axial electron and ion flow between the two source plasmas and the target plasma can be controlled by independently biased and closely spaced grid pairs on both sides of the target plasma and also by source biases, show indeed a very strong dependence on the electric B.C. (Leung et al., 1980; Baker et al., 1981; Hershkowitz et al., 1981).

Varying also the effective size L/λ_D of the target plasma by changing either the Debye length λ_D or the distance L between the grid pairs, shows that multiple double layers can exist in sufficiently long systems, $L/\lambda_D \geqslant 300$. They evolve into a single double layer and then into a sheath like structure as L/λ_D is decreased to $\leqslant 100$ (Chan and Hershkowitz, 1982). In a similar long device, multiple potential steps formed for sufficiently large applied potential differences, but an almost uniform potential drop developed instead, if the low potential (cathode) plasma source was turned off (Guyot and Hollenstein, 1983). Moreover, in contrast to other experiments and

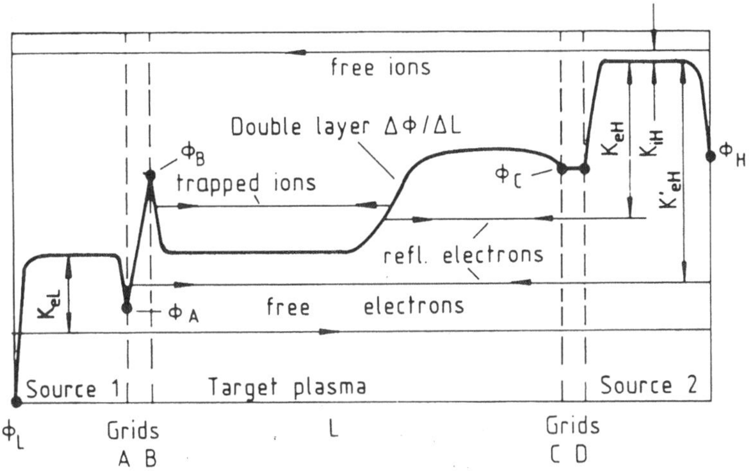

Figure 1 Schematic Representation of a. Triple plasma device (after Hershkowitz et al., 1981). Plasma flow between the sources and the target plasma is controlled by applying potentials Φ to the closely spaced grid pairs A-B. C-D, and the anodes (Φ_L,Φ_H). Walls are at ground potential. Depending on these electric boundary conditions and the ratio of plasma length L to Debye length, sheaths, single or multiple double layers (D.L.) are formed. b. Laminar particle motion is controlled by $W=K_j+e_j\Phi=\text{const.}$, where $K_j=m_j v^2/2$ is the kinetic energy. Collisions create ions trapped between the D.L. and grid B. Particle populations are also merged by scattering from turbulence.

laminar theory, localized potential jumps occured for electron drift velocities u_e substantially below the electron thermal velocity v_e. Multiple double layers solutions corresponding to the observed potential structures do not seem to exist at all in the context of the laminar and stationary Bernstein Greene Kruskal (BGK) model with plausible particle distribution functions (Chan and Hershkowitz, 1981). Charge exchange and ionizing collisions with neutral atoms in concert with radial losses depending on the lateral wall potential and on the confining magnetic field may be significant in some experiments (Leung et al., 1980), at least in forming the trapped particle populations (Baker et al., 1981; Hershkowitz et al., 1981).

Although this is still a controversial subject within the "double layer community", the very high levels of turbulence that are observed in many experiments (Leung et al., 1981; Hollenstein and Guyot, 1983) strongly suggest an important role of wave-particle scattering and of the ponderomotive force connected with the large gradient of the high frequency fluctuation intensity.

The potential step roughly separates the regions of high and low frequency turbulence. Turbulence with frequencies near the electron plasma frequency appears in a narrow region on the high potential, side as expected from plasma wave excitation by an accelerated electron beam which is subsequently flattened by the wave-particle interaction. Current (or ion beam) driven ion acoustic waves have substantial fluctuation levels on the low potential side of the potential step. For large drifts, wave frequencies corresponding to the Buneman instability are also observed in this region.

Clearly, if effective mean free paths for wave-particle scattering become small compared to the scale lengths of the potential structure, the purely ballistic description which distinguishes between free and trapped particles and retains a perfect memory of global boundary conditions should be abandoned in favor of a diffusive theory which relates fluxes to local gradients. Ballistic effects, however, may still be important for limited regions in phase space because of the usually strong velocity (speed and angle) dependence of turbulent scattering and the spatial or temporal variation of turbulence levels (Dum, 1981). In turbulent heating experiments, turbulence levels are fairly uniform along the axis for most experimental conditions. Potential drops and drift velocities are then clearly related by anomalous resistivity, due to essentially the same waves as observed in double layer experiments. A tail of runaway electrons, for which ballistic effects and thus boundary conditions are important, appears nevertheless. Plasma waves are observed, and X-ray production when these electrons strike the anode or the walls indicates energies sometimes even in excess of the applied potential difference. This flux naturally has the strongest effect on axial electron heat loss. The contribution to the total electric current may be small. The "problem of excess local power deposition" with anomalous resistivity, usually mentioned by double layer advocates for potential drops e.g. along auroral field lines, obviously is nonexistent in these experiments. Just the contrary, heat flux to the anode and not the potential drop that can be achieved, is a major limiting factor.

The boundary value problems discussed here will usually require a full kinetic treatment. Heat flux is then automatically included, because it is simply related to the third (and lower) moments of the particle distribution functions. Separate consideration of heat flux B.C. is required in a fluid model. The effect of electric fields on heat flux is also included in a kinetic treatment. Heat flux inhibition by double layers has been studied experimentally (Chan et al., 1983). Trapping of hot electrons on the high potential side is just one effect that may be important in a thermal sheath. Intense turbulence and related "thermalization" of transiting and reflected particle populations is also observed.

Electron heat flux is also crucial in the dynamics of ablation and implosion of laser heated targets (Zimmermann and Kruer, 1975; Kruer 1979). Coulomb collisions can be important, at least in the dense regions inside the heated corona. As gradient scales become comparable to mean free paths, however, heat flux is no longer dependent solely on local plasma parameters and their gradients but directly on conditions over an extended region that may include the boundaries and the heat source of these tiny systems. This nonlocal behavior becomes especially significant for suprathermal electrons produced by laser absorption, because of their much larger mean free paths. Heat flux limitation by Coulomb collisions, which becomes nonlocal as the ratio of the mean free path to scale length increases much above $\lambda/L \simeq 10^{-2}$, has been demonstrated recently in a number of simulations using either a spherical harmonics expansion for the solution of the Fokker-Planck equation (Bell et al., 1981; Matte and Virmont, 1982) or Monte Carlo methods (Khan and Rognlien, 1981; Mason, 1981a). Instabilities and self-generated magnetic fields are likely also important in heat flux limitations depending on experimental conditions (Zimmermann and Kruer, 1975). Just as for double layers, the debate on this subject is not closed (Kruer,1979).

So far only quasi-stationary states of a current carrying plasma were considered. For not very different B.C., however, impulsive phenomena may occur, without any approach to a steady state. If the amplitude of the applied voltage pulse is raised, double layers may move away from the cathode, dissolve at the anode and then reform again near the cathode (Leung et al., 1980; Iizuka et al., 1982). The period of the corresponding large spikes in the current and other quantities is related to the ion transit time through the system. A very similar behavior was seen in early particle simulations of a low pressure cesium thermionic converter, (Burger, 1965). The model assumes emission of both electrons and ions with half-Maxwellian (v>0) velocity distributions and densities $n_i \gtrsim n_e$ towards a positively biased cold plate. The plates are absorbing, i.e.all particles that reach either plate are collected and lost for the system. The results (Figure 2) demonstrate a subtile dependence on B.C. for particles and electric fields near emitting electrons. Oscillations occur because ions respond only within their transit time to much faster electron rearrangements. From this basic mechanism, Silevitch (1981) inferred a negative dynamic resistance of double layers. His effort was to explain flickering auroras. Moving and recurring double layer formation for other B.C. and varying system length has been studied in a Vlasov simulation by Singh (1980) and Singh and Schunk (1982a).

Again the transit time between the two emitting electrodes has a decisive effect on the dynamics. Relaxation oscillations of the current and associated turbulence levels are also possible for the more uniform electric fields related to anomalous resistivity (Dum, 1981).

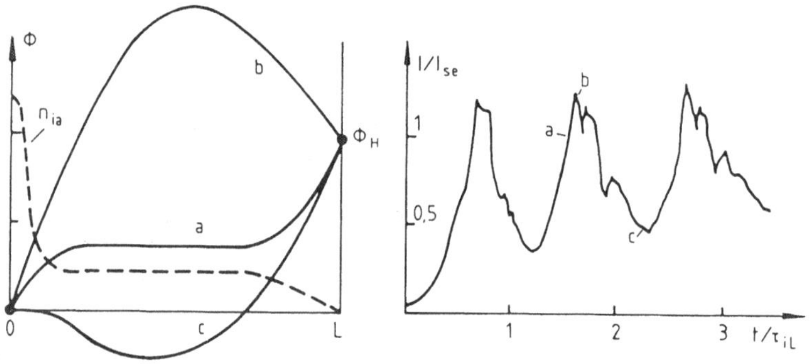

Figure 2 Fluctuating potential distribution and current in a plasma diode (simulation, Burger, 1965) with emission from (Half-) Maxwellian electron and ion sources at x=0, towards an anode at potential $Φ_H$. Both electrodes absorb exiting particles. The period of current fluctuations about the saturation current I_{se} corresponds to the ion transit time $τ_{iL}$. Electron rearrangement occurs on the much shorter time scale of electron transit through the system and leads to the potential distribution (b) consistent with the steady electron flow and the fixed ion distribution n_{ia}. An undershoot (c) and current disruption results as ions try to re-establish potential (a) which is also consistent with steady ion flow.

The external electric circuit is an important B.C. for all these impulsive phenomena, as may be demonstrated directly by simulations. External circuit elements such as resistors, inductances or capacitors allow for a feedback interaction between a battery and the plasma that is between the two extremes of constant applied voltage and constant current. It is also possible to use injection from plasma reservoirs with drifting Maxwellians (Goertz and Joyce, 1975; Joyce and Hubbard, 1978; Singh 1980) or any combination of these B.C. with an electric circuit (Sato and Okuda, 1981; Smith, 1982). It must be noted, however, that some of these B.C. create singularities and prescribe potential structures (Borovsky and Joyce, 1983) with very little room left for the plasma to respond. This is especially true for small system size and "laminar" double layer simulations (see also Sec.3.2). Coupling of an external electric circuit to a homogeneous plasma or a simulation with periodic B.C. is more straightforward (Sec.4) and may also produce interesting effects, such as short repeated bursts of anomalous resistivity (Mondelli and Ott, 1974).

2.2 Electrodynamic Phenomena: Shocks, Reconnection and Tearing

An experimental study by Stenzel et al. (1983) on repeated double layer formation and associated current disruption in magnetic reconnection is particularly interesting in the context of Sec.2.1 and the electrodynamic phenomena to be discussed now (Figure 3). The

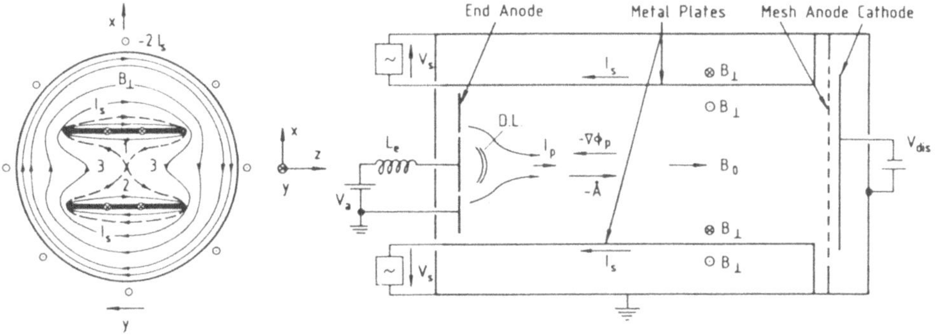

Figure 3 Schematic circuits for reconnection and current disruption (after Stenzel et al., 1982). Magnetic flux transfer involves the private fluxes (cells 1,2) encircling the plates with pulsed current I_S and their common flux (cell 3). Shown is the X-type configuration before the flow of the inductively driven plasma current I_P. Disruption of this current is triggered by applying a positive bias to the central anode and involves lateral deflection of the plasma current by the double layer, as well as inductive electric fields due to plasma and external electric circuit (L_e).

process is triggered if the current density in the center of the current sheet is raised by applying a potential difference between a large anode and its separated central section. During current disruption connected with double layer formation near this anode, the potential difference rises substantially above the applied voltage, as expected from the inductance of the external electric circuit. Alfvén and Carlquist (1967) discussed this familiar effect in connection with current disruption in solar flares. The most interesting point, although somewhat distressing for simulation attempts, is, however, that this experiment shows double layer formation and current disruption as an inherently three-dimensional dynamic phenomenon that is also affected by the magnetic field topology and the inductive electric field due to magnetic flux changes inside the plasma. The total current, being inductively driven by the electric circuit which is also responsible for the pulsed X-type vacuum magnetic field, actually stays nearly constant during the disruption. The current is only diverted from the center anode. Because ions are essentially unmagnetized, direct expulsion of ions by the high central electric potential is possible in transverse directions. The cross section of the region of large potential and related significant density depression is somewhat smaller than the center anode plate. No current disruptions are observed for a floating large anode plate, even if the total (electron) current that is now collected by the center anode alone is doubled. The transverse magnetic field configuration for this concentrated current corresponds to an O type (magnetic island) topology. The different behavior may arise from the transverse jetting (tangential to the current sheet) of plasma in a reconnecting X-type magnetic field topology as opposed to pinching in an O-type topology (Syrovatskii, 1981).

Spontaneous redistribution of the axial electric field E_Z into a double layer structure was also observed for the simplest O-type topology, an azimuthal magnetic field $B_\theta(r)$ produced by an axial current where, however, a strong external axial magnetic field $B_Z \gg B_\theta$ prevents pinching (Kalinin et al., 1971). It occurred in the second phase of turbulent heating and apparently resulted from a macroscopic instability of the turbulently resistive column, and not from a sheath effect as in laminar double layer theory. The growth rate is related to the transverse gradient of anomalous resistivity and is inversely proportional to the axial magnetic field. Rapid redistribution of the current density with a significant current increase near the axis, generation of azimuthal electric field $E_\theta \gg E_Z$, and an almost complete balance of turbulent heating by energy transport, mostly to the anode, was observed. Examination of X-rays from the anode shows that only a small fraction of this energy is transported by freely accelerated electrons, although the intensities of X-ray emission and especially of radio emission increase considerably during the disruption, compared to the earlier turbulent heating phase. The peak plasma density decreases markedly in the second phase.

These two experiments demonstrate that not every structure in an electric field component that resembles the distribution of a double layer should be interpretated uncritically in terms of the usual laminar and one-dimensional theories. Poisson's equation is not questioned, of course, but the electric field may not only be multidimensional but may have a substantial inductive component E^t in addition to the electrostatic field,

$$\underline{E} = -\nabla\phi + \underline{E}^t; \quad \nabla \times \underline{E}^t = -\frac{1}{c}\frac{\partial \underline{B}}{\partial t}; \quad E^t = -\frac{1}{c}\frac{\partial \underline{A}}{\partial t}; \ \nabla \cdot \underline{E}^t = 0 \qquad (1)$$

where $\underline{A}$ is the vector potential $\underline{B}=\nabla\times\underline{A}$ (in the Coulomb gauge $\nabla\cdot\underline{A}=0$). Furthermore, as just seen, particle motion may be non-laminar, i.e. subject to scattering by waves.

As also demonstrated by these experiments, B.C. for the magnetic field and B.C. related to the external electric circuit for the currents are important for current disruption and reconnection. Syrovatskii (1981) points out that it may be more appropriate (certainly for the experiments described by him and Figure 3) to treat reconnection as an initial-boundary value problem for current sheet formation, rather than as stationary reconnection models suggest, impose antiparallel magnetic fields and inward plasma flow as B.C. The magnetic fields in the vicinity of the current sheet and the sheet dimensions are then found from a solution of this problem and the superimposed external field $\nabla\times\underline{B}_0=0$. A vacuum field $\underline{B}_0$ with X-point separatix favors formation of an elongated current sheet. In contrast to the stationary reconnection models, a highly dynamic behavior may result, as evidenced by experiments (Baum and Bratenahl, 1980; Syrovatskii, 1981) and space observations (Haerendel and Paschmann, 1982). Dependence of the merging rate (plasma influx velocity) on microscopic dissipation, which is itself nonstationary, is strongly affected by the B.C. for plasma flow and magnetic fields. Compared to laboratory plasmas with clearly defined electric current paths and wall boundaries (see Figure 3), their proper definition in space plasmas is more difficult (see Sec.3).

Usually, computational constraints will allow to treat only part of the entire system. The global aspect of reconnection, emphasized by Baum and Bratenahl (1980) is the root of artificial "wall" problems when a small volume about the magnetic separator (neutral line) is excised for simulation purposes. The important questions on how the magnetic energy was built up by the global system, the nature of the trigger for the release of this energy by reconnection and of the feedback from this release to the global system will necessarily remain mostly unanswered by such simulations for small excised systems. Electrodynamic coupling to the rest of the system engaged in the flux transfer should also be simulated somehow. These problems are not faced by theoreticians who simply declare a steady state for the excised system. The inductive electric field $\underline{E}^t$ in (1), related to the release of magnetic energy from a finite reservoir and to changes of the magnetic flux through some surface (cf. Figure 3)

$$\Phi_m = \frac{\partial}{\partial t} \int \underline{B}\cdot d\underline{S} = - c \int \underline{E}^t\cdot d\underline{l} = \frac{\partial}{\partial t} \int \underline{A}\cdot d\underline{l} \qquad (2)$$

in particular, is then not treated self-consistently.

B.C. in simulations that seek to treat stationary reconnection models as an initial-boundary value problem are in fact quite arbitrary. In <u>driven</u> reconnection models, the reconnection rate and the location of the neutral line is determined by the imposed influx profile. The size of the diffusion region adjusts itself to the imposed influx in a magnetohydrodynamic (MHD) model with constant resistivity and incompressible flow by Biskamp (1982), becoming a thin layer of increasing length (Sweet-Parker layer) as the inflow velocity increases or the resistivity η decreases. The initial equilibrium configuration is a plane current sheet $\underline{B}=(0,B_y,0)$ with $B_y(x)$ reversing at $x=0$. A peaked influx profile, $u_x=\pm u[1+\varepsilon(t)\cos(\pi\, y/L_y)]$ is used only during an initial period to set up the X-type profile with neutral line $(x=0,y=0)$ and concave magnetic field lines. It then becomes homogeneous, $\varepsilon\to 0$. Although the assumptions of the classical

Petcheck model are followed most closely, a considerably different and more complex behavior is revealed by the simulation (Figure 4). Upstream (ideal) flow region and diffusion region, in particular, are strongly and discontinously coupled. This leads to the η dependence of the diffusion region and an upstream flux pile up for too small η=const. By contrast, in the MHD model of Sato and Hayashi (1979) anomalous resistivity adjusts itself to this influx. The evolution of the outflux or the neutral line electric field and of the neutral line temperature are strongly dependent on the influx but not on details of the anomalous resistivity law,

$$\underline{E} + (\underline{u} / c) \times \underline{B} = \eta j, \tag{3}$$

e.g. the parameters α and j_C and in $\eta=\alpha(j-j_C)^2$ for $|j|>j_C$ and zero otherwise, for current density j.

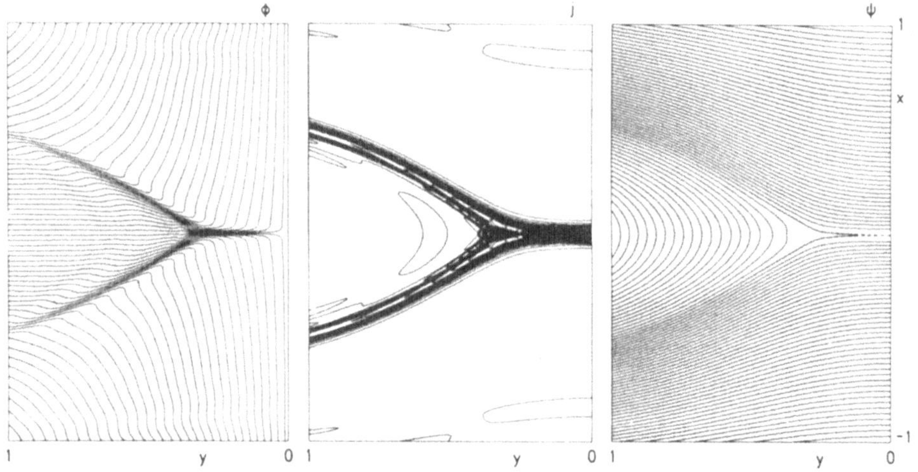

Figure 4 Stream lines (Φ) current density (j_Z) and magnetic field lines (Ψ) for reconnection driven by plasma inflow at the x boundaries (courtesy, D. Biskamp, 1984). Only half of the system is shown. In contrast to the predictions following Petschek, the central diffusion region is strongly coupled to the upstream ideal flow region, increasing in particular its length with decreasing resistivity or increasing influx. Outflow in the y direction is also reduced by a "plug" of $j_Z B_X<0$, just outside the diffusion region (x=0).

In the MHD models of Ugai (1982) plasma is "sucked" in by a sudden increase in anomalous resistivity that is highly localized near the neutral line, $\eta=S(t)\exp[-(r/R_K)^3]$, $r=(x^2+y^2)^{1/2}$. The time dependence of $0<S(t)<1$ may be chosen e.g. to correspond to exponential growth or decay depending on wether the relative electron-ion drift velocity $-j/ne$ is above or below a critical velocity. The size of the diffusion region increases and the current density decreases with increasing resistivity η. The electric field along the neutral line (flux transfer rate) thus depends more weakly on η. The ultimate cause for reconnection is the ad hoc localized resisitivity increase. No influx at $x=\pm L_x$ needs to be imposed in addition, these boundaries may even be closed to the flow $u_x=0$. (Plasma is still sucked in through the y boundaries, however.) All models, however, require open boundaries at $y=\pm L_y$ through which the plasma may freely enter or exit, in order that nonlinear plasma motion can set up a Petchek type configuration with two shock pairs attached to a diffusion region in which dissipation dominates over convection.

Open boundaries may be defined by the condition that derivatives normal to the boundary surface vanish

$$\frac{\partial \rho}{\partial n} = \frac{\partial P}{\partial n} = \frac{\partial \underline{u}}{\partial n} = \frac{\partial \underline{B}_t}{\partial n} = 0 \tag{4}$$

for the mass density, the pressure, the flow velocity, and the tangential magnetic field component (Ugai and Tsuda, 1977). In order to reduce wave reflection from such a boundary, Sato and Hayashi (1979) include an buffer zone in which a friction term $\nu(S-S_o)$ is added to the time derivative for any variable S with intial value S_o. If y boundaries are closed or periodic, a closed convection pattern of the tearing type is set up (Ugai, 1982). Standard B.C. for tearing mode calculations assume periodicity in the y direction, with alternating X-and O-type neutral points, and conducting walls at which the tangential electric field (E_z) and the normal flow velocity (u_x) vanish are imposed at $x=\pm L_x$. Still other combinations of resistivity laws and ad hoc boundary conditions for reconnection and tearing are possible, but it is not clear exactly which additional questions are answered by such models.

Not only the nonlinear behavior is affected by the B.C. which describe sources and sinks for energy flow etc., but also the linear mode structure, $A(x)\exp(ik_y y)$. For large wavelengths ($k_y L_x \ll 1$), details of the x boundaries naturally enter, which makes consideration of their artificial effects important for astrophysical applications such as solar flare energy releases. Steinolfson and Van Hoven (1983) compute the linear resistive tearing mode structure for a force free initial magnetic field($\underline{j}\times\underline{B}=0$)

$$\underline{B}_o(x) = B_o[0,\tanh(x/a),\text{sech}(x/a)], \tag{5}$$

using either conducting walls at $x=\pm L_x$ or extrapolations to the gridpoints just beyond the computational domain, $x=\pm L_x$, by an ideal asymptotic solution corresponding to free space B.C. The critical surface for tearing modes in the case of (5) is defined by the condition

$$k_{\parallel}(x) = (k_y B_{oy} + k_z B_{oz})/B = 0. \tag{6}$$

Note that plane current sheet models frequently use an initial magnetic field that correspond only to the y component of the force free field. The axial current and equilibrium pressure profiles are then

$$j_z(x) = (c/4\pi a) B_0 \mathrm{sech}^2(x/a) \; ; \; P(x) = (B_0^2/8\pi) \mathrm{sech}^2(x/a) \quad , \qquad (7)$$

as is easily verified from the properties of the hyperbolic functions (e.g. Sato and Hayashi, 1979).

It is to be expected that the box length (or periodicity length) L_y in numerical simulations also affects the evolution of long wavelength tearing modes. The fact that, depending on box length L_y, single or multiple tearing modes are observed must be taken into account in the physical interpretation of these results, especially for space plasmas (Sec.3). In laboratory devices such as tokamaks mode numbers are actually restricted by the requirement of poloidal (y) and toroidal (z) periodicity. These devices also have a (non-force free) magnetic field with a substantial component $B_z >> |B_y|$ in the direction of current flow, which leads to incompressible flow.

The resistivity used in the MHD models of reconnection and tearing may correspond to classical resistivity, but at least in aplications to space plasmas it must be interpretated as anomalous, being caused by some instability. Collisionless reconnection and tearing is also possible, and has been simulated by particle codes. The particle simulation model of Leboeuf et al. (1982) resembles the Stenzel et al. (1983) reconnection experiment in that the X-type magnetic field configuration is induced by two external sheets in which the current rises in time. No dependence in the current direction is included, however, in this 2 D model and the use of periodic B. C. implies interaction of the system with its periodic continuations (2 D refers here to two space dimensions and generally three velocity dimensions). The very small simulation box (64×32 Debye lengths) makes electrostatic interactions much more important than in an actual physical system. Still, another price was paid for this self-consistent simulation: the ion-electron mass ratio is only 10, which is likely to modify the microscopic dynamics. Similarities, but also interesting differences, arise between MHD models, the kinetic model, and the experiments.

Lin (1978) uses such a code with the standard B.C. for tearing modes, i.e. conducting walls in x and periodicity in y. The strong external magnetic field is slightly tilted in the y-z plane, $B_{oy} << B_{oz}$, in order to allow for kinetic effects due to electron motion along $\underline{B}_o$, in a 2D (x,y) calculation. The coupled drift-shear alfvén wave $k_\parallel << k_\perp$ is excited by these effects and tearing results from current flow.

A similar particle code has been used by Katanuma and Kamimura (1980) to study collisionless tearing modes, with the standard B.C. mentioned above. The strong axial external magnetic field $B_{oz} >> |B_y|$ is now perpendicular to the plane of computation. By turning off the electrostatic (space charge) fields, which for the given simulation box size and other parameters must be large, it could be shown that these fields enhance tearing mode growth. Burn (1970) reduced dominance of electrostatic forces over magnetic forces which are proportional to v/c, by introducing a low "bogus" speed of light c/v=O(1). This early simulation uses free space electric B.C. in x and periodicity in y (see Sec.4).

Space charge effects are not included in the MHD models and in simulation models that consider only the ion particle dynamics with electrons as a neutralizing background (Dickman et al., 1969; Teresawa, 1981; Swift 1983). The electric field in the latter models is purely inductive $E_z=-(1/c)\partial A_z/\partial t$, see (1). However, using a hybrid description in which the electric field satisfies the inertialess equation of motion for fluid electrons of pressure p_e, cf. (3)

$$\underline{E} + (\underline{u}_i/c)\times \underline{B} = - (j/n_e)\nabla p_e + \eta \underline{j} + (\underline{j}/nec)\times\underline{B} \quad (8)$$

along with quasi-neutrality ($\sigma/ne \ll 1$)

$$n_e = n_i = n \; ; \qquad \underline{u}_e = \underline{u}_i - (j/ne) \quad (9,ab)$$

modifies the collisionless "ion tearing" mode by adding electric fields (E_x,E_z) and a Hall current in the plane of computation (x,y), as well as a magnetic field B_z (Sonnerup, 1979).

The lower hybrid drift instability of a current sheet is often invoked for providing the anomalous resistivity for tearing or reconnections. 2 D simulations, unfortunately, neatly separate these two modes and thus eliminate the dynamics that would result from their interaction. The same profiles and B.C. in x may be used, but for the lower hybrid drift mode it is essential to include the space dependence along the current direction (z), generally with periodic B.C. In a neutral sheet with magnetic field (B_y) reversal, the predominantly electrostatic lower hybrid instability is excited only in the regions of steep density gradients, away from the neutral line. An electromagnetic mode in the magnetic reversal region develops at a later stage. The resulting anomalous resistivity in the reversal region is small (Tanaka and Sato, 1981a; Winske, 1981). B.C. are important in these simulations. The electromagnetic mode prefers long wavelengths and thus is affected by too small box length L_z. Penetration of the lower hybrid mode will be affected by the profile in x and the box width L_x. But perhaps even more important is to consider B.C., for external sources that can maintain the instability beyond a short transient stage. Tanaka and Sato (1981b) included an external electric field $E_z(x)$ which drives particles towards the neutral line. The result is recurrent bursting of the instability and thus effectively larger anomalous resisitivity (Figure 5). The period of excitation appears to be related to the particle transit time across the neutral sheet. Their external electric field E_z may be identified as the inductive electric field in a reconnecting plasma (Figure 3), but it then clearly requires consideration of the global

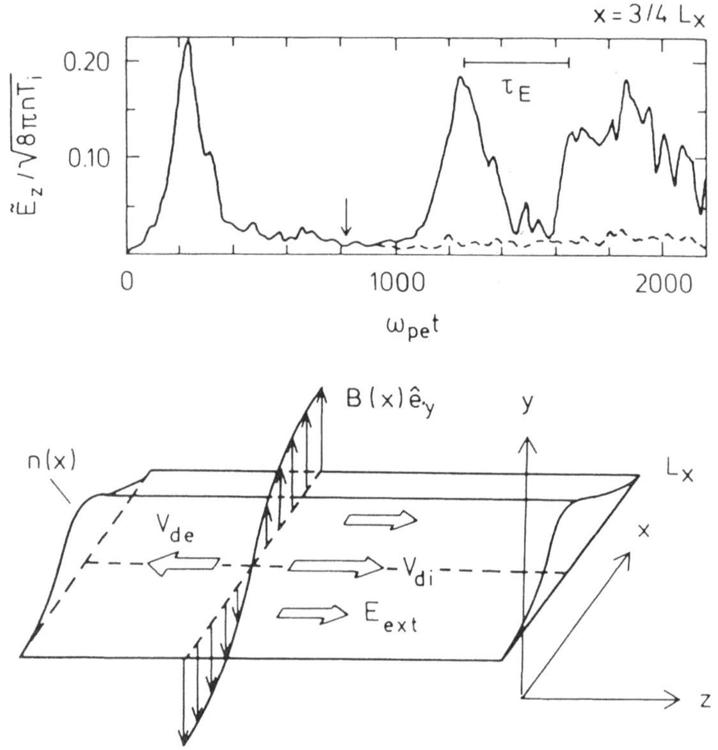

Figure 5 Multiple Excitation of lower-hybrid drift waves in the neutral sheet (particle simulation by Tanaka and Sato, 1981). Shown is the (initial) configuration with magnetic field reversal, and re-excitation of the instability after an external electric field is turned on at $\omega_{pe}t=800$. The period of re-excitation corresponds to the $E_{ext}xB$ transit time across the sheet.

system and its B.C.

Skin current penetration in a z-pinch or in a turbulent heating device (axial current), and shocks driven into a theta pinch (azimuthal current) by a magnetic piston, i.e. a rapidly rising magnetic field at the plasma boundary, are outstanding examples for strongly coupled electric, magnetic, and plasma processes in a simple geometry, a cylinder for example. The coupling between macroscopic and microscopic dynamics is studied more easily than for the complex reconnection geometries. Again, the waveform of the total plasma current $I_p(t)$ is determined by the electrodynamic coupling of the plasma to the external electric circuit and has a decisive effect, not only on the macroscopic plasma dynamics, but also on the dynamics of the microscopic processes responsible for anomalous transport (Dum, 1981). From $\nabla\times\underline{B}=(4\pi/c)\underline{j}$, and assuming a cylindrical cross section of radius R_p, we obtain the magnetic B.C.

$$B_\theta(R\ ,t) = (2/cR_\theta)\ I_{pz}(t)\ ;$$
$$B_z(R_p,t) - B_z(0) = -\ (4\pi/cR_p)\ I_{p\theta}(t), \qquad (10a,b)$$

respectively for the z and theta pinch of length $L_z >> R_p$. Both current systems and corresponding magnetic fields are employed in the more general screw pinch configuration (Sgro & Nielson, 1976).

The balance between rising plasma current $I_p(t)$ and diffusion due to anomalous resistivity, anomalous thermoelectric effects, and anomalous viscosity determines the magnetic and current density profile inside the plasma. Anomalous heating and anomalous heat conduction determine the temperature profile. Convection also contributes in the case of pinching (Dum, 1978). Considering then these space and time varying profiles of current density etc. as background for the growth of the microscopic instabilities, closes the loop. The macroscopic dynamics affects wave growth, often by not leaving enough time for the waves to grow to (nonlinear) saturation or by causing repeated growth and quenching. A stationary turbulence level would be exceptional. Time scales may even be short enough for some instabilities, e.g. the lower hybrid drift instability, to become important only at later (post-implosion) phases.

These details of microscopic dynamics could be found in particle simulations with appropriate, generally time dependent, B.C. for a slice of the plasma volume, but they may not matter too much for macroscopic evolution, if the required dissipation is provided on the average by a rapid increase in anomalous transport above some instability threshold. A marginal stability approach which, however, accounts for the interrelationship among the various anomalous transport processes and with wave growth, leads then to a simplified, but usually adequate code for the macroscopic dynamics with anomalous transport (Dum, 1978, 1981, 1983a). A hybrid description with fluid electrons and kinetic ions (see Equs. 8-9) may be used with advantage for macroscopic scales comparable to ion scales, such as the Larmor radius. This is the usual situation in pinch devices (Chodura, 1975; Chodura et al., 1975; Sgro and Nielson, 1976; Chodura et al., 1977). In addition to B.C. for the magnetic field and the ion particles, B.C. for electron heat conduction must be specified unless a simple adiabatic law is used for the electron pressure in (8). Symmetry relations define B.C. on the axis of a cylindrical device, with important consequences e.g. for ion reflection (conservation of angular momentum). The plasma wall boundary problem (see above) is generally treated in a highly simplified manner, with the primary aim of avoiding numerical difficulties which arise for the near vacuum conditions (except for particle emission and reflection by the wall) created by pinching (Hewett, 1980; Horned, 1982). A free plasma boundary may be defined by some cutoff density, and owing to the great uncertainty about electron heat flux B.C., the initial electron temperature often serves as a B.C. (Sgro and Nielson, 1976). Two dimensional particle simulations (electrons and ions) of skin current penetration have been carried out by Lin and Dawson (1978). The choice of the computation plane in these two-dimensional simulations determines, of course, which microscopic processes survive in the restricted geometry. The system size is, however, only 64x32 Debye lengths and the mass ratio is 25. Recent particle code advances allow an easing of these restrictions, as shown by 2 D simulations of the earth's bow shock (Forslund et al., 1983; Quest et al., 1984). For computational convenience, a magnetic piston is still used to produce the shock.

The magnetic fields (10a,b) may be programmed by the external electric circuit, such as to control or even drive the reconnection processes discussed previously. This is not only of great current importance for the fusion community, but also provides for interesting simulation models with self-consistent B.C. A reversed magnetic field configuration may be created by first establishing a bias field $B_Z(0)<0$, followed by quick field reversal, $B_Z(R_P)>0$, and plasma compression. Small scale reconnection (island formation and coalescence) along the central neutral layer starts during the early implosion phase, as also shown by hybrid 2D(r,z) simulations (Hewett and Seyler, 1981). The Hall term in (8) creates an azimuthal magnetic field B_θ. If it is turned off, reconnection is delayed considerably. After the main implosion phase large scale reconnection starts from the open ends of the theta pinch and then the configuration with closed field lines contracts axially towards an elongated equilibrium toroid. Plasma on open field lines drifts out the ends. Similarities to the Sato-Hayashi-model (1979) of driven reconnection exist in a MHD simulation model of this process (Milroy and Brackbill, 1982), including a weak dependence on the assumed anomalous resistivity law. Important differences include, however, cylindrical geometry, limited reverse bias flux and more complicated B.C. Reconnection is aided by end magnetic mirrors and delayed by end cusp fields. All these magnetic fields can be programmed to produce various effects. Although this leads beyond the scope of this article, reversed field pinches should be briefly mentioned. They have azimuthal (poloidal) fields B_θ comparable to the average axial (toroidal) field, $B_\theta > \langle Bz \rangle$, and Bz reversal may even occur spontaneously, as observed already in early z pinch experiments. The confinement properties excite the fusion community and astrophysical interest stems from the fact that nearly force free magnetic fields, cf. (5), are created and maintained, apparently by reconnection. However, B.C. for a plasma torus with enclosing, conducting walls (review by Bodin and Newton, 1980; Caramana et al., 1983) and a coronal loop with ends in the photosphere and no walls (e.g. Priest, 1983) are obviously quite different and likely affect phenomena such as field reversal.

3. SIMULATION MODELS AND BOUNDARY CONDITIONS SPECIFIC OF SPACE PLASMAS

We have discussed laboratory experiments in some detail, not only because B.C. are more obvious and more easily explored, but also for pointing to the benefits and perils of using these experiments and related simulations as an explanation for space phenomena. Some emphasis has been put on double layers and anomalous resistivity, not only because microscopic processes for sustaining electric fields are of fundamental improtance in space plasma physics as well, but also because double layers which, as just shown, are an outstanding example for a very strong dependence on B.C., have recently been invoked quite uncritically for almost everything under the sun and beyond. Similar comments apply to reconnection. Appropriate B.C. are most certainly quite different in these cases as space plasmas have no electrodes or external electric circuits, nor enclosing walls. Only bodies immersed in the plasma, like the sun, planets, comets or spacecrafts, provide reasonably well defined boundary surfaces. Because computational domains are finite, additional boundary surfaces will generally have to be defined for computational purposes. Analogs such as grids and external electric circuits (e.g. Baum und Bratenahl, 1980; Spicer, 1982) in laboratory devices (Figs.1,3) may be very helpful in constructing appropriate B.C. on these <u>artificial</u> surfaces, but they are only analogs which cannot deal with <u>dynamically distributed</u> and often <u>nonlinear</u> effects in a plasma.

3.1 Solar and Polar Winds (Exospheric Theory)

It is appropriate to begin considerations of B.C. in space plasmas with the sun as the major source of the interplanetary plasma. A detailed discussion of its many fascinating phenomena (e.g. Priest, 1983: Spicer 1982 and references therein) can be circumvented by assuming that the properties of the solar wind are known on some surface in the outer corona (Figure 6). Approximating the rapid change of the expanding solar wind from a collision dominated to a (nearly) collisionless regime by a sharply defined <u>exobase</u> (baropause), exospheric theory describes the further spatial (but stationary) evolution of the distribution functions, using Liouville's theorem, conservation of energy and conservation of magnetic moment

$$W = m_j(v_\parallel^2/2) + V = \text{const.} \; ; \; \mu = m_j v_\perp^2/2B = \text{const.} \qquad \text{(11a,b)}$$

Particle motion along the magnetic field $\underline{B}$ is determined by the effective potential

$$V(\underline{x}) = e_j\Phi(\underline{x}) + \mu\, B(x) - GMm_j/r \;, \qquad \text{(11c)}$$

where the last term is the gravitational potential at heliocentric distance r. Some particles can be reflected near maxima of V or trapped near its minima. Distribution functions for outgoing and reflected particles are specified by B.C. for $v_\parallel > 0$, at the exobase. The distributions of incoming, $v_\parallel < 0$, and of trapped particles must

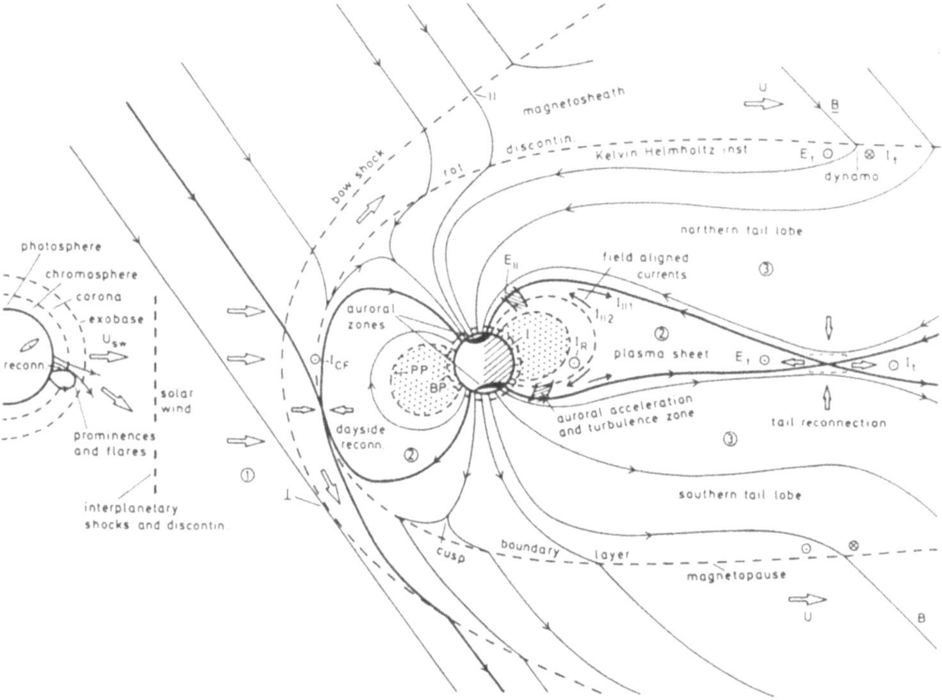

Figure 6 Coupled solar wind-magnetosphere-ionosphere system (obviously schematic and not to scale). In addition to global MHD models of electrodynamic coupling, local simulations of greatly varying detail have been carried out for the processes indicated in this map. Linking of these elements within the global system remains one of the most challenging tasks. The plasma may be considered collisionless (but not laminar!) between the solar exosphere and the ionospheric baropause (BP). Magnetic flux transfer involves the cells 1-3 in day-side and tail reconnection. Primary dynamos for magnetospheric convection and currents are located at the magnetopause. Secondary dynamos in the plasma sheet play an important role in the dynamics of substorms and aurorae. Mangetosphere-Ionosphere coupling drives an ionospheric current system and involves field aligned currents and particle acceleration in auroral flux tubes.

also be specified. Particle distribution functions for outgoing or incoming particles at different locations along a magnetic field line are related by Lioville's theorem.

$$f_{\pm}(\underline{x}, v_{\perp}, v_{\parallel} \gtrless 0) = f_{\pm}(\underline{x}', v_{\perp}', v_{\parallel}' \gtrless 0) \tag{11d}$$

provided the conservation laws (11a-c), which relate $\underline{v}$ and $\underline{v}'$, allow an orbit with no turning point $v_{\parallel}=0$ between $\underline{x}$ and $\underline{x}'$. Reflected particles and trapped particles with two turning points satisfy the symmetry relation

$$f_{t}(\underline{x}, v_{\perp}, v_{\parallel} > 0) = f_{t}(\underline{x}, v_{\perp}, v_{\parallel} < 0) \ . \tag{11e}$$

Since expansion of the solar wind is essentially into a vacuum and the plasma is assumed to be collisionless, incoming particles reaching the exobase are neglected. In order to avoid discontinuities in the distribution functions, it is assumed, however, that weak populations of trapped particles and of reflected particles with energies slightly below the escape energy are established by residual collisions (Jockers, 1970). An analogous procedure may be followed for the polar wind on high latitude open magnetic field lines of planetary exospheres (Lemaire and Scherer, 1971). The electric potential $\Phi(x)$ in the solar or polar wind is determined self-consistently from the conditions of quasi-neutrality, $n_e=n_i$, and zero net current.The total potential drop is in the kV and volt range respectively and depends mainly on the assumed exobase electron temperature.

The role of B.C. and of turbulence on potential structures and collisionless flow was already discussed above for laboratory and simulation experiments. The much larger uncertainty of B.C. in the case of solar and planetary exospheres has been fully exploited in attempts to produce the desired results from laminar theory. The fact that boundaries and points of observation are separated by scaled distances L/λ_D many orders of magnitude larger than the total size of largest laboratory or simulation experiments requires at least considerably more courage in assuming that particle motion is laminar, i.e. unaffected by collisional or turbulent scattering over the entire path, and that fields are static. Nevertheless, these theories correctly describe boundary effects that in the absence of other mechanisms would become fully effective. It remains to investigate these additional mechanisms.

As discussed for laboratory experiments and related simulations e.g. of laser heated plasmas, it is not necessary that effective collision frequencies are high enough to determine the entire distribution function by a local diffusive process, in order to have a profound effect. A cumulative effect of scattering and of convection in the global fields over appreciable distances from the point of observation is sufficient and, as demonstrated in some detail for the interaction of observed heat flux driven ion acoustic like turbulence with observed solar wind electron distributions, also much more likely (Dum, 1983b). Scattering of energetic (suprathermal) electrons is weakest in this case and, indeed, only they show some remnants of exospheric behavior, e.g. a strong focussing by the decreasing magnetic field (μ=const!). The macroscopic electric field acting on these particles may, however, be considerably enhanced by anomalous momentum transfer between ions and low energy electrons. Anisotropies are in general much less extreme, and for protons often also of opposite sign to that predicted by exosheric theory for scatter-free flow along a strongly diverging magnetic field. The simultaneous observation of turbulence throughout the interplanetary solar wind makes wave-particle scattering the most plausible explanation for these discrepancies. Transport in this case may still be nonlocal, but the direct dependence on distant microscopic B.C. implied by exospheric theory may be eliminated or at least considerably weakened.

The consideration of plasma flow, heat flow and of instabilities in the diverging geometry of the collisionless solar wind should provide for many interesting simulation models. So far there are any number of classical fluid calculations which use ad hoc heat flux limitation, similar to fluid models for laser heated plasmas (Zimmerman and Kruer, 1975), in order to avoid a catastrophic breakdown for plasmas that actually are nearly collisionless (Hundhausen, 1972). A technically interesting point is that integration of steady state fluid equations starting from asymptotic B.C. at large heliocentric distances, rather than from coromal B.C., is more efficient (Cuperman et al., 1975; Nerney and Barnes, 1977).

3.2 Plasma Flow, Field Aligned Currents and Electric Fields in Auroral Flux tubes

With respect to the effects of B.C. and of turbulence on plasma flow and electric fields along auroral field lines, the situation is still somewhat similar to that described by Hundhausen (1972, p. 81) for solar wind expansion: "It would appear that unique conclusions regarding the physical processes that have dominant effects on coronal expansion can be deduced only by true believers, usually with a parental relationship to one of the competing ideas or models. It should not, in fact, be surprising that the incorporation of several different physical mechanisms can lead to similar modifications of the basic models. Introduction of each new mechanism or assumption is akin to the inclusion of several free parameters in the system of equations, greatly enhancing the possibility of finding a solution that can reasonably connect the observed state at two positions." The number of positions and the quality of observations is now rapidly increasing, but it is still desirable to connect these by more realistic simulations and to be open minded about mechanisms for electric fields and particle acceleration. In particular, as we have seen, localized electric fields and turbulent scattering can very well coexist (Dum, 1981). Although this is more difficult, theoretical progress based on this more flexible approach appears to be in sight. Admitted, the situation is likely quite different from the solar wind or from laboratory experiments. For one thing, electrostatic ion cyclotron waves, which may people believe to play an important role in the auroral acceleration zones, leave electrons tied to magnetic field lines. Electron plateau formation, in accordance with quasilinear theory, quickly stabilizes these current driven waves in a homogeneous plasma at a negligible level (Okuda et al., 1981; Pritchett et al., 1981). To match theory with observations of very large fluctuation levels, it is therefore imperative to consider not only particle sources and sinks (Dum and Dupree, 1970; Dum, 1981; Okuda and Ashour-Abdalla, 1983), but also macroscopic fields and the question of electrodynamic coupling to the rest of the system (Dum, 1981, 1983a).

The full details of the particle distribution functions at the distant magnetospheric (plasma sheet) and ionospheric (baropause) boundaries and the exact location of these artificial boundaries are directly reflected in the results of laminar, collisionless theories of auroral flux tubes, and hence are a subject of intense debate. Just as for the solar or polar wind, particle distribution functions and the potential distribution along the auroral flux tube in these theories are determined from the B.C., using the conservation laws (11) for adiabatic motion and the condition of quasi-neutrality (review by Kan, 1982). The latter condition for extended potential drops, $L_Z \gg \lambda_D$, replaces Poisson's equation which has to be used for localized potential drops such as double layers. This is, however, not the only difference from the standard double layer models. Drawing a large field aligned upward current (precipitating electrons) requires a sufficiently large electric field to overcome magnetic mirroring of electrons in the converging magnetic field. Other electrons are reflected by this mirror force and return to the magnetospheric source. Still other electrons could by fully trapped in the magnetosphere, with turning points due to reflection by the potential and magnetic mirror. This region of phase space (pitch angles centered about 90^0), however, has to be populated by effects such as wave-particle scattering and residual collisions which are not included in the models. Kan and Lee (1980) exploit the freedom in choosing trapped electron populations for the construction of potential drops of assumed shape and amplitude within laminar theory. Chiu and Schulz (1978) allow for different anisotropies of electrons and ions at the magnetospheric source. Backscattered primary and secondary electrons resulting from the interaction of precipitating electrons with the atmosphere have been included in the ionospheric source, in addition to the cold ambient ionospheric electrons and ions (Figure 6).

Height dependent (100 km<h<400 km) downward and upward electron fluxes, ionization and optical emission for various incident electron energies were computed by Banks et al. (1974). The full pitch angle distribution of electron fluxes was included by Strickland et al. (1976). In the model of Pulliam et al. (1981), secondary electrons and primary electrons that have been backscattered and degraded in energy are reflected again by the potential drop which was assumed to have accelerated the incident primary electrons, as suggested by Evans (1974). About half of the electrons may be recycled in this way between the acceleration region and the atmosphere. Prasad et al. (1983) compute also pitch angle distributions from this model. Electrons which have been backscattered by the conjugate ionosphere with energies and pitch angles appropriate for escaping from the conjugate potential drop according to (11), are added to the incident flux. It is thus not only assumed that the region of field aligned electric fields is well above the collisional atmosphere, but also that electrons suffer no scattering, e.g. in pitch angle, or encounter time varying electric fields on their way between the two conjugate ionospheres. Anyway, it is seen that the particle flux entering the auroral flux tube at some assumed inner boundary, a baropause for example, has a very significant dependence on the distribution of the incoming particle flux, i.e. on the solution of the boundary value problem within the chosen flux tube volume. As emphasized by Chiu and Schulz (1978), latitudinal variations of the ionosphere are also significant. Depletions (troughs) in the ionosphere along with anisotropic distribution functions at the equatorial symmetry plane and inclusion of ionospheric backscatter allowed them to crank up the self-consistent potential drop between a baropause at h=2000 km and the equator to typically 1 kV as compared to a few volts obtained for other B.C. (Lemaire and Scherer, 1973,). B.C. at the chosen outer boundary of an auroral flux tube are by no means obvious, although collisions are not important there. Even if symmetry relation at the equatorial plane are used for closed magnetic field lines, it is still necessary to consider the sources that maintain the distribution functions at this boundary.

It is expected that more detailed observations will restrict the freedom in choosing B.C. for auroral flux tubes. Similar to the solar wind, where the search for a quiet state proved to be unsuccessful (Hundhausen, 1972), dynamic processes which are not included in these laminar models will however leave plenty of variety in the particle distribution functions. The same is true for the electric field. Available electric field measurements are highly localized and are a far cry from the mapping of potential distribution that has been achieved in some laboratory experiments. Moreover, electric field measurements mostly refer to the dominant component perpendicular to the magnetic field. Field aligned electric fields are then inferred from particle measurements, assuming laminar acceleration. These measurements indicate a limited acceleration region for precipitating electrons and upgoing ions that is centered at typically 6000 km ($1R_E$) altitude. Particle measurements at low altitudes which are apparently in conflict with the assumption of purely laminar high altitude acceleration were, however, already discussed by Evans (1976), see also Kaufmann et al. (1976), Hoffman and Lin (1981).

The acceleration region at 1 R_E also happens to be a region of intense plasma turbulence, corresponding to electrostatic ion cyclotron waves in particular. Embedded in this turbulence region are highly localized parallel electric fields $E_z<15$ mV/m which recently could be directly measured (Temerin et al., 1982). Each event consisted of from 10 to 400 separate spikes. Enhanced downward electron fluxes and upgoing ion beams indicated the presence of an upward directed average electric field above and below the spacecraft. Although there is considerable uncertainty about the velocity, even its direction, at which these spikes passed the spacecraft, the inferred length of perhaps $40\lambda_D$, the spacing of approximately $1000\lambda_D$ and the amplitude are in many ways similar to the results of <u>one-dimensional</u> particle simulations and corresponding theories of ion acoustic double layers and solitons (Sato and Okuda, 1981; Okuda and Ashour-Abdalla, 1982; Hudson et al., 1983). What tends to be ignored, however, in addition to the presence of intense ion cyclotron turbulence, is that the measurements also indicate that scale lengths perpendicular to the magnetic field of an individual structure may be comparable to the parallel scale length. Also, it is not in any way established that fields are electrostatic (see Sec.2.2).

Taking λ_D=10 m (n=5 cm^{-3}, T_e=10 eV) as a typical Debye length in the acceleration zone, its height above the ionosphere corresponds to $6\times10^5\lambda_D$. Periodic B.C. for a sufficiently large box thus appear to be most appropriate for a particle simulation of electric field structures with scale sizes several orders of magnitude smaller than the distance to the ionospheric and magnetospheric boundaries. Indeed, randomly localized potential spikes are readily formed out of current driven ion acoustic waves and anomalous resisitivity in <u>one-dimensional</u> systems of sufficient length $L_z \geqslant 512\lambda_D$ and mass ratios $M/m \leqslant 100$. For slightly larger initial drifts $u=-j/n(e)>v_e=(T_e/m)^{1/2}$ or by recycling electrons at a small rate such as to nearly maintain the initial drifting Maxwellian distribution (Okuda and Ashour-Abdalla, 1982), these spikes are also formed in somewhat smaller systems. No spikes are found, however, in a periodic 2 D system, unless it is initialized with a large and very localized 1 D density depression As expected, in 2 D simulations with a very strong magnetic field, where at least electrons cannot as easily move sideways, other than in the homogeneous and thus irrelevant $\underline{E}\times\underline{B}$ direction, potential spikes are readily formed. However, their random locations never line up in a direction perpendicular to the magnetic field such as to form a 1 D structure (Kindel et al., 1981). Potential spikes and associated particle trapping were also seen in particle simulations in which an additional uniform electric field maintained a constant drift and which started in the reactive (Buneman) regime $u>>v_{eo}$ (Dum and Chodura, 1979). Spikes appeared at random locations shortly after the (sudden) transition to the ion sound regime, $u<v_e$. Because oblique wave modes and electron pitch angle scattering are very essential features of the instability and of anomalous resisitivity in both regimes (Dum, 1981), they were, however, considered an <u>artifact</u> of 1 D simulations. Indeed, no spikes or trapping were noticeable in (generally somewhat smaller) 2 D systems with the same physical parameters. (No recycling or initial density depressions were imposed.) Turbulence and anomalous resisitivity remained very homogenous in this case. As shown by turbulent heating experiments, and contrary to the arguments for 1D simulations, the multidimensional behavior remains essentially intact even for the largest ratios

Ω_{ce}/ω_{pe} of gyro-and plasma frequency to be expected in the auroral zone. Perpendicular scattering is related to cyclotron resonances (anomalous Doppler effect).

It is not concluded that either coherent (laminar) effects or turbulence are all-important with the other effect secondary or nonexistent, but that factors such as dimensionality, mass ratio, box size, and initial conditions strongly affect the properties of a simulated plasma "conductor", in addition to details of the B.C. Simulation result thus should be examined with the greatest caution and open-mindedness, keeping in mind the parameters and B.C. for the real physical system.

Periodic B.C. have the undesirable feature that simulations essentially reduce to an initial value problem in which free energy sources cannot be maintained, because neither energy and particle inflow nor losses by convection or heat conduction are included. Applied uniform electric fields and homogenous recycling of particles are only a very crude approximation to these processes. The alternative are B.C. for particle inflow and the electric potential at some boundary surfaces. The trouble in applying such B.C. to a system like the auroral acceleration region is that the artificial boundary surfaces will always be recognized by the system as emitting electrodes, which are connected by an external electric circuit as specified by the electric B.C. These models thus represent more closely very schematic laboratory experiments, provided B.C. are at least self-consistent. For these boundary effects to disappear, it would be necessary that inflow B.C. and electric B.C. at the artificial boundary surfaces exactly match the unknown processes within the volume. Taking, e.g. a half-Maxwellian or the positive part $v>0$ of a drifting Maxwellian for an inflow distribution does not meet this criterion, because in general it will not be smoothly joined by an outflow distribution $v<0$. It would have to be the positive part of some unknown skewed distribution which describes the current and associated heat flow, etc.

Although these difficulties are probably recognized by most computational physicists, in the strict geophysical context, to state it bluntly, some "auroral double layer" simulations must be considered exercises on the propagation of singularities into a plasma. They may still provide food for thought. System size is, of course, a key factor for the truth of this statement. It is thus especially troublesome for 2 D, not to speak of 3 D simulations. On the other hand, multi-dimensional simulations allow for more turbulence and scattering which may iron out singularities created by B.C., although this may not have been intended in simulating "laminar" double layers. Other B.C. may, however, result in the numerical instabilities, probably experienced by most computational physicists when trying to find a compromise between reality and numerics.

B.C. which are a compromise between the physical ionospheric and plasma sheet B.C. discussed before in some detail and the numerics, have been incorporated in the particle simulation models of Wagner et al. (1981), and of Singh et al. (1983), Thiemann et al. (1983). We will call them model I and II, respectively. The authors realize, of course, that the scaled dimensions of their systems, $L_x \times L_z = 256\lambda_D \times 512\lambda_D$ for model I, are negligible compared with the auroral system. The ionosphere (anode) is treated as an emitting plate. Energy degraded backscattered electrons, along with a converging magnetic field of mirror ratio $B_z(I)/B_z(M)=1.1$ between "ionosphere" and "plasma sheet" (cathode) are included in modelI. These two features proved to be essential in sustaining an extended potential drop that otherwise would be concentrated into a narrow anode sheath. In model II this is achieved by emitting ionospheric ions from a drifting Maxwellian rather than using a half-Maxwellian reservoir In both models the flux of upgoing electrons from the ionosphere is adjusted such as to preserve overall charge neutrality (but not particle number) in the plasma volume. The flux tube is initially filled with a cool and uniform background plasma. Model II has hot plasma reservoirs extending over the entire width L_x of the magnetospheric boundary z=0. A narrow unneutralized electron beam is injected at the center of this boundary. In model I the reservoirs are restricted to a narrow plasma sheet ($\Delta x/L_x=6/128$). Half-Maxwellian reservoirs are used except for the plasma sheet electrons which may also be drifting. In both models it is recognized that suitable wires for connecting a battery with the plasma sheet and the ionosphere are not available, hence no potential is prescribed at the magnetospheric boundary. The Neumann and Dirichlet B.C.

$$E_z = -\partial\Phi/\partial z = 0 \quad , z = 0 \quad ; \quad \Phi = 0 \ , z = L_z \ , \qquad (12a,b)$$

respectively are used. The "ionosphere" (anode) is thus a conducting and grounded plate to which charged particles are attracted by their oppositely charged mirror images. The prescribed boundary condition at z=0 is numerically convenient, but actually $E_z(0)$ would be determined self-consistently by the magnetospheric emitter and the charge flow into the system, cf. Figure 1,2.

B.C. for the particles and the potential dominate over initial conditions in the control of potential formation and particle acceleration. Intense turbulence develops and in a very dynamic fashion strongly interacts with the laminar acceleration process, again depending on details of the B.C. (Thiemann et al., 1983). For model I it was noted that the potential drop is slightly less than the thermal energy of plasma sheet ions $e\Delta\Phi=0.8\ T_i$ (compare Chiu and Schulz, 1978), but it is increased sixfold if plasma sheet electrons are drifting and consequently lead to intense turbulence and corresponding anomalous resistivity . Except for the importance of space charge effects, which compared with the physical system are grossly exaggerated by the small simulation box, double layers in these bounded models bear little resemblance to the potential spikes discussed above for periodic B.C.

Injection of an energetic beam can directly drive a return current in the ambient plasma. In contrast to the analytic models and simulations just discussed for injection from the plasma sheet, which allow only for electrostatic interactions, this process involves in general also an inductive electric field created by the beam head. According to Lenz's rule this field will try to compensate the beam current j_b by creating a plasma return current j_p. Indeed from Maxwell's equations we have the wave equation

$$\nabla\times\nabla\times\underline{E} + \frac{1}{c^2}\frac{\partial^2 \underline{E}}{\partial t^2} + \frac{4\pi}{c}\frac{\partial}{\partial t}\underline{j}_p = -\frac{4\pi}{c}\frac{\partial}{\partial t}\underline{j}_b \tag{13}$$

The equations of motion determine the conductivity relation between the current j_p in the ambient plasma and the electric field. It may involve anomalous resistivity. The electric field can be split into an inductive and electrostatic part, according to (1). Poisson's equation may then be used to derive an equation for charge neutralization of the beam. The underlying wave mode in the process of return current generation and beam neutralization depends on the beam diameter, the beam profile (rise time and speed), and the parameters of the ambient plasma. For the injection of relativistic beams into laboratory plasmas where this boundary value problem has been studied in considerable detail, it is the electromagnetic electron plasma wave that initially sets up return current such as to cancel the beam current, $j_p=-j_b$. The relevance at these results to solar flare injections has been pointed out by Spicer and Sudan (1983). It is not difficult to see that more gentle injection may involve other electromagnetic wave modes, Alfvén waves for example. An interesting simulation of such a case has been carried out by Byers et al. (1978), using a hybrid code with resistive electrons (equ.(8)). An ion beam is injected perpendicular to an ambient magnetic field in order to create larger mirror ratios or even reverse the magnetic field (Sec.2.2) through the oppositely directed beam self-magnetic field. This information, however, is carried to the ambient plasma by compressional Alfvén waves which reflect at the plasma boundaries. The plasma reacts with a return current which subsides only after many Alfvén transit times. These processes are eliminated from the outset in purely electrostatic analytic models or simulations.

3.3 Electrodynamic Magnetosphere-Ionosphere Coupling

A great variety of electromagnetic waves is observed in the magnetosphere. Some are directly associated with magnetospheric substorm and auroral activity. They include in particular shear Alfvén waves which are guided by the ambient magnetic field because of their dispersion relation.

$$\omega = k_\parallel V_A \ ; \qquad V_A = (B^2/4\pi\rho)^{1/2} \tag{14}$$

(see e.g. reviews by Southwood and Hughes, 1983; Baumjohann and Glaßmeier, 1983). These waves, in contrast to the fast isotropic mode $\omega=kV_A$, can carry a field aligned current. Any predominantly electrostatic fields on auroral flux tubes have to be built up starting from the boundaries. Obviously, the auroral system has no electrodes to which a potential difference could be applied in order to draw the field aligned current, although such laboratory models have often been invoked in "explaining" auroral particle acceleration, by double layers in particular. Propagation of shear Alfvén waves from a distant magnetosheric or ionospheric generator is then the only plausible mechanism by which a field aligned current can be established and which as a limit may also include a stationary state with electrostatic fields. In most studies of magnetosphere-ionosphere coupling the electrostatic nature of the electric field is never questioned, although measurements are highly localized and certainly have not mapped out the assumed V or S shaped potential structures. Dissipation in the ionospheres or along the field lines is required in order to reach a steady state but because of wave reflection it may take many bounces between the ionospheres and the magnetosphere. Inhomogeneities along the flux tube, such as a localized dissipation region, also produce reflections (Lysak and Dum, 1983). Any (sudden) change in the B.C. or along the field lines will result in renewed shear Alfvén wave generation. The close parallels to electrical transmission lines should be obvious.

Even if (quasi-) stationary states with predominantly electrostatic fields are established, it is still of importance for the consideration of microscopic processes which locally must sustain these fields, to know how the current or electric field is maintained by electrodynamic coupling to other parts of the system (Dum, 1983a). As also discussed above, simulations of double layers or anomalous resistivity never result in a universal (quasi-) stationary state which is independent of these electric B.C. System size is certainly a factor, but laboratory experiment have also never established universal laws for laminar potential drops or anomalous resistivity, independent of these B.C.

The fundamental question on how a current or electric field is applied to an extended system, such as auroral flux tubes, and the observation of shear Alfvén waves and of small scale turbulence on these flux tubes provided a major motivation for the development of a 2 D electrodynamic coupling model with anomalous resistivity in a localized acceleration region (Lysak and Dum, 1983). The resistivity law is derived from nonlinear saturation of ion cyclotron turbulence (Dum and Dupree, 1970; Dum 1981). Just as for the reconnection models mentioned in Sec. II.2, details of this law are not important, however. Partially laminar acceleration of particles may also provide an energy sink, although the usual laminar models have no obvious generalization to non-stationary states. The important point is that the dissipation process is self-limiting, by enhanced reflection back to the generator, for example. Construction of appropriate B.C. for the magnetospheric generator thus turns out to be a somewhat vexing, but very important problem. Whether or not a stationary electrostatic structure is established strongly depends on these B.C., in addition to dissipation in the localized acceleration region. For typical values of the height integrated Pedersen conductivity Σ_P, the ionosphere is an almost perfect reflector for the waves which by itself would allow for many wave bounces, before any stationary state may be established. The amplitudes of reflected and incident electric field are related by

$$\frac{E_\perp^r}{E_\perp^i} \equiv R = -\frac{\Sigma_P-\Sigma_A}{\Sigma_P+\Sigma_A} \quad ; \quad \Sigma_A = \frac{c^2}{4\pi V_A} \quad , \tag{15}$$

Note that the magnetic field and the field aligned current retain their direction in the reflected wave. It can also be concluded from (15) that an anomalous resistivity region corresponding to a very poor conductor would have the opposite reflection properties, i.e. returns $E_\perp$ with the same sign and reverse the field aligned current. The actual dynamics of this process is much more complex, however (Lysak and Dum, 1983).

It is also possible for the ionosphere to become a generator of Alfvén waves if the conductivity which is proportional to the electron density is modulated by appropriate feedback from electron-precipitation (Sato, 1978; Miura and Sato, 1980; Sato, 1982). The continuity equation for the height integrated density N may be written as (Sato, 1982)

$$\frac{\partial N}{\partial t} + \frac{c}{B_I^2}(\underline{E} \times \underline{B}_I)\cdot\nabla N = -\frac{1}{e}(j_\parallel+\gamma j_h) - \alpha(N^2-N_O^2) \tag{16}$$

where $j_h<0$ is the component of the field aligned current carried by energetic precipitating electrons which give rise to the production of secondary electrons, α is the recombination rate and N_O is the background density. It may be assumed that the height integrated conductivity satisfies $\Sigma=\Sigma_O N/N_O$. The convecting and current carrying ionosphere basically acts for small scale perturbations as a negative impedance which is coupled to the distributed impedance of the auroral flux tubes (Tamao and Miura, 1982; Miura et al., 1982). A standing Alfvén wave pattern is set up between the two conjugate ionospheres. Field aligned electric fields are not included in these models. The Sato & Miura model treats auroral field lines as uniform electrical transmission lines, the other model allows for a nonuniform (cold) magnetosphere but considers only small amplitude linear wave motion. Geomagnetic pulsations are usually also treated in this way (Southwood and Hughes, 1983). For the magnetic field of these very low frequency waves to leak through the ionosphere, which is treated as an infinitely thin conducting sheet, it is essential to include the height integrated Hall conductivity Σ_H. The Pedersen conductivity is responsible for wave reflection, but in a nonuniform ionosphere Σ_H also contributes to this process. This may be seen from the continuity relation between the field aligned current above the ionosphere I and the height integrated ionospheric current

$$j_\parallel(I) = \nabla_\perp\cdot\underline{J}_I = \Sigma_P\nabla_I\cdot\underline{E}_I + \underline{E}_I\cdot\nabla\Sigma_P + \underline{B}_I\cdot(\underline{E}_I\times\nabla\Sigma_H)/\underline{B}_I \quad , \tag{17}$$

Nonuniformities in the height integrated conductivities also strongly affect the structure of induced ionospheric electric fields $\underline{E}_I$ and change the phase shift of the magnetic field in the atmosphere below the ionosphere from 90^0 in the uniform case (Glaßmeier, 1984). The earth's electrical conductivity also influences the signal in the atmosphere, depending on its wavelength and frequency (e.g. Miura et al., 1982, Appendix; Southwood and Hughes, 1983). For other purposes, the lumped description of the ionosphere by a boundary surface has to be replaced by numerical models which describe the actual height dependent structure and the dynamic response of the polar ionosphere to, for instance, changing convection patterns and changing precipitation energy fluxes of auroral electrons associated with magnetospheric storms. These storm variations are considered as a given magnetospheric input in the numerical simulation model of Sojka and Schunk (1983), which includes ionospheric electrodynamic and thermospheric convection, diffusion, and a variety of chemical reactions. Chemical equilibrium at an altitude of 120 km and zero plasma flow at 800 km are considered as B.C.

3.4 Dynamos for small Scale Field Aligned Currents and Dynamics of the Plasma Sheet

Magnetospheric dynamos for field aligned currents or plasma flow were repeatedly mentioned above, but mostly as a troublesome B.C. which is necessary for the termination of computations at some, in simulations actually not so distant, boundary surface. Especially for discrete auroral arcs there is no universal agreement on the detailed nature and location of these dynamos. The answer to the important question as to what determines the narrow, east-west-aligned and often multiple structure that is characteristic of discrete auroral arcs, for example, should then not be prejudiced by arbitrarily imposed B.C. for describing the dynamo. It is quite possible that the structure is mostly determined at lower altitudes, in the acceleration zone or in the ionosphere itself. A model for multiple discrete arcs based on a number of observational facts and on the simulations of Lysak and Dum (1983) has been proposed by Haerendel (1983). It relates the arc structure to an interference pattern arising from multiple alfvén wave reflection between the acceleration zone and the ionosphere, for a plasma which convects, e.g. equatorward with respect to the dynamo. Because interaction with the dynamo on a given magnetic field line is now limited in time, one major boundary problem, feedback to the generator, is actually eased considerably in these new simulations (Christiansen et al., 1983). B.C. on the north-south walls of the simulation box require increased attention, however, (Figure 7).

Generation of field aligned currects can be related to plasma convection and pressure effects. The equation of motion determines the polarization (inertial) and diamagnetic currents.

$$\underline{j} = \underline{j}_{in} + \underline{j}_d = \frac{c\underline{B}}{B^2} \times \left(\rho \frac{\partial \underline{u}}{\partial t} + \nabla P \right) \qquad (18)$$

The continuity equation yields

$$\nabla \cdot \underline{j}_{\parallel} = B \frac{\partial}{\partial S} \frac{j_n}{B} - \nabla \cdot \underline{j}_{\perp} \qquad (19)$$

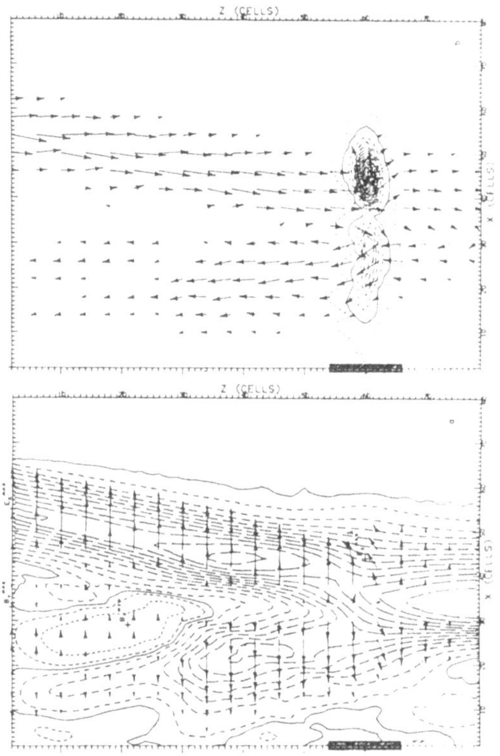

Figure 7 Simulation of Magnetosphere-Ionosphere Coupling (Christiansen and Dum, 1984).

a. Shear Alfvén wave field produced by a magnetospheric dynamo for field aligned currents, after 1.66 Alfvén wave travel times to the conducting ionosphere. The contours correspond to the magnetic field B_y (---<0 ; -.->0) and the current stream lines. The vectors represent (E_x,2000 E_z). Significant electric fields E_z parallel to the ambient magnetic field occur only due to anomalous resistivity in the (shaded) acceleration zone. Feedback between the magnetospheric dynamo (moving relative to the plasma)and its loads is by wave reflection from both the ionosphere and the acceleration region. The electric field E_x is reversed by an ionospheric reflection but retains its sign for reflections from the poorly conducting acceleration region (center). Some reflection from the artificial magnetospheric boundary with imposed B_y is just beginning.

b. Poynting flux and contours of dissipation. The left zone is largely fed by reflections from the ionosphere, while input from the moving dynamo dominates in the right zone. Selflimiting of dissipation and acceleration, and magnetosphere-ionosphere decoupling is related to wave reflection from the acceleration zone.

with s the distance along the magnetic field line. It can be seen that a self-consistent treatment of field aligned current generation requires a 3D computation. In the 2D calculations of reconnection and tearing modes, discussed in Sec.2.2, the current flows out of the plane of computation, thus (19) is ineffective. Extensions to 3 D have recently been carried out, which contain generation of field aligned currents. They are primarily intended for the description of dynamic processes in a plasma sheet volume of the magnetospheric tail. The tearing mode calculation of Birn and Hones (1981) is initialized from an ideal MHD equilibrium stretched magnetotail configuration by suddenly turning on a constant and uniform resistivity. The temperature is constrained to its initial uniform value. Symmetry relations are used at the z=0 (equatorial) and y=0 (noon-midnight) planes of the conventional coordinate system for the magnetospheric tail with the x axis pointing earth-and sunward. (The other coordinate system for the purposes of Sec.II is almost equally conventional. It is hoped therefore that the least confusion is caused by applying them both in their usual way.) At the other boundaries of the 3 D rectangular box the B.C.

$$\underline{u} = 0, \; B_n = B_n(t) \qquad , \; \frac{\partial \rho}{\partial n} = \frac{\partial p}{\partial n} = \frac{\partial \underline{B}_t}{\partial n} = 0 \qquad \text{(20a-e)}$$

are used for the velocity, normal magnetic field component and the normal derivatives of the mass density, pressure and tangential magnetic field component. These B.C. for closed surfaces are to contrasted with (4) for open boundaries. In a corresponding 2 D calculation in the x, z (noon-midnight meridian) plane also other pressure laws (adiabatic indices) and less restrictive B.C. for the flow

$$\underline{u}_t = 0, \; \frac{\partial u_n}{\partial n} = 0, \text{ or } u_n = 0, \; \frac{\partial \underline{u}_t}{\partial n} = 0 \qquad \text{(21a-b)}$$

have been employed (Birn, 1980). As expected, flow near the boundaries is increased by the B.C. (21). This is more important for the earthward, x=0, boundary, because the X-type neutral lines forms closer to this boundary than to the tailward boundary surface. The location of the neutral line is not imposed by the inflow pattern or the resistivity law as in the models of Sec.2.2, but is related to the asymmetry of the initial tail configuration. However, if in the 2D calculation the plasma is allowed to move parallel to the earthward boundary x=0, corresponding to B.C. (21b), the neutral line forms closer to this boundary. Moreover, in contrast to 2D calculations with other combinations of flow B.C., for B.C. (21b) at x=0, the flow on the x-axis is tailwards $u_x < 0$, in the entire range $0>x>-L_x$, although u=0, B.C. (20) is imposed at the tailward boundary $x=-L_x$. Flow patterns on the tailward side of the X-type neutral point remain essentially similar. Provided L_x is sufficiently large, B.C. at the far earth boundary $x=-L_x$ have comparatively little influence on the overall flow and magnetic field pattern, except, of course, in the vicinity of this boundary.

Forbes and Priest (1983) on the other hand assume line tying at the x=0 boundary, corresponding to the conditions (20) for the flow and the magnetic field. The "free-floating" B.C. (4) are imposed at the $x=-L_x$ and $z=\pm L_z$ boundaries of this 2 D calculation. Symmetry is used at the x-axis z=0. The initial configuration is a uniform current sheet with $\underline{B}=\underline{B}_0(z)$ parallel to the x-axis and $B_0(z=0)=0$. The asymmetry is introduced by line tying at the x=0 surface. At a small distance from this surface, of the order of the sheet width, the main X-type neutral point is formed which then moves tailward. Other X- and O-type neutral points are formed closer to the tailward x boundary. There number might increase if the box were bigger. Secondary tearing is observed in the vicinity of the main X-type neutral point. It gives rise to plasmoids which move towards the x=0 surface, coalescing in its vicinity. This behavior of plasmoids or magnetic islands is not seen in the calculations by Birn (1980) and Birn and Hones (1981) where periodically formed X-and O-points move tailwards. These calculations, however, differ in the B.C. as discussed above, and also in the numerical algorithm. More details can be found in solar flare related calculations (Forbes and Priest, 1982, 1983a) which, except for some parameter changes and the coordinate system of Sec.2, use an identical model. Line tying at x=0 is assumed to be due to the ionosphere in the case of the magnetospheric tail and due to the high density photosphere for solar flare case. The underlying wave mode in 2D reconnection is the compressive. It is clear, however, that between plasma sheet and ionosphere, auroral flux tubes along which information is transmitted chiefly by shear Alfvén waves, should be included in the tail model. The same transmission mechanism is likely also of importance in solar flares (Dum, 1983a).

An alternative simulation model for plasma sheet dynamics and generation of field aligned currents has been presented by Sato et al. (1983). It is a 3 D extension of the driven reconnection model discussed in Sec.2.2. Plasma inflow (from the tail lobes) is imposed at the $z=\pm L_z$ boundaries of the rectangular volume, cf. Figure 6. The flow pattern forces the initially uniform current sheet into an X-type topology. The other boundaries, $x=\pm L_x$, $y=\pm L_y$ are open to the flow. Symmetry about the equator, z=0, the noon-midnight meridian, y=0, and the neutral line x=0 is used. The dawn-dusk cross tail current ($j_y>0$) in the vicinity of the x-point is diverted earth-and tailward. To find the field aligned current component, it is necessary, however, to also take into account the change in the magnetic field configuration implied by Ampere's law. The field aligned currents are formed in narrow regions of strong pressure gradients, near the outer edges of the plasma sheet. They are directed earthward on the dawn side (y<0) and tailward on the dusk side (y>0), consistent with the simple current wedge model for tail current diversion. Without any specific physical mechanism this model has long been used for the interpretation of magnetotail phenomena and the enhancements of ionospheric currents (of the westward auroral electrojet in particular) after substorm onset (Kamide, 1982, Baumjohann, 1983). On the other hand, it is understood that the expansive (as seen in the ionosphere) phase after substorm onset involves, on a large scale, relaxation to a more dipolar magnetic field, plasma flow and motion of neutral lines Hones, 1979), which to be fair, cannot be fully explained by a simulation for a small volume, excised in the vicinity of a fixed X-type neutral line. Moreover, the circuit for the electrical current is not completed. In particular, the transmission of current and power along magnetic field lines that connect with the ionospheric load is not included. It should be noted that for transient events, current closure is not necessarily in the ionosphere alone, but involves polarization currents, j_{in} in (18), across the magnetic field lines and significant wave reflection with feedback to the dynamo (Lysak and Dum, 1983).

Electrodynamic coupling to the ionosphere and other parts of the magnetotail is not included either in the tearing mode simulation by Birn and Hones (1981) for a more realistic 3 D plasma sheet configuration which also allows for the motion of multiple neutral lines. The imposed B.C. (20), however, do not allow for the plasma flow and Poynting flux to enter or leave the system. The latter may be seen by using Ohm's law (3),

$$\underline{S}^{EM} = \frac{c}{4\pi} (\underline{E}\times\underline{B}) = \frac{1}{4\pi} (B^2\underline{u} - \underline{u}\cdot\underline{B} + c\eta\underline{j}\times\underline{B}) \quad . \tag{22}$$

Using the equation of motion

$$\rho\frac{d\underline{u}}{dt} + \nabla P = \frac{\underline{j}}{c} \times \underline{B} \tag{23}$$

along with the B.C. (20), it follows that also the normal component of the last term in (22) vanishes. The displacement current can be neglected

$$\nabla \times \underline{B} = \frac{4\pi}{c} \underline{j} \quad , \tag{24}$$

because phenomena are slow and the stored electric energy is negligible compared to the magnetic energy. Poynting's theorem for the latter therefore becomes

$$\frac{\partial}{\partial t} \frac{B^2}{8\pi} + \nabla\cdot\underline{S}^{EM} = - \underline{E}\cdot\underline{j} \tag{25}$$

With vanishing Poynting flux into the volume, only the <u>stored</u> magnetic energy can be tapped for acceleration and dissipation.

$$\underline{E}\cdot\underline{j} = \underline{u}\cdot(\frac{\underline{j}}{c}\times\underline{B}) + \eta j^2 \tag{26}$$

It is very instructive to display these terms (integrated over the volume) separately, along with input and output Poynting fluxes, as done by Sato and Hayashi (1979) and Sato (1979) for driven reconnection. In the latter paper it is argued that during the early phase of reconnection in the plasma sheet the earthward and tailward sides $x=\pm L_x$ may be closed. Therefore an O-type (magnetic island or

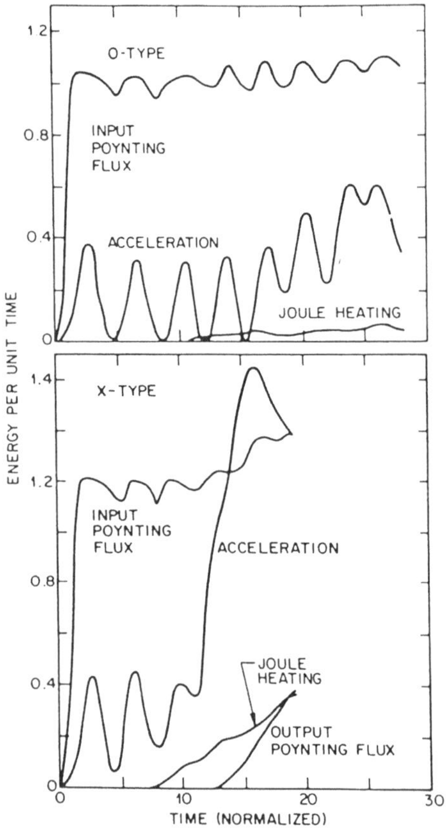

Figure 8 Energy balance for driven reconnection in closed (O-type) and open (X-type) configurations (Sato, 1979). Note the sudden ncrease in acceleration of bulk flow by the j×B force which takes place after some internal reorganization of the open ended (free outflow) configuration. Output thus does not simply follow the input, even in driven systems.

plasmoid) configuration is also simulated (in 2 D) by closing the x boundaries, $u_X=B_Z=0$. Also, the chosen z inflow reaches its maximum near these boundaries and vanishes in the central region about x=0 corresponding to the O-type neutral point. Acceleration, i.e. the first term in (26), stays much smaller in this case, with most of the input flux (there is no output Poynting flux) going into stored magnetic and thermal energy, see Figure 8). With the heat equation

$$\frac{dP}{dt} + \gamma P\nabla\cdot\underline{u} = (\gamma-1)\eta j^2 \quad , \tag{27}$$

where γ is the (adjustable) polytropic index, one derives the conservation law

$$\frac{\partial}{\partial t}\left(\frac{\rho u^2}{2}+\frac{P}{\gamma-1}+\frac{B^2}{8\pi}\right)+\nabla\cdot\left[\left(\frac{\rho u^2}{2}+\frac{\gamma}{\gamma-1}P\right)\underline{u}+\underline{S}^{EM}\right]=0 \qquad (28)$$

for the total energy. For isothermal compression, $\gamma=1$, conservation does not hold as an external heat bath is implied for the maintenance of the temperature. Birn and Hones (1981) assume a fixed uniform temperature. Choosing instead $\gamma=5/3$, was found to have a small effect (Birn, 1980). Sato and Hayashi (1979) and Sato (1979) consider $\gamma=2$ more appropriate for a 2 D system.

Birn and Hones (1981) find a field aligned current system that, in contrast to Sato et al. (1983) has the opposite flow directions towards the earth and is located close to the center of the plasma sheet. Although no coupling or mapping to the ionosphere is included, this pattern corresponds to the so-called region 2 Birkeland currents observed above the ionosphere at lower latitudes than the oppositely directed region 1 currents. These differences in the current pattern are perhaps not very surprising, considering the vast differences in the initial configuration, the resistivity law and especially in the B.C.

We have concentrated on dynamos for transient, substorm related, field aligned currents which should provide the proper B.C. for auroral flux tubes (transmission lines). The global average field aligned current system and its relation to magnetospheric dynamos has been reviewed recently by Sato and Iijima (1979) and Sato (1982). We also discussed in Sec.3.3 the important role of the ionospheric B.C. and its modification by auroral precipitation and by convection. These effects may be associated with substorms and therefore be transient. They are then capable of launching secondary Alfvén waves (Maltsev et al., 1984). Mallinckrodt and Carlson (1978) who include plasma convection across dynamo regions for Alfvén waves, point out an important difference between magnetospheric and ionospheric dynamos. In the first case the net current (almost field aligned within the ratio $u_D/V_A \ll 1$ of convection speed to Alfvén velocity) delivered to the ionosphere is enhanced by strong reflection, while there is essentially only an outgoing wave for ionospheric generators.

Actually, the plasma sheet is not a resistive fluid, but is hot and collisionless. A kinetic treatment in which electrons and ions can behave quite differently is thus required in general (review by Galeev, 1982). The analytic study of Goldstein and Schindler (1978) is interesting in the context of field aligned current generation and coupling to the ionosphere. They consider the (linear) growth of the ion tearing mode with magnetized, isothermal electrons and unmagnetized isothermal ions ($T_i \gg T_e$), in a plasma sheet with the normal magnetic field $B_z \gg |B_x|$ inside the singular layer and $B_z \ll |B_x|$ outside. A significant potential electric field is generated in this case, in addition to the inductive electric field (in the y direction) considered in the MHD models. It is suggested that if the conducting ionosphere can essentially shorten out the potential electric field across the magnetic field lines, a pattern of field aligned currents should result. It is assumed that the pattern arises from mapping along field lines to multiple neutral lines in the plasma sheet dynamo. No self-consistent treatment of this effect, of the auroral transmission line, nor an explicit consideration of proper B.C. is given, however. Note also that this explanation for multiple structures of field aligned current in the north-south direction (east-west aligned) differs significantly from the model of Haerendel (1983) in which the structure is determined not by the dynamo but by the acceleration zone at appoximately 1 R_E altitude. Moreover, a key result of the Lysak-Dum (1983) model for the auroral transmission line is partial <u>decoupling</u> of magnetosphere and ionosphere by the acceleration region which becomes most effective for structures with small scale across the magnetic field or intense field aligned currents. Nevertheless, it would be interesting to combine the dynamo for field aligned currents proposed by Goldstein and Schindler (1978) with a magnetosphere-ionosphere coupling model.

It may be recalled from Sec.2.2 that 2D Hybrid models of tearing with collisionless ions and a resistive electron fluid (8) do indeed yield currents and electric fields in the (x, z) computation plane, in addition to the current j_y and the inductive electric field E_y contained in MHD models. A magnetic field B_y is also generated (Hewett and Seyler, 1981). As $\partial B_y/\partial t=c(\partial E_z/\partial x-\partial E_x/\partial z)$ this indicates also a nonpotential electric field component in the (x,z) plane. The additional current system is a Hall current due to negligible electron inertia. It also has a y component. As described in Sec. II. 2, the model of Hewett and Seyler (1981) corresponds to "driven" reconnection in an imploding reversed field theta pinch. The significance of this Hall current effect for reconnection in the magnetiotail was discussed by Sonnerup (1979).

It is important to investigate the stabilizing effect of the normal magnetic field component B_z in the magnetotail configuration (Galeev, 1982). The model of Hamilton and Eastwood (1982) includes a specified (external) magnetic field B_{oz}. The self-consistent magnetic field is calculated from the ion current, with electrons neglected (A background of thermal electrons is included in some runs). In addition to the self-consistent inductive electric field $E_y=-(1/c)\,\partial A_y/\partial t$, a uniform convection electric field E_{oy} is also included which for northward B_z drives particles earthward. The chosen B.C. are interesting. The boundaries $z=\pm L_z$ are conducting, $E_y=0$, and absorbing to outgoing particles, but particles are also injected into the plasma at these boundaries. Those leaving the box in x are reinserted at the opposite x boundary (Periodic B.C.). To simulate the finite sheet width in y, particles with $|y|>L_y$ are removed. The y coordinate is not otherwise used in the 2 D calculation. It turns out that stabilization against tearing occurs if the typical radius of ion gyration in the (x,y) plane becomes smaller than the width L_y. This would indicate stabilization for arbitrary small normal magnetic field B_{oz} if the sheet width L_y is infinite. Obviously, the y boundary requires more attention. It must also be noted that this effect differs considerably from stabilization in an infinite medium, which also involves electron kinetic effects (Galeev, 1982). A similar 2 D simulation model for magnetotail dynamics has been devised by Swift (1983). It differs by the B.C. in z and the initial particle loading. Hot plasma sheet ions are loaded for z<32 (grid points) and cold tail lobe ions are loaded for $z<L_Z=128$. The axis z=0 is a symmetry axis. Particles leaving the range $z<L_Z$ are reinserted as cold tail lobe particles. The conducting plates at $z=\pm(L_Z+D)$ are separated from the plasma by a vacuum region (D=128). Periodicity in x is used, as in the other simulation models. Due to the presence of the external magnetic field B_{OZ} no neutral lines are formed, however. The periodic B.C. in x allows only for circulation of particles as in a torus, and excludes any real earthward or tailward acceleration (injection) which is of special interest in connection with substorms. Another problem with the system length L_X in kinetic models is that it may severely restrict the number of modes, often to a single fixed neutral point. An alternative is to calculate particle orbits from the electric and magnetic fields obtained in MHD simulations (Sato et al., 1982). The particles are treated as test particles, that is their self-consistent fields are not included in the orbit calculations but the dynamics of the MHD fields adds an important element of realism to analytic calculations with static fields.

Earthward convecting and accelerated plasma sheet particles are a source for injection into auroral field lines, thus provide the magnetospheric B.C. for the flux tube models discussed in Sec.3.2. A free energy source for instabilities may be contained in certain features in the distribution functions. Convection is thus coupled with diffusion, leading to precipitation into the atmosphere for particles with appropriate energies and pitch angles. The (energy dependent) inner i.e. earthward edge of the plasma sheet results from a balance of convection and loss rate. The zone of diffuse aurorae magnetically maps to this edge. The evolution of instabilities is coupled with the distribution functions and thus also with particle convection and losses in the bounded system (Kennel and Ashour-Abdalla, 1982). It should provide for many interesting simulations.

3.5 Magnetospheric Substorm: A Problem involving Global Magnetospheric Convection and its Coupling to the Solar Wind and the Ionosphere

"A magnetospheric substorm is a transient process initiated on the night side of the earth in which a significant amount of energy derived from the solar wind-magnetosphere interaction is deposited in the auroral ionosphere and magnetosphere." This is a consensus definition quoted in the review by McPherron (1979), who also noted that general consensus does not extent much beyond this statement.

We have just described simulation models for such an energy deposition. These models, however, cannot be used directly for exploring a number of other important questions attached to this substorm definition, because they describe only excised parts of the global system and the answers to these question are then to a large extent already implied by the assumed B.C. and initial conditions. A very fundamental question is whether the sudden departure from an undisturbed ground state with low level magnetospheric convection and precipitation is a driven response to increased energy input from the primary solar wind-magnetosphere dynamo due to changing solar wind parameters (Akasofu, 1980), or whether it is the result of an explosive instability inherent in the plasmasheet that is related to plasma parameters in the magnetotail but not directly to changing solar wind input (Schindler, 1980; Galeev, 1982).

Because solar wind flow and the interplanetary magnetic field embedded in this flow are subject to many transient and irregular effects related to solar activity, the first concept appears to be more straight-forward and extensive experimental evidence in its favor has been presented. For example, Akasofu (1980) finds a close correlation between an energy input function $\varepsilon(t)$ that depends on solar wind parameters, the orientation of the interplanetary magnetic field in particular, and an index AE(t) which characterizes dissipation in the auroral ionosphere. The correlation applies to the three phases of a substorm, growth phase, expansive phase and recovery phase with the latter related to declining energy input from the solar wind-magnetosphere dynamo. It is also very interesting to note that correlation ceases for the largest energy inputs, $\varepsilon(t) \geqslant 10^{19}$ ergs/sec (or 10^{12} Watts), indicating partial decoupling of the ionosphere from the magnetosphere. This process is exactly one of the principal features in the dynamic magnetosphere-ionosphere coupling model of Lysak and Dum (1983) which involves electric fields related to enhanced field aligned currents. It must be admitted, however, that not only is the characterization of the complicated dynamic processes at the magnetospheric boundary (magneto pause) and the ionosphere by single indices subject to many questions, but the long chain from the primary solar wind-magnetosphere dynamos to global mangetospheric convection and to secondary dynamos for field aligned currents must also be studied. Direct space observations and the models of Sec.3.4 show that the plasma sheet is a very important link in this chain, especially for the sudden onset of the expansive phase. Also, in the ionosphere this phase begins with a sudden brightening of an auroral arc just poleward of the zone of diffuse aurorae (Sec.3.4) and near local midnight, and then spreads in a characteristic way. The poleward boundary of the zones of discrete nightside aurorae magnetically maps to the outer plasma sheet boundary layer(see Figure 6). No detailed discussion of auroral morphology is intended here. Returning then to the plasma sheet, disruption of the cross-tail current and its diversion to the ionosphere appears to be a logical description of substorm onset. Akasofu (1980) relates this onset to critical solar wind input powers $\varepsilon(t) \gtrsim 10^{18}$ erg/sec. Below this level, the magnetosphere is assumed to respond to increasing $\varepsilon(t)$ with enhanced convection and field aligned currents, without major changes in the convection pattern. No viable physical mechanism for this threshold or for current interruption is offered, however. It is possible that the current increases, due to the increased driving by the solar wind dynamo, and is disrupted by some instability above a certain current density. According to Galeev (1982), no viable current driven instabilities with reasonably low threshold for the assumed plasma sheet parameters have been proposed however, other than tearing modes which depend on the special magnetic field topology of the magnetotail. Before discussing these models in the global context, it is recalled from Sec.2.2 that disruption of sheet currents is common in reconnection experiments. In the experiment of Stenzel et al. (1983) it could be systematically triggered by raising the potential of the central anode (Figure 3). The double layer, however, formed near this anode and thus the process cannot by directly employed for the plasma sheet. This study shares however, with the phenomelogical current disruption model for the plasma sheet, the realization that the global circuit for current closure is very important. The cross-tail plasma sheet current closes in the outer regions of the tail lobes where dynamo action $\underline{E}\cdot\underline{j}<0$ is also assumed to take place. This circuit is ignored in all the models of reconnection and tearing described in the previous sections. The cross-tail current can adjust freely to changes in the magnetic field without any feedback from the exterior or the dynamos for this current. The existing kinetic models are not capable of also including the y dimension which must be involved in important microscopic processes such as anomalous resistivity, by lower hybrid waves for example, or double layer formation (Sec. 2. 2). The MHD models use a resistivity, which for the collisionless plasma sheet must be interpretated as anomalous. As already discussed in some detail, anomalous resistivity is, however, strongly dependent on B.C., although to a lesser extent for current flow predominanty perpendicular to the magnetic field. It is expected that electrodynamic coupling processes affect the structure of driven reconnection models, in addition to the plasma flow which also should be coupled to global magnetospheric convection.

One important point should be noted for driven reconnection. Although the output strongly depends on the energy inflow, it is not an exact replica. Figure 8 shows that acceleration increases drastically after some period of internal reorganization in the system and that some release of previously stored energy is also involved. It is likely that given an appropriate resistivity law and appropriate B.C., describing the coupling to the global system, repeated bursts of acceleration could be produced with a period and intensity depending on the input flow. A model that is between the two extremes of a strictly driven response and explosive release of previously stored energy would result. McPherron (1979) discusses experimental evidence for such substorm sequences or multiple onset substorms. The implication of periodic bursting has often been used as an argument against instability (energy unloading) models for substorms. These models indeed tend to ignore the recovery phase of a substorm. After this phase a energy reservoir for explosive release could be built up again in the magnetotail. Provided other conditions do not change, the process would indeed repeat itself periodically. These models, by starting with an instability analysis, also tend to bypass the growth phase in which such an unstable configuration must be built up. It is usually assumed that reconnection on the day and nightside of the magnetosphere transfers magnetic flux and energy between the interplanetary space and the region of closed magnetospheric field lines, with the open field lines serving as an intermediary. There is no reason to assume that these processes always exactly balance. Thus magnetic flux and energy can be built up in the magnetotail. Again experimental evidence in support of this picture for the growth phase of magnetospheric substorms, such as dependence on the orientation (southward component) of the interplanetary magnetic field, expansion of the polar cap region, and evolution to an increasingly tail like magnetic field (implying increased cross-tail current density) can be found.

Models involving a global instability of the magnetotail for explosive release of energy in a substrom assume that somehow a sufficient amount of free energy can be stored in the magnetotail, without premature release during the growth phase, although they may admit that a finite external disturbance may trigger such a release. It is implied, however, that in contrast to a driven response, the energy in the initiating perturbation is small compared to the total energy release (Schindler, 1980). This concept of an instability has a number of familiar mechanical representations, like a ball on top of a hill, water dripping from a leaky faucet, or a ball in a dip on top of a hill for a metastable state that requires a small triggering energy input for the sudden release of stored energy. However no explanation of plasma effects is provided by such models. A detailed examination of plasma instabilities, actually reveals in general a soft onset of instabilities for parameters exceeding some marginal instability thresholds. Amplitudes increase continously beyond this threshold, but without external driving the system would quickly return to marginal stability. Sufficiently strong driving is also required in the first place to get the system past marginal instability before the instability can grow enough to stop this process. As descrived in Sec.2.2 relaxation oscillations about marginal instability will result in this case, with dissipation on the average related to energy input. Anomalous resistivity always shows this behavior. This represents a severe problem for tearing mode calculations in closed system such as the model of Birn (1980) and Birn and Hones (1981) which rely on triggering of the stored energy release by a sudden increase in (anomalous) resistivity. Certainly, the nature of this trigger needs to be explored. It should also be noted that additional driving by plasma inflow enhances tearing mode growth (Sato and Hasegawa, 1982), expecially if the flow is parallel to the plasma sheet (Sato and Walker, 1982).

Collisionless tearing modes as a mechanism for the explosive release of stored energy suffer from the same problem: Driving by the external system is required to produce an unstable state in the first place (growth phase) and this system cannot be ignored during the further evolution in the expansive and recovery phases. The proposal by Galeev and Zeleny (1976) that the plasma sheet evolves during the substorm growth phase to a metastable state from which it can be lifted into an instability gap by a small increase in the normal magnetic field component B_{OZ} (trigger) is interesting. However, in addition to a more detailed (numerical) investigation of the instability mechanism, it would again require the investigation of the global system, in order to establish such a trigger and to make certain that the system does not prematurely fall back to stability.

It is seen that in addition to the many interesting investigations of the various physical mechanisms involved in magnetospheric substorms and aurorae a large task in coupling this elements remains. To some extent, this can be done for a particular link in the chain by appropriate dynamic B.C. The obvious way out of this difficult problem, however, appears to be a global simulation of the magnetosphere and its coupling with the solar wind and the ionosphere. It is expected that conflicts between the various theories and simulation models discussed above would then mostly disappear. Such global simulations are clearly also very desirable for connecting the very localized spacecraft observation in a way other than by cartoons of various degrees of ingenuity or by highly simplified theories, which usually entails new controversies. The computational effort is enormous, however. Because solar wind flow, interplanetary magnetic field and the intrinsic field of planets are not aligned, the problem is inherently three-dimensional. This fact and the size of the system allow only one fluid MHD simulations at this time.

In addition to 2 D models for the noon-midnight meridian etc. (Leboeuf et al., 1978), 1979; Lyon et al., 1980, 1981), fully three-dimensional global MHD simulation models have recently become possible (Leboeuf et al., 1981; Brecht et al. 1981, 1982; Wu et al. 1981) B.C. are comparatively simple. On the sunward side of the box a supersonic solar wind enters with prescribed plasma parameters and interplanetary magnetic field. These parameters could be time dependent, corresponding to interplanetary observations. All other boundaries are open for plasma outflow. Wu. et al. (1981) for example, use linear extrapolation of fluid variables and magnetic field from the computational domain to implement these open B.C., choosing a direction of 45^0 from the x axis (sun-earth line) for the y and z boundaries, because this direction is roughly tangent to the bow shock. Lyon et al. (1980) use absorbing z walls. Leboeuf et al. (1978, 1979) use periodic field variables. Use of "particles" to convect fluid properties, however, allows to include re-injection of exiting flow with its original solar wind velocity, in a way that is familiar from particle codes (Sec.4). A magnetic dipole is placed at the origin of the coordinate system, which generally is not coincident with the center of the box, in order to account for the long magnetotail. Typical box sizes may be as large $30\ R_E > x > -100\ R_E$, $|y| < 40\ R_E$ and $|z| < 40\ R_E$ (Brecht et al., 1982). The number of mesh points in this case is 29x21x21, but somewhat higher in other calculations.

Grid spacing, on the average (grids can be nonuniform), thus corresponds to several R_E in 3D and about 1 R_E in 2D. This not only indicates severe problems with numerical resolution and numerical dissipation, but also that details of the ionospheric B.C. and magnetosphere-ionosphere coupling hardly matter for the present models, at least not in the distant magnetotail. The near earth B.C. is usually handled by holding plasma parameters fixed at their initial values. We know, of course, that in reality magnetosphere-ionosphere coupling is very important for such dynamic events as magnetospheric substorms. Nevertheless, very interesting results on magnetotail dynamics and its interaction with the solar wind have emerged, which could serve in defining B.C. for studies on a smaller scale and hence with better resolution. Reconnection in these global models is generally due to numerical resistivity alone, but plasma parameter dependent ad hoc resistivity models can be added. It was found e.g. in the 2D calculations of Lyons et al. (1981) that repeated formation of neutral lines in the tail ceased if Joule heating was included in the MHD equations, indicating that anomalous resistivity models for fluid calculations require more attention. Artificial viscosity it used to handle discontinuities such as the bow shock and the magnetopause.

There are a number of important phenomena taking place in the various plasma regions and boundary layers of the inner magnetosphere, that are not even discernable in these global MHD calculations,nor for being inherently kinetic can be treated by MHD models. Besides the various simulation models for parts of the magnetospheric system that were already discussed, the global single particle simulation model for the inner magnetosphere (L<10) developed by the Rice group (e.g. Harel et al., 1981a,b; Spiro et al., 1981) should be mentioned here, because the philosophy in treating B.C. is quite different from other models. Namely, experimental input is used as much as is possible, based on availability and numerical restrictions. The model finds hot particle distributions in the magnetosphere, starting from an assumed plasma sheet source. Particles are convected earthward using drift motion in a given magnetic field and an electric field that is determined from magnetosphere-ionosphere mapping, neglecting field aligned potential drops. Some effects of an inductive electric field are included by allowing equatorial crossing points of magnetic field lines to move. Field aligned currents are calculated from the magnetospheric source, using continuity. The ionospheric potential distribution can then be calculated (Kamide, 1982), using an empiral conductivity model and the potential distribution at the polar boundary of the model as an input. Zero current flow across the equatorward boundary at latitude 21^0 is also assumed. Mapping this potential distribution back to the magnetosphere closes the loop. The basic features of this chain were described by Vasyliunas (1970) but the model has now advanced to a dynamical description of substorm related phenomena, starting from experimental time dependent input at the boundaries. A number of features, such as injection of ions from the ionosphere etc., still need to be included. The effort in modeling is continuing.

3.6. Magnetospheric Boundaries

Global MHD models, as we have seen, can with a minimum of ad hoc assumptions, describe the transfer of energy, magnetic flux and plasma between interplanetary space and magnetosphere. The boundaries across which this transfer takes place, however, appear as rather sharp discontinuities which can only be handled by artificially increasing dissipation in their vicinity. The balance between numerical oscillations for too little dissipation and an inaccurate transition for too much dissipation is rather delicate. One expects at least that not only the location of the discontinities is correctly given, but also that the (Rankine-Hugeniot) jump conditions, expressing conservation of mass, momentum, and energy as well as continuity of normal magnetic field and tangential electric field, are satisfied. Transfer from the interplanetary medium plays a key role in magnetospheric dynamics. Intrinsic theoretical and obervational interest is also attached to the structures of bow shock magnetosheath, magnetopause and boundary layers which effect this transfer (Haerendel and Paschmann, 1982). More detailed simulations than are provided by global MHD models are thus very desirable, despite the increased problems attached to the introduction of artificial computational boundaries with which we are by now quite familiar. Local MHD models (Sato, 1984) can also be used to study magnetic field and flow charcteristics near the magnetopause. Day side reconnection models (Hoshino and Nishida, 1983; Sato, 1984) differ from the models we have already discussed by the pronounced asymmetry of the regions representing magnetosheath and magnetosphere. Not only are the magnitudes of density, pressure and magnetic field quite different in these regions, but the magnetic field must also make a rotation from its rather arbitrary interplanetary orientation to the well defined magnetospheric direction.

The shear flow along the magnetopause gives rise to the Kelvin-Helmhotz instability. Its nonlinear state can provide a turbulent viscosity for momentum transfer to the magnetosphere. This fundamental mechanism can be investigated in 2 D simulaiton models with an initial shear flow

$$u_y(x) = u_o \tanh(x/a) \quad , \qquad (29)$$

a magnetic field $\underline{B}_o$ in the y,z plane, and periodicity in y. The ideal MHD calculation of Miura (1984) assumes that u_x, B_x and the normal gradients of the remaining variables vanish at the x boundary surfaces $x=\pm x_b$ (=10a). These B.C. in x, corresponding to the set (20b-e) and (21b), along with periodicity in y imply that there is no net energy flow in or out of the simulation box, see (22-28) with $\eta=0$. The same configuration (for normal magnetic field B_{oz}) has also been studied in the electrostatic limit, using particle simulation (Pritchett and Coroniti, 1984). Kinetic effects, which result in particular in a reduction of wave growth with increasing Larmor radius can be examined in this model. The initial distributions are drifting Maxwellians. The nonuniform E×B drift (29) implies a nonuniform electric field E_{ox} and a net charge in the system. Ions are loaded uniformly while electrons are nonuniform in order to meet these initial conditions. The implementation of B.C. in x also differs from the MHD model. Particles were reflected and either the condition of vanishing perturbations of the normal electric field, $E_{1x}=0$, or of the electrostatic potential, $\Phi_1=0$, i.e. $E_{1y}=0$ at the boundary, was imposed. The results were very similar as long as the boundaries were sufficiently far from the shear region, $x_b \geqslant 8a$. The shear length a was then rather unimportant in the nonlinear stage of the instability, which was, as is so often the case (see Sec.4), dominated by the longest wavelength mode allowed by the periodicity length L_y. Because of the rather stringent requirements on box size ($128\times128\ \lambda_{De}^2$) and particle number (N=130,000), a guiding-center code (Lee and Okuda, 1978) was also used. The full ion motion is retained, but the electron motion across the magnetic field is restricted to the E×B drift.

Box size and B.C. in these local models are chosen such as to allow the study of dynamic processes within a limited region of the magnetosphere, with the least interference from B.C. What is totally missing is thus, once again, the electrodynamic coupling to the inner magnetosphere and the ionosphere. Vorticity motion induced by the Kelvin-Helmholtz instability at the boundary layer may not only encompass large portions of the plasma sheet (Hones et al., 1981) but can also be responsible for the generation of region 1 field aligned currents (Sato, 1982). There is also strong evidence that solar wind driven surface waves at the magnetospheric boundary are a major source of geomagnetic pulsations. The coupling to the inner regions of the magnetosphere involves field line resonances for shear Alfvén waves (Review by Southwood and Hughes, 1983). A numerical investigation of this process was recently carried out by Junginger (1984). Waves in the (cold) plasma are driven by a surface wave imposed at the magnetopause boundary. The increase of density by up to two order of magnitude at the plasmapause acts as a solid wall, $u_n=0$. The other B.C. are provided by the conducting ionospheres, cf. Figure 6.

The fact that magnetic transitions occur in regions of the collisionless magnetospheric or interplanetary plasma that may have a thickness of only a few gyroradii suggests an investigation by hybrid codes which retain the full ion particle dynamics and use a fluid description only for electrons, see Sec.2.2. The transitions in this case are initialized using the Rankine Hugeniot jump condition. B.C. are designed to maintain these conditions at the upstream and downstream boundaries. This is reasonably consistent, as long as the perturbed ion fluxes and the wave activity do not reach these boundaries. Electron heat flux out of the transition region, especially along the magnetic field, however, presents a problem that is ignored in these models. Wave excitation involving electron kinetic effects, naturally, is also missing. Heating by instabilities is actually enhanced for quasi-perpendicular shocks, and heat conduction, it turns out, is also important, even there is only a small gradient component along the magnetic field (Forslund et al., 1984). Investigations of strictly perpendicular shocks in theta pinches also showed the importance of heating by instabilities that are actually largely heat flux driven (Dum, 1978, Chodura et al., 1977). Simpler codes which ignore these electron effects can nevertheless be used with advantage in studying various fundamental aspects of ion dynamics. The rotational discontinuity at the magnetopause has been investigated by Swift and Lee (1983), assuming cold massless electrons. The same 1D code was used to also simulate slow switch-off shocks (Swift, 1983) and the bow shock (Kan and Swift, 1983). A constant phenomenological resistivity, the Joule and adiabatic electron heating terms, but not heat flow, are included in the bow shock simulations of Leroy et al. (1982) and of Leroy and Winske (1983). A detailed description of the simulation technique is given in this volume (Winske and Leroy, 1984).

4. BOUNDARY CONDITIONS AND THE SOLUTION OF FIELD EQUATIONS

4.1 Periodic Boundary Conditions and Fast Fourier Transforms

Periodic boundary conditions in which the system is extended to points outside the computational domain by conditions of the form

$$A(x + L) = A(x) \tag{30}$$

where L is the system or periodicity length in a given direction, are the easiest to implement and thus are used in simulations whenever this appears to be possible. A Fourier series

$$A(x) = \sum_k A_k e^{ikx}, \qquad A_k = \frac{1}{L}\int_0^L dx A(x) e^{-ikx} \tag{31}$$

with wavenumbers $k=(2\pi/L)n$, $n=0,\pm1,\pm2,\pm3\cdots\cdots$ is naturally connected with such a periodic system. It or its multidimensional extension affords a convenient method of solution, at least for linear equations with constant coefficients such as Poisson's equation

$$\nabla^2\Phi(\underline{x}) = -4\pi\sigma(\underline{x}) \tag{32}$$

for the electrostatic potential, and a variety of other field equations. In practice, the number of modes must be limited to some maximum $n \leqslant N$. Such a truncation arises naturally if the field variables are defined neonly at grid points with spacing Δx. This is the case for example in simulations by the particle in cell (PIC) method where the charge density from particles distributed throughout the cells is accumulated at the grid points delineating these cells. As there are $L/\Delta x=N$ grid points $x=j\Delta x$, there must be an equal number of independent Fourier modes that provide the mapping between real space and wavenumber space, $|k|\leqslant\pi/L$. In lieu of (31) one obtains the amplitudes

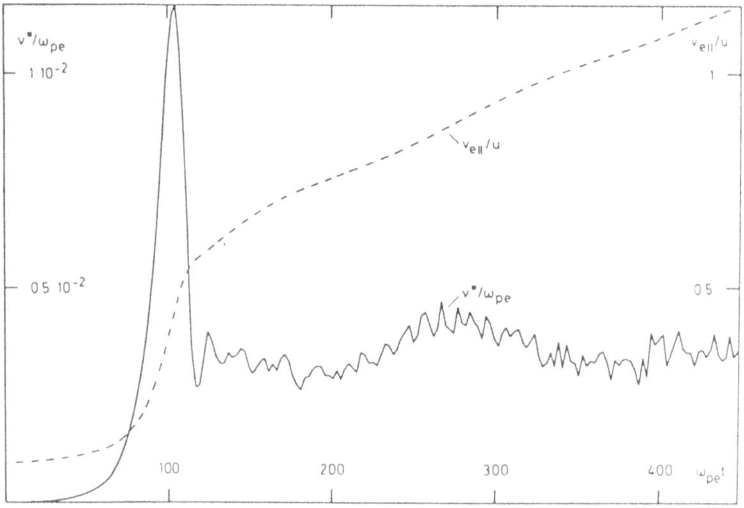

Figure 9 Transition from the reactive Bunemann instability ($u \gg v_e$) to the kinetic ion sound instability ($u < v_e$) for constant drift u. (From 2D simulations of Dum and Chodura, 1979, mass ratio M/m= 1600). The applied electric field and the energy input are proportional to the effective collision frequency ν^*. The electron thermal velocity v_e continues its steady increase during the ion sound regime which lasts for several $10^3 \omega_{pe} t$.

$$A_n = \frac{1}{N} \sum_{j=0}^{N-1} A(j) \exp(-2\pi i jn/N) \tag{33}$$

where $A_{N-n} = A_{-n}$. At the grid points (2a) becomes then

$$A(j) = \sum_{n=0}^{N-1} A_n \exp(2\pi i jn/N) = A_o + (-1)^j A_{N/2} \tag{34a}$$

$$+ \sum_{n=1}^{N/2-1} [\, A_n^e \cos(2\pi nj/N) + A_n^o \sin(2\pi nj/N) \,]$$

$$A_n^e = A_n + A_{-n} = \frac{2}{N} \sum_{j=0}^{N-1} A(j) \cos(2\pi nj/N) \tag{34b}$$

$$A_n^o = i(A_n - A_{-n}) = \frac{2}{N} \sum_{j=1}^{N-1} A(j) \sin(2\pi nj/N) \tag{34c}$$

$$A_o = \frac{1}{N} \sum_{j=0}^{N-1} A(j) \quad ; \quad A_{N/2} = \frac{1}{N} \sum_{j=0}^{N-1} (-1)^j A(j) \tag{34d}$$

As the basis functions exp $(2\pi i jn/N)$ are eigenfunctions of linear difference operators, the discretized versions of (32) etc. are solved

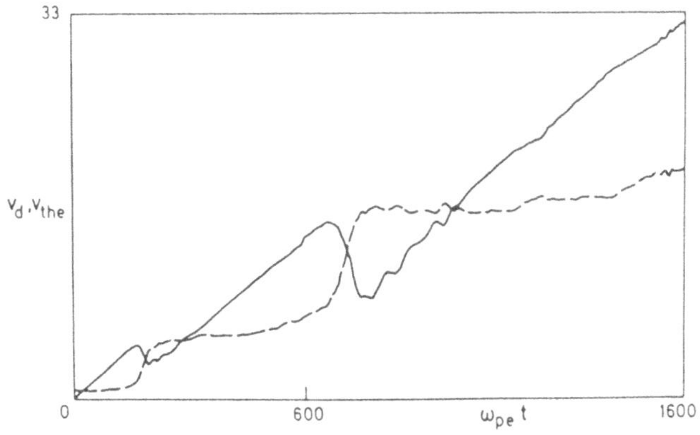

Figure 10 Two-stream instability for constant applied electric field E_0 (1 D simulation by Biskamp and Chodura, 1973; $eE_0/m\omega_{pe}v_e$ =0.03, mass ratio M/m=1000). Drift velocity v_d (solid line) and thermal velocity v_{the} (dashed line) increase linearly, apart from oscillations about $v_d \cong v_{the}$, until the wave modes outgrow the simulation box ($L=128\lambda_{De}$). Joining two such systems at $\omega_{pe}t=600$ restarts this process, but the ion sound regime $v_d < v_{the}$ is never entered.

directly by the expansion (34). Assuming not only that N is even but that $N=2^M$, where M is integer, the transformations in (33) or (34) can be carried out by the very efficient Fast Fourier Transform algorithm (FFT). The correspondence of this procedure for $L \to \infty$ to the important notions of wave modes and fluctuation spectra in an infinite homogeneous medium is also obvious and is often a strong reason for using it. The art in approximating this case numerically by periodic B.C. consists then in choosing simulation parameters such that the restriction in wave numbers, imposed by system size, grid spacing and accuracy requirements which imply usable $|n|<N/2$ or $|k|<\pi/\Delta x$, is tolerable and remains so in the course of the simulation. Especially the latter is often not the case as microscopic scale lengths such as the Debye length or the average gyroradius increase as a result of plasma heating. For instabilities with a very narrow range of unstable wave numbers, such as the Buneman instability, $ku/\omega_{pe} \simeq 1$, $\Delta(ku/\omega_{pe}) \simeq 3/2(m/M)^{1/3}$ for drifts $u \gg v_e$, one must make certain that a reasonably large number of wave modes are unstable initially (Dum and Chodura, 1979). It turns out that for simulations in which the current is held constant by an applied uniform electric field $E_0=\eta j$, where η is the anomalous resistivity, the linear growth phase is terminated suddenly with a spread of the wave spectrum to smaller wavenumbers (Figure 9). Again, this decay of the wave spectrum must be made possible by appropriatly chosen periodic B.C. New conditions for the wave number range apply in the ion sound regime $u<v_e$ that follows. As mentioned in Sec.3.2, restrictions on wave vector direction imposed by dimensionality also very strongly affect results especially in the kinetic (ion sound) regime, and for strong magnetic fields where oblique modes become increasingly important. An entirely different picture of the physical processes has emerged from previous (1D) simulations in which the applied electric field is held constant. The drift velocity u and the electron thermal velocity v_e increase linearly with time, apart form oscillations about the marginal stability condition $u \simeq v_e$. The ion sound regime $u<v_e$ thus is never entered. Biskamp and Chodura (1973) show by combining simulation systems that the evolution may be visualized as a sequence of instability bursts, each one doubling the dominant wavelength by coalescence of Bernstein Greene Kruskal nonlinear modes (Figure 10). At late times nearly all of the energy resides in the largest wavelength mode, $k=2\pi/L$, allowed by the size of the simulation system. Note that the energy input is much larger in the case of constant applied electric field and that in contrast to the real physical situation there are no energy losses. The energy content of the system and its characteristic scales thus must continue to increase. Other instabilities, e.g. the Weibel instability for temperature anisotropies in an unmagnetized plasma can also outgrow the simulation box in the course of their evolution, even though there is no external energy input (Brackbill and Forslund, 1982). The reason in the case of the Weibel instability is that relevant wavenumbers $0<(kc/\omega_{pe})<A^{1/2}$ and also growth rates become smaller as the anisotropy, $A=(T_x/T_y)-1$, decreases. The residual anisotropy and the time dependence of the fluctuation level thus depend on the box

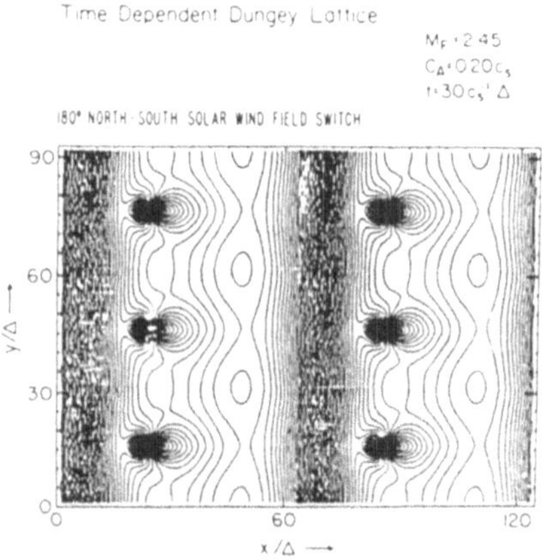

Figure 11 Lattice of magnetospheres (Leboeuf et al., 1979). Interaction with repeated images is due to periodic boundary conditions for the fields and leads to the island-like downstream structure. The effect of periodic field boundary conditions is very slight for other runs.

size. Such statements as that dominance of a single large amplitude wave is due to nonlinear suppression of other initially also unstable modes have often arisen from neglecting the effects of too small a box size. Restrictions on dimensionality and mass ratios also contribute to such unrealistic simulations.

Periodic B.C. and FFT are also used for phenomena which are not all wave like, such as the interaction of the solar wind with the earth's dipoe field (Leboeuf et al., 1979), see Figure 11. The effect of such B.C. may be interpretated as the interaction of the computational domain with its repeated images. Difficulties in the physical interpretation arise if this effect does not remain localized near the boundaries. These difficulties are, however, not unique to periodic B.C. but may arise any time B.C. are chosen for mathematical convenience in cases other than the one for which they derive on physical grounds.

4.2 Nonperiodic Boundary Conditions

In addition to pure initial value problems and the cases of constant current j_p or constant applied electric field E_0 that were just discussed, it is also possible to allow for a more general time dependence. The electric field E_0 may be applied by conducting plates at $x = 0, L$ with potentials $\Phi_{L,R}$ such that $E_0=(\Phi_L-\Phi_R)/L$. The surface charges on these plates are then

$$Q_L = (E_0A/4\pi) - \sum_j q_j(1 - x_j/L) \; ; \; Q_R = - (E_0A/4\pi) - \Sigma q_j x_j/L \qquad (35a,b)$$

where A is the cross section and the second terms are image charges induced by charges in the interior $0<x<L$ of this plasma capacitor. The continuity equation for the external current is therefore

$$j_{ext} = \frac{1}{4\pi}\frac{\partial E_0}{\partial t} - \frac{1}{V}\sum_j q_j v_j \qquad (36)$$

where the last term is the volume averaged plasma current j_p for charges of velocity v_j. The external electric circuit connected to the capacitor determines then the relation between j_{ext} and the potential difference $\Phi_L-\Phi_R$ or E_0. It may be possible to simulate a more realistic dynamic behavior in this way. An external inductance, e.g. may restrict the current to slow variations or may oppose current disruption by increasing the plasma potential drop over the applied potential. The difference between j_{ext} and j_p in (36) is often neglected in these electric B.C. If E_0 is supported by anomalous plasma resistivity, $E_0=\eta j_p$, $\eta=4\pi\nu/\omega_{pe}^2$ and is slowly varying with timescale T, this difference is of order $(\nu/\omega_{pe})(1/\omega_{pe}T)$ and thus is usually negligible. However, the remaining B.C. for the high frequency component

$$\frac{1}{4\pi}\frac{\partial \tilde{E}_0}{\partial t} + \tilde{j}_p = 0 \qquad (37)$$

in contrast to the B.C. $j_p=0$, allows the excitation of the k=0 plasma mode with the plasma frequency ω_{pe}. This sometimes makes a difference in the nonlinear phase of evolution.

An alternative to the Dirichlet B.C. of prescribed potentials $\Phi_{L,R}$ at x=0, L is the Neumann B.C. of prescribed electric fields. Poisson's equation in 1D implies, however, that the fields at the boundaries x=0, L must be the same unless there exists a homogenous net charge component σ_0 in the plasma slab, $E_R-E_L=4\pi\sigma_0L$. The electric fields may e.g. be imposed by surface charges of a plasma slab in vacuum. At the surface x=0 e.g., the electric field jumps then from the constant vacuum field E(0-) to

$$E_L = E(0+) = E(0-) + 4\pi Q_L \ . \tag{38}$$

It is seen (Decyk and Dawson, 1979) that in 1D the general B.C.

$$\alpha(x)\Phi(x) + \beta(x)d\Phi/dx = \gamma(x) \quad ; \ x=0,L \tag{39}$$

can be accomodated by adding an appropriate solution of the form

$$\Phi^h(x) = \Phi_0 - E_0x - 2\pi\sigma_0x^2 \tag{40}$$

to the periodic solution of the inhomogenous equation (32). The most common 2 D extension of this plasma slab model consists in assuming periodicity in y,

$$\Phi(x,y) = \sum_m \Phi_m(x)\ \exp(ik_my) \tag{41}$$

An appropriate homogenous solution

$$\Phi_m^h(x) = A_me^{|k_m|x} + B_me^{-|k_m|x} \tag{42}$$

may then be added to the doubly periodic inhomogenous solution $\Phi_m{}^P$ in order to meet B.C. of the form (39) also for the $k_m\neq0$ components. For instance, at a vacuum interface x=0 with surface charge Q_m^L, continuity with the decaying vacuum solution

$$\Phi_m(x) = \Phi_m(0-)e^{|k_m|x} \qquad ; \quad x < 0 \tag{43}$$

requires

$$B_m = -\frac{1}{2|k_m|}\ \{\ E_m^P(0) + k_m[\Phi_m^P(0)-4\pi Q_m^L]\ \}\ . \tag{44}$$

The periodic solution Φ_m^P of (32) may be found by Fourier transform or its discretized FFT approximation. For the Dirichlet B.C. $\Phi_m=0$, x=0,L and the Neumann B.C. $\Phi_m'=0$, Fourier half series expansions in $\sin\pi$ x/L and $\cos\pi$ x/L, respectively, may be also used. The combination $\Phi'(0)=0$, $\Phi(L)=0$ may be implemented by even-odd continuation of the computational domain, corresponding to a period 4L in a cos expansion.

The options of Dirichlet, Neumann, periodic B.C. (30) or the mixed B.C. (39) are also available for other direct field solvers which may be faster than 2D FFT. Sets of tridiagonal equations of the general from

$$a_j\Phi_{j+1} + b_j\Phi_j + c_j\Phi_{j-1} = d_j \qquad (45)$$

result from lowest order central differencing $\Phi'(x)=[\Phi(x+\Delta x)-\Phi(x-\Delta x)]/2\Delta x$ and after applying separation or FFT (or cyclic reduction) in the case of separable 2D differancial operators. They may be solved by the ansatz $\Phi_j=e_j\Phi_{j+1}+f_j$. The B.C. are used to terminate the resulting recursion relations. The generalization to vector fields is straightforward. It is not difficult so see that for an intervall L with $N=L/\Delta x$ and the interior grid points, $x_j=j\Delta x$, the solution of (32) requires the charges at grid points $j=1,2,\cdots,N-1$ in the case of the Dirichlet B.C., with given $\Phi(0)$, $\Phi(N)$. Neumann B.C. with given $\Phi'(0)=(\Phi_1-\Phi_{-1})/2\Delta x$, $\Phi'(N)=(\Phi_{N+1}-\Phi_{N-1})/2\Delta x$ or the mixed B.C. (39) are used to determine Φ_{-1}, Φ_{N+1} just outside the computational domain. The charges at $j=0,1,2,\cdots,N$ are required in this case. Periodic B.C., $\Phi_o=\Phi_N$, require the charges at $j=0,1,2.\cdots,N-1$ or $(1,2,\cdots.N)$ and the continuation $\Phi_{-1}=\Phi_{N-1}$ or $\Phi_{N+1}=\Phi_1$. The discretized version of (43) is required in the case of a vacuum boundary at x=0, in order to find a relation for Φ_{-1}, see also Langdon and Lashinski (1976). Finite difference methods for the solution of field equations that include also nonseparable cases and complex boundary surfaces are surveyed by Hockney and Eastwood (1981, Chapter 6).

The methods of magnetostatics may be used in a similar manner for the implementation of magnetic B.C. The equation for the vector potential $\underline{A}$,$\underline{B}=\nabla\times\underline{A}$, indeed is just the vectorial version of (32) in terms of the current density

$$\nabla^2\underline{A} = -\frac{4\pi}{c}\,\underline{j} \qquad (46)$$

In order to guarantee charge continuity, the Darwin approximation (Nielson & Lewis, 1976; Busnardo-Neto et al. 1977) does not simply neglect the displacement current but only its transverse component. Thus, only the transverse current $\nabla\cdot j^t=0$, is used in (46). The longitudinal components j^l satisfies

$$\frac{4\pi}{c}\,\underline{j}^l + \frac{1}{c}\frac{\partial E^l}{\partial t} = 0\ , \qquad (47)$$

c.f. with (37), and the transverse electric field is determined from

$$-\nabla\times(\nabla\times\underline{E}) = \nabla^2\underline{E}^t = \frac{4\pi}{c}\frac{\partial \underline{j}^t}{\partial t} \qquad (48)$$

Simulation models with conducting, current carrying plates at x=0,L have been developed in this way by Lin and Dawson (1978) for skin current penetration and by Leboeuf et al. (1982) for reconnection.

The case of fully electromagnetic B.C., required e.g. for the irradiation of a plasma slab, is complicated by spurious wave reflection from the computational boundaries. Methods for reducing reflections by improved matching of fields in the computational domain and the outside (vacuum) region can be quite elaborate (Langdon and Lashinski, 1976). An absorbing boundary with tolerable reflection can be provided in some cases by simple masking of some field components, i.e. by gradually reducing their amplitudes by $E(x) \to f(x)E(x)$, within a buffer zone of width ΔL. (Tajima and Lee, 1981). If particles are prevented by reflecting B.C. from entering these zones, they are not affected by the field modifications. Long wavelength modes, $\lambda > \Delta L$, however, also do not sample the ramp structure and thus contribute to residual reflection. Instead of this primarily reactive modification of wave propagation near the boundary, the computational domain may also be extended by a sufficiently large region of damping (sponge layer), before applying simple B.C. such as Neumann B.C. This may work not only for electromagnetic waves (Hashimoto et al., 1983) but <u>open</u> boundaries which allow MHD flow to leave the computational domain proper without significant distortions can also be constructed (Sato and Hayashi, 1979). Making direct use of the wave (hyperbolic) character in constructing such <u>outflow</u> B.C., is more complicated, but may save computation costs. In 1 D this amounts to some discretization of radiation conditions for outgoing waves such as

$$\frac{\partial A}{\partial t} + c \frac{\partial A}{\partial x} = 0 \ , \tag{49}$$

where the local and instantaneous phase speed c is evaluated from the grid points closest to the open boundary. It is possible to discretize (49) such that no reflection occurs for a purely harmonic outgoing wave (Orlanski, 1976). The scheme, however, is nonlinear and not very simply implemented. Simplifications are possible, but as pointed out by Tajima and Lee (1981), Simple spatial extrapolation, e.g. $A_N=3(A_{N-1}-A_{N-2})+A_{N-3}$, gives 100 % reflection of purely harmonic outgoing waves. One verifies also very easily that for damped waves, reflection may be reduced but remains wave number dependent in magnitude and phase. Combinations of radiation conditions such as (49) with damping or wave speed modifications in a buffer zone (sponge layer or filter) can be advantageous (Israeli and Orzag, 1981). Characteristics analysis of boundaries may also be helpful. Physical B.C. can be prescribed only for ingoing characteristics, while the additional relations needed to determine dependent variables at the boundary are <u>boundary approximations</u> (Blottner, 1982) obtained by extrapolations such as (49), characteristic compatibility conditions, or difference relations based on the governing equations. For subsonic 1D entry flow of an isotropic fluid, the path line u and Mach line $u+C_s$ are entering, while the characteristic corresponding to $u-C_s$ is leaving the computational domain. To determine the fluid variables ρ,u,P at the boundary, two B.C. and one boundary approximation (B.A.) are thus needed. Subsonic exit flow requires 1 B.C., 2 B.A., while no B.C. can be prescribed for supersonic exit flow, etc. The extension of this analysis to multidimensional MHD flows, describing e.g. perturbed solar wind flow from the corona, is very difficult (Nakagawa, 1981). Minimizing the effects of artificial boundaries remains a rich field for analytical and numerical experimentation, as no method is perfect or universally applicable.

5. BOUNDARY CONDITIONS FOR CHARGED PARTICLES

In principle, since particles are handled individually in particle codes, as it were by Maxwell's demon, B.C. for particles may be implemented in any way this is desired. The only caveat is the danger of creating unphsical boundary sheaths and numerical instabilities that may also spread to the interior of the system, thereby obscuring if not dominating the physical behavior of interest. Although it is plausible that B.C. are less critical at distant boundaries with low plasma density, this case is often not realized in practice.

To simulate an infinite homogeneous plasma, periodic B.C. are also employed for particles. Particles leaving the simulation box at a given time step are re-inserted with the same velocity at the opposite end. This B.C. presents no numerical problems and may be used e.g. for the study of velocity space instabilities. Drift waves and other instabilities that are excited in regions of local plasma gradients, however, already require a finite slab model, if interference from positive and negative slopes, which would arise from periodic continuation, is to be excluded. Moreover, if there is actually a symmetry axis, half of the computational effort may be saved by using symmetry relations as B.C.

Reflecting B.C. are used to confine the plasma within a slab. The most obvious way is to use the scheme

$$x \to 2x_w - x \ ; \qquad v_x \to -v_x \tag{50}$$

corresponding to elastic reflection from a wall at $x=x_w$. Some complications arise, however, for this B.C. It is clearly inconsistent with a streaming distribution $v_x \neq 0$ and may then result in a two stream instability with the reflected distribution. From its physical interpretation it is clear that (50) is consistent with equilibrium distribution functions. Initial particle loading, however, should also be consistent with the desired distribution. In the case of a magnetic field tangential to the boundary surface this requires that guiding center loading, which is appropriate for magnetic confinement, be modified near a reflecting boundary where we have, in effect, the limiting case of electrostatic confinement by a short range potential. The shift of guiding centers by reflection may otherwise result in surface currents, density gradients and space charge electric fields in a sheath with a width of the order of the Larmor radius. Drift wave like instabilities excited by such sheaths, however, spread into the interior of the system (Naitou et al., 1979). Potential fluctuations along the boundary may be smoothened to reduce this spreading (Lee and Okuda, 1978). In magnetostatic or electromagnetic codes, however, the magnetic field is also modified by numerical diamagnetic currents. It is possible to use modifications of the reflection scheme in order to reduce numerical instabilities, but any resemblance to physical B.C. may be lost in this way. Consistent initial particle loading certainly is preferable. For uniform density near a wall, say at $x_w=0$, this can be done (Naitou et al., 1979) by first loading a uniform guiding center distribution for $X=x+v_y/\Omega_j>0$, and then returning out of bonds particles $x<0$ by

$$x \to -x\ , \quad v_y \to -v_y \tag{51}$$

where v_y is the velocity component in the boundary plane and normal to the magnetic field, and $\Omega_j=e_jB/m_jc$ is the cyclotron frequency. Using (49) also for this initial operation (Lee and Okuda, 1978) is inconsistent with the sense of particle gyration.

It may also be verified from the equations of motions that the reflecting B.C. (50) is equivalent to a symmetry condition with reversal of the tangential magnetic field and normal electric field across the boundary. This B.C. may thus be used for the mid-plane of a reconnection geometry in restricting computation to one half of the system (Swift, 1983). If, however, the magnetic field remains symmetric across the mid-plane, inversion symmetry (Nevins et al., 1981) $\underline{E}(-\underline{x}_j)=-\underline{E}(-\underline{x}_j)$; $\underline{v}(-\underline{x}_j)=-\underline{v}(\underline{x}_j)$, is easiest to implement. Again, computation may be restricted to $x>0$, but the regions $y>0$ and $y<0$ are coupled by this B.C., with actual consequences depending on how close to the axis the phenomena of interest are located. Moreover, for both cases, physical symmetry applies, strictly speaking, in a statistical sense and not to individual particles, as implied by these B.C.

Both, the periodic and reflecting B.C. are designed to maintain a constant number of particles in the computational domain. No mechanical energy is tranferred across such boundaries. If, however, electromagnetic B.C. are non-periodic, transfer of electromagnetic energy, as described by Poynting's theorem, its still possible. The purely electrostatic version of energy conservation is given by (Decyk, 1982)

$$\frac{\partial}{\partial t}\left[\ \frac{(E^l)^2}{8\pi} + \sum_\alpha n_\alpha m_\alpha \frac{\langle v^2\rangle_\alpha}{2}\ \right] + \nabla\cdot\left[\ j^t\Phi + \sum_\alpha n_\alpha m_\alpha \frac{\langle v^2 \underline{v}\rangle_\alpha}{2}\ \right\} = 0 \tag{52}$$

where $\underline{E}^l=-\nabla\Phi$ is the electric field, and $\underline{j}^t= \underline{j}+(1/4\pi)\partial\underline{E}^l/\partial t$ is the transverse component of the conduction current $\underline{j}=\sum_\alpha n_\alpha e_\alpha\langle\underline{v}\rangle_\alpha$. For reflecting B.C., $\underline{j}=0$, but electrical energy is still transferred by the electrostatic displacement current across a boundary with nonzero potential and electric field. For periodic particle B.C. in general at the boundary and a transfer of electrical energy is possible also for the Neumann B.C. $\underline{E}=-\nabla\Phi=0$, with floating potential. Swift and Ambrosiano (1981) have shown that for the electric B.C. $\Phi(0)=0$, $\Phi'(L)=0$ and periodic particle B.C., the current across the boundaries becomes correlated with potential fluctuations so that the term $\tilde{J}\tilde{\Phi}$ may have a non-zero time average and contribute to a numerical instabilitiy by buildup of wave energy.

Mechanical energy may be transferred across boundaries with inelastic or random reflection, while still preserving the number of particles in the system. In random reflection, particles are re-introduced with a velocity drawn at random from a distribution function with unit flux. Taking

$$g_\alpha(v_x) = (v_x/v_\alpha^2)\exp(-v_x^2/2v_\alpha^2)\ \mathrm{sgn}\ v_x \tag{53}$$

where $\mathrm{sgn}v_x \gtrless 0$ for the left and right boundary, respectively, simulates a <u>heat bath</u> with temperature $T_\alpha = mv_\alpha^2$.

Particles may also be <u>emitted</u> independently of their exit from the system, by drawing particle velocities from a flux distribution at a specified rate. A neighbouring plasma <u>reservoir</u> with given density, mean velocity and temperature may be simulated in this way. For a Maxwellian reservoir $nv_e/(2\pi)^{1/2}$ particles/sec are drawn at random from (53). At an <u>absorbing</u> boundary, exiting particles are simply deleted. Applying this B.C. to a boundary that is at the same time emitting, means that the reservoir is not affected by particles entering it. Combinations of these basic particle B.C. are used in the various simulation models discussed in Sec.2 and 3. Some modifications also occur. A simulation of ionospheric backscatter (Sec.3.2) e.g. requires that a fraction ($\simeq$50 %) of the exiting particles be randomly reflected, using non Maxwellian fluxes that depend on the energy of the exiting particles. Swift and Lee (1983) use injection of particles into homogeneous buffer zones without an electric field, in order to reduce boundary effects for oblique magnetic fields. Pritchett and Coroniti (1984), on the other hand, find that energy conservation (52) in an electrostatic model, with guiding center approximation for electrons, is markedly improved if reflection is used only for ions. Electrons are then confined by an electrostatic boundary sheath. Experimentation with B.C. is thus always advisable.

Particle injection is used not only to simulate actual sources, including sources within the plasma volume, but injection rates may also be adjusted such as to reduce undesirable space and time dependences that are not related to the physical effect of interest, but to transit times of particles across a too small simulation box. Initial and B.C. must also be matched for this purpose. Loading e.g. a uniform Maxwellian distribution with density n_0 in a box of length L and injecting particles from a thermal reservoir at x=0 with density n_I creates a nonuniform density distribution,

$$n(x,t) = \frac{n_I}{2} + \frac{n_0 - n_I}{2}\,\mathrm{erf}\frac{x/\lambda_{De}}{\sqrt{2}\,\omega_{pe}t} + \frac{n_0}{2}\,\mathrm{erf}\frac{L-(x/\lambda_{De})}{\sqrt{2}\,\omega_{pe}t} \tag{54}$$

assuming free streaming and that all exiting particles are absorbed. (Electron plasma frequency ω_{pe} and Debye length are introduced only for normalization purposes.) With absorption, the initial particle number $N_0=n_0L$ would decay at a rate

$$\frac{1}{N_0}\frac{\partial N}{\partial t} = 2\,\frac{\omega_{pe}}{(2\pi)^{1/2}}\,\frac{\lambda_{De}}{L}\left\{1 - \exp\left[-\frac{(L/\lambda_{De})^2}{2(\omega_{pe}t)^2}\right]\right\} \tag{55}$$

For typical system lengths of some $100\lambda_{De}$ ($\lambda_{De}=v_e/\omega_{pe}$) this is usually intolerable. Changing the x=L boundary from absorbing to reflecting, effectively doubles the system length and after some transit times $\tau_{eL}=L/v_e\sqrt{2}$ establishes a Maxwellian injected distribution rather than the Half-Maxwellian $v_x>0$ that is created for (54). The equivalent Vlasov representation of (53), i.e. precribing $f^{\pm}(v)$ for $v_x \gtrless 0$ at the left or right boundary respectively, has been used for this simple exercise. In an 1D Vlasov simulation code f(v,x) is actually transported from the B.C. and initial conditions, using Liouville's theorem and the characteristics of the Vlasov equation which just correspond to self-consistent particle trajectories. The distribution function is evaluated at (x,v) grid points by an area weighting scheme which is quite analogous to a 2 D particle in cell scheme (Sakanaka et al., 1971). Although there is still grid noise, shot noise due to particle discreteness is avoided.

The self-consistent electric field of a plasma modifies, of course, the simple relations (54-55) which are based on free streaming. Within the body of the plasma, quasi-neutrality will be maintained by electrons dragging along ions, provided time scales are also appropriate, i.e. $O(\omega_{pi})$. Studies of sudden plasma release into a vacuum show indeed that the characteristic expansion speed is reduced from v_e to $c_s=(T_e/M)^{1/2}$, but (linearly) increasing towards an ion front which is continuously accelerated by precursor electrons (Denavit, 1979; Singh and Schunk, 1982b). A rarefaction wave with velocity -c propagates into the undisturbed plasma. B.C. are not very critical as long as the ion front and the head of the rarefraction wave are far from the boundaries. A plausible choice are absorbing boundaries, but with injection from a thermal reservoir, in order to maintain an undisturbed plasma near the left boundary (Denavit, 1979). In 1 D, the electric field at x is simply determined by the net charge in x'<x,

$$E(x) = E_L + 4\pi \int_{x_L}^{x} dx'\sigma(x') \tag{56}$$

where E_L is the electric field at the left boundary x_L. It is plausible to assume that the electric field is negligible deep inside the undisturbed plasma, $E_L\simeq 0$. It also vanishes at a boundary corresponding to a symmetry plane. (Random-) Reflection of particles exiting at this plane is then an appropriate particle B.C. (Mason, 1971).

The (steady) flow of a plasma towards an absorbing wall is a closely related problem. However, if one is interested in a simulation of the sheath structure near this wall, it is uneconomical if not infeasible to also simulate the largely undisturbed pre-sheath plasma, although some preacceleration by a weak electric field must occur in this region. Injection of particles from a <u>drifting</u> distribution at the (poorly defined) sheath boundary and absorption of all exiting particles is a plausible alternative. The ion drift velocity must exceed some critical speed ($\simeq c_s$) in order to get a smooth transition, $E\simeq 0$, $E'=4\pi\sigma\simeq 0$ to the presheath plasma. Chodura

(1982) extends this Bohm criterion to flow in a magnetic field which is at an angle to the boundary. In his simulaiton the electric field at the boundary is related to the charge imbalance that will develop in the initialy charge neutral presheath plasma if there is a net current outflow,

$$E(x=0,t) = 4\pi Q_L = - \int_0^t dt' \; j(x=0,t') \tag{57}$$

In contrast to (56) this is a B.C., as the presheath plasma is not simulated but replaced by a reservoir with fixed emission rate. If e.g. the ion emission rate is too small for maintaining charge neutrality, an electric field develops near the boundary that will increase the ion drift within the simulation box to the Bohm limit. The bulk of the electrons is reflected by the space charge potential that develops within the plasma sheath and returns thus to the reservoir.

A stationary double layer is a limited region of charge imbalance, with vanishing electric fields outside. In addition to the Bohm criterion for the ion drift at the high potential side, a similar criterion $u_e > v_e$ must be satisfied in general for the electrons injected at the low potential side (Bloch, 1972). The electron and ion fluxes should also satisfy the Langmuir condition, $j_e/j_i = (m_i/m_e)^{1/2}$, in order to have overall charge neutrality within the layer. Strong instabilities develop on the high potential side of the layer if the Langmuir condition is violated (Singh, 1980). Injecting particles from boundaries with prescibed potentials is a consistent choice of B.C. The electric feild at these electrodes is then related to the induced self-consistent surface charges, cf. (35a,b). It may undergo large fluctuations, depending on processes in the plasma volume. The diode model of Burger (1965) is a drastic example (Figure 2). An additional control of the plasma flow and electric fields near the boundaries of the target plasma can be exercised by applying potentials to narrowly spaced grids (Figure 1). The potential within the body of the plasma reservoirs adjusts self-consistently in this case, however (Hershkowitz et al., 1981). The Langmuir criterion need not be satisfied in this case, with the exception of very strong double layers $e\Delta\Phi/T_e >> 1$. Weak double layers depend strongly on the B.C. Ions created by collisional processes and trapped between the high potential side and a biased grid (B in Figure 1) also have an important effect on the potential structure. Similarly, the Bohm-Bloch criterion for electrons may be relaxed if there is a sufficient number of electrons trapped between the low potential side and a magnetic mirror (Kan and Lee, 1980).

These considerations on sheaths and stationary double layers may be tested in time dependent simulations or used for their design. The study of sheaths near or between emitting and absorbing electrodes is generally not the object of space plasma simulations. As discussed in Sec.3.2, a simulation of potential drops on auroral field lines, for example, requires however two artificial boundary surfaces at same distance from the ionospheric and magnetospheric plasma sources, respectively. For short simulation boxes one is thus in practice faced with the same boundary problems. Even if one succeeded, by appropriate particle injection and electric B.C., in avoiding large gradients in the density and electric field near the artificial boundaries and is able to control the current (net fluxes), the microscopic particle distribution functions still will not be the same as in the non-excised system. The response of this system to processes within the simulated region cannot be modeled by injection from sources with fixed (Maxwellian)particle distributions. The same comment applies to electric B.C., which moreover, by internal reflection, may also modify the small scale turbulence. Convective instabilities due to the accelerated electron beam may become absolute. An enhanced turbulence level at the high potential boundary was noted by Joyce and Hubbard (1978). A restriction to 1D further modifies potential structures and turbulence. In 2D, the design of electric B.C. consistent with plasma flow is, however, an even more subtile problem (see Sec.3.2).

6. COUPLING IF LOCAL AND GLOBAL DYNAMICS

A simulation code which calculates particle orbits in self-consistent electromagnetic fields provides, in principle, the most complete description of plasmas. Numerical stability and energy conservation in conventional particle codes actually demand the resolution of small scale phenomena, i.e. the time and space steps are restricted by the scales of electron motion, $\omega_{pe}\Delta t \lesssim o(1)$, $\Delta x/\lambda_{De} \lesssim O(1)$, $v_e \Delta t \lesssim O(\Delta x)$. Explicit solution of field equations furthermore imposes the Courant condition $v_{ph}\Delta t \lesssim O(1)$, corresponding to the largest phase velocity of plasma modes, e.g. the speed of light. A sufficiently large number of particles must also be present in the computational cells, in order to make statistical noise tolerable (e.g. Hockney and Eastwood, 1983, Ch. 9). The resulting practical restrictions on system size and the range of time scales can be very severe, especially for low frequency phenomena, and much of the potential realism of particle simulations may actually be lost. The number of time steps or the number of particles may be reduced if unwanted high frequency are eliminated by some combination of filtering, implicit methods, and orbit averaging (Mason, 1981; Denavit, 1981; Cohen et al., 1982, Brackbill and Forslund, 1982). An intriguing aspect of implicit schemes is that they also make contact with the fluid continuity and momentum equations, although only in an intermediary fashion, when advancing fields. Practical constraints on the design of simulation experiments, certainly, will be eased by these very promising developments and other efforts in increasing computing power.

A basically different approach is to use some kind of hybrid model which eliminates a detailed simulation of what is already reasonably understood, and thus allows to focus more strongly on new features. In the conventional sense (cf. Sec.2.2, 3.6) this means that a fluid description is used for some plasma components, while others are described by particles. The incorporation of anomalous transport is a strong and continuing challenge for this approach. Space physics, especially, as we have seen, presents another, perhaps even bigger challenge, i.e. the coupling of local (kinetic) phenomena within a large global system.

6.1 Fluid and Hybrid models

A fluid description in the classical sense applies to plasmas in which collisions are sufficiently frequent to maintain the near Maxwellian distributions of classical transport theory. Conditions for the validity of this theory are rather stringent, especially for heat flux, cf. Sec.2.1. As there are many levels of (ad hoc) idealization in fluid models, implementation of fluid B.C. should be based on the actual governing equations for the field variables, following the procedures of Sec.4. Physical intuition is still helpful, of course. It is always instructive, if not necessary, to mentally follow the finite difference algorithm for a few steps near the boundaries of the space and time domain, in order to see the actual workings of B.C. and initial conditions. Numerical stability may impose additional compatibility conditions between the algorithm for the volume and numerical B.C.

By drawing exospheres around the sun and planets (Figure 6), we have excluded collision dominated plasmas from our consideration, yet found (Sec.3) many simulation models which use an MHD description, either in its ideal form, or with some phenomenological dissipation terms. Although these models are generally introduced in an ad hoc manner, without any particular justification, they can usually be considered as the first significant step towards understanding the flows of mass, momentum, energy, and especially the interchange between mechanical and magnetic energy. Despite the model idealizations, results for global dynamics often are more realistic than attempts at a full kinetic simulation, because the computational domain can be increased dramatically. Of course, there are kinetic effects which have no fluid counterparts, but dependence on other kinetic effects induced e.g. by turbulence, may be tested by varying the parameters or structure of transport relations within reasonable limits. What is reasonable depends on the understanding of kinetic effects. Conservation laws must be satisfied, of course, but kinetic effects may also open up new avenues for the flows, corresponding to effective transport relations which differ, even in structure, very much from the classical relations (Dum, 1978). Anomalous transport is related to the turbulence level and wave mode structure, and thus may have a much stronger space and time dependence than classical transport. This dependence is related to the growth of microinstabilities which in turn, as we have seen (Sec.2.2), depends on the macroscopic dynamics. A reasonable incorporation of anomalous transport effects into fluid or hybrid models is thus a difficult, but not impossible, task which requires close contact with kinetic theory and simulations. Collisionless heat flux, especially, represent a challenge which has not been accepted to any extent for space plasma simulations, although electron heat flux largely controls solar wind flow, for example (Sec.3.1).

Even highly idealized fluid models can have a very useful function in advancing our understanding. It is necessary, however, to examine this question not only in the design phase of a simulation, but it needs to be constantly re-examined, in order to determine the appropriate moment for taking the next step towards a more complete model. Such an analysis (e.g. Dum, 1984) may show that it has become essential to use multifluid models or hybrid models with a full kinetic description for some plasma components. B.C., obviously, should also progress step by step from convenient ad hoc conditions towards more realistic B.C. The variety of models and possibilities for extensions become still larger in this case. The full reasons for the particular design of a model are generally only known to its builder. From some papers it is even impossible to make a reasonably certain guess about key features, involving B.C., in particular. It may be permitted therefore, to return briefly to electrodynamic magnetosphere ionosphere coupling (Sec.3.3). Major aspects of the philosophy on the construction and analysis of simulation models, as expressed by this paper, may then also be summarized. This philosophy, not too surprisingly, is largely formed by the sum of the authors own work and experience.

The simulation model of Lysak and Dum (1983) for auroral flux tubes differs in a number of aspects from the MHD models for global or local phenomena that were described in the present paper. Rather than trying to work with one of these very advanced and fairly general MHD codes, we decided on a "home-made" code for our (time-)limited purposes. It performs without problems, which should greatly encourage newcomers to the field. We saved efforts by not inluding pressure terms, which limits us to $\beta_e=8\pi nT_e/B^2=2(v_e/v_A)^2 m/M\cong\beta_i<2m/M$, or heights below a few R_E, although propagation of shear Alfvén waves is not drastically modified for somewhat larger β. Wave dispersion in the cold plasma is due to electron inertia and friction,

$$\omega = k_{\parallel}V_{AK} \quad ; \quad V_{AK} = V_A[\ 1 + (k_{\perp}c/\omega_{pe})^2\ (1+i\nu/\omega)\ ]^{-1/2} \qquad (58)$$

as compared to (14) for ideal MHD. Our code thus determines the parallel electric field E_z from the full nonlinear equation of electron motion along the ambient magnetic field, rather than using an Ohm's law (3) with $\eta=m\nu_o/ne^2$. In addition to electron inertia, the rate of momentum transfer R_{ez} from saturated ion cyclotron turbulence to electrons (Dum, 1981) is included,

$$\frac{1}{ne} R_{ez} = \eta_z^* \ (j_z - n_e u_c \mathrm{sgn} j_z); \ \eta_z^* = \frac{m\nu^*}{ne^2} \ , \ \nu = \nu_1 + \nu_2\left(\frac{|u_z|-u_c}{u_c}\right)^{\alpha}, \quad (59)$$

where $\alpha=2$. As the self-consistent evolution of the temperatures and of the non-Maxwellian shapes of the distribution functions is not computed, all parameters in (59), except for the drift $u_z=-j_z/ne$ which is related to the self-consistent field aligned current, are varied within reasonable limits, in order to test the dependence of global processes on the microscopic dissipation mechanism. The height profile of the critical drift velocity $u_c \simeq 13\ v_i\ T_i/T_e$, is such that instability, $|u_z|>u_c$, can occur in a region centered about 1 R_E, cf.Figs.6, 7.

The perpendicular electric field is related to the (ion) polarization current by the ion equation of motion, cf. (18) with $\underline{u}_\perp = c\underline{E}\times B/B^2$. Self-consistent pressure terms are excluded not only by our low β assumption, but also by our restriction to 2 D and shear Alfvén waves. An ad hoc pressure gradient normal to the plane of computation is one of the methods for introducing a dynamo, which cannot be done self-consistently in a 2D model, cf. Sec. 3.4. Other methods we use are to prescribe the magnetic field (parallel current injection) or the perpendicular electric field at the magnetospheric boundary z=0. We can experiment with different B.C. in order to see the effect of the dynamo on structures in and below the acceleration (dissipation) zone. The buffer zone between the dynamo and the acceleration region corresponds to several R_E, thus is many orders of magnitude larger than in corresponding particle simulations, $1R_e \cong 10^6 \lambda_{De}$. The same applies to the distance, $\Delta L \sim 1R_E$ from the ionospheric boundary at which the field aligned current j_z is channeled into the horizontal ionospheric current sheet, cf. (17). Continuity requires that the relations

$$E_{xk}^{(i,r)} = \pm \frac{V_A}{V_{Ak}} \frac{V_A}{c} B_{yk}^{(i,r)} \quad ; \quad E_{xk}^{I} = \frac{c}{4\pi\Sigma_P} B_{yk}^{I} \tag{60a,b}$$

for the field of incident (upper sign) and reflected waves

$$E_x = [\ E_{xk}(z-V_{Ak}t) + E_{xk}(z+V_{Ak}t)\ ]\ e^{ikx} \tag{61}$$

and the fields on top of the ionosphere be matched. B.C. (60b) can be obtained in several ways, e.g. by equating dissipation in the ionosphere with the Poynting flux into its surface. Integral conservation laws for boundary cells are always useful in deriving B.C. A full solution of the wave equation for ionosphere, atmosphere and the conducting earth (review by Southwood and Hughes, 1983) is important for the interpretation of ground observations i.e. geomagnetic pulsations, see Sec.III.3, but in our case gives nearly the same result (60b), i.e. $B_y=0$ below the ionosphere and practically 100% reflection of the incident shear Alfvén wave, $E^{(r)} \approx -E^{(i)}$, cf. (15) with admittance $\Sigma_A \rightarrow \Sigma_{Ak} = (c^2/4\pi V_A)\ V_{Ak}/V_A$ for the currents

$$j_{zk}^{(i,r)} = ikB_{yk}^{(i,r)} \frac{c}{4\pi} = \pm \Sigma_{Ak}\ ikE_{xk}^{(i,r)} \ ; \ j_{xk}^{(i,r)} = \mp \Sigma_{Ak} \frac{\partial E_{xk}^{(i,r)}}{\partial z} \tag{62a,b}$$

The fields are advanced by relations of the form

$$[\ 1 - \frac{\partial}{\partial x}(\frac{c}{\omega_{pe}})^2 \frac{\partial}{\partial x}\ ] \frac{\partial B_y}{\partial t} = -\ c \frac{\partial E_x}{\partial z} + \frac{\partial}{\partial x} \frac{cR_{ez}}{ne} , \tag{63}$$

leaving out nonlinear convection terms which along with the

calculation of density modulations (depletions) are retained in the code. The parallel electric field $|E_z|<<|E_x|$ is computed from $j_z=c\partial B_y/\partial x\ 4\pi$, ($j_x=-c\ \partial B_y/\partial z\ 4\pi$) and contributes diffusion-like terms to Faraday's law (63), due to electron inertia (l.h.s.) and momentum transfer (r.h.s.). An implicit solution in x is required in order to stably resolve structures of $O(c/\omega_{pe})$. Narrow north south (east-west aligned) structures are an outstanding characteristic of discrete auroral arcs. Explicit stepping $\Delta z>V_A\Delta t$ is used along the ambient magnetic field $B_o>>|B_y|$. A staggered mesh is used for central differencing. E_x and B_y thus are not defined at the same grid points and some experimenting was necessary in order to implement the ionospheric B.C. in a numerically stable manner: (60b) is used to write (64) as an equation for the advance of B_y^I. If L_x is large enough, simple B.C. of the Dirichlet or Neumann type suffice at the $x=0$, L_x boundaries. Open boundaries could be constructed, following the methods of Sec.4.

The significance of wave reflection by the acceleration region (Figure 7) is one of those typical discoveries by simulation of a basically simple process that apparently was not realized by anyone, although it should have been quite obvious, especially to people familiar with electrical transmission lines. Depending on the parameters of acceleration region and dynamo, ionosphere and magnetospheric dynamo can be effectively <u>decoupled</u>. In addition to the physical reflections at the ionosphere ($R=E^{(r)}/E^{(i)}\cong-1$) and acceleration region ($R\cong+1$), wave reflection from the artificial boundaries at $z=0$ and $x=0$, L_x also occurs. If B_y $(x,z=0,t)$ is prescribed (current injection), waves that were reflected by either the ionosphere or acceleration region are reflected again at $z=0$ with $R=1$ (B_y reversal). For prescribed E_x, clearly $R=-1$. A conservation theorem for the total energy associated with the wave motion (cf. 28) is very useful for the study of electrodynamic coupling,

$$\frac{\partial}{\partial t}\left\{\frac{B_y^2}{8\pi}+\left(\frac{c}{v_A}\right)^2\frac{E_x^2}{8\pi}+\frac{1}{2}nmu_z^2\right]+\frac{\partial}{\partial x}\left[-\frac{c}{4\pi}E_zB_y+\left(\frac{c}{v_A}\right)^2\frac{E_x^2}{8\pi}u_x\right]$$
$$+\frac{\partial}{\partial z}\left[\frac{c}{4\pi}E_xB_y+\frac{1}{2}nmu_z^3\right]=-Q=-\frac{1}{ne}R_{ez}j_z \tag{65}$$

where $u_x = j_x/ne$.

We have purposely constructed the simplest possible model for the study of some very fundamental processes. A number of extensions can be made. The B.C. at $z=0$ can be improved by introducing some kind of dynamo impedance for the response to the ionospheric and magnetospheric (acceleration region) loads which is mediated by the reflected waves. Sato (1978) used the impedance of a lossless and inertialess auroral transmission line, c.f. (60-64), for a standing wave pattern (pulsations), in order to describe the load of an ionospheric dynamo. A proper description of this wave pattern requires the extension to an inhomogeneous configuration with dipole-like ambient magnetic field (Sec.3.3). It has been carried out for our magnetosphere-ionosphere coupling model (Dum and Lysak, to be published, 1984), but the uniform magnetic field model is still used for other fundamental studies, among which the electron acceleration process plays an important role. In addition to bulk acceleration which appears explicitely in (65), a large part of the dissipation term Q can be channeled into acceleration of runaway particles, rather than electron or ion bulk heating (Sec.2.1). The fluid model is used to provide the appropriate macroscopic conditions for particle simulations of these kinetic processes (cf. also Sec.4.2). It can be seen from (63) that microscopic processes in turn affect global fields only through the momentum transfer to electrons. The implicit solution of (63) implies that only the R_{ez}/ne averaged over the macroscopic time step Δt and an imposed current that rises linearly between j_z^t and $j_z^{t+\Delta t}$, i.e. practically constant, is needed. The time and space averages correspond to scales $\Delta t \simeq 2\Omega_{ci}^{-1} \simeq 1000\omega_{pe}^{-1}$, $\Delta x \simeq 100\lambda_{De}$, $\Delta z \simeq 10^5 \lambda_{De}$, that is they are much larger than typical particle simulation domains or microscopic scale sizes.

6.2 Concluding Remarks

The preceding discussion of fluid and hybrid models illuminates a central theme of this article on the design and analysis of simulation models. It is the coupling of microscopic and macroscopic effects, and in space plasmas especially, the electrodynamic and mechanical (heat flux) coupling of local phenomena within a global system. It is safe to assume that this theme will remain a major, but exciting, challenge for the design of simulation models. Certainly, dramatic advances in computing power will further ease many of the boundary "woes" we have encountered, by allowing larger systems. A fully kinetic and realistic particle simulation of say magnetosphere-ionosphere coupling, not to speak of solar wind-magnetosphere coupling, however, seems inconceivable. Certainly, no one would want to analyze this mass of microscopic data. What is needed in this case is some kind of "patching", of parts of our map, Figure 6, which require a particle simulation of kinetic effects, with other parts for which a fluid model is adequate to describe bulk flows.

Another conclusion that emerges from this study, is that simulations should not necessarily be judged by the description they provide of the real physical world. It is one of the big strengths of simulations that, by eliminating some of the complexities of the real world at will, modern day "Gedanken experiments" can be carried out which advance our understanding of basic processes, step by step. Publications, however, should provide more "consumer" information on this point and avoid excessive claims of realism. Close contact with the real world, of course, should not be lost. Inappropriate simulations have also led to side tracking into rather elaborate, but practically useless theories. Simulations, in principle, allow unlimited diagnostics and thus can be complementary to physical observations. Simulation models that have been tested against detailed measurements in the near earth environment can be especially valuable in solving the puzzles of astrophysics.

REFERENCES

Abarbanel, S.S., and E.M. Murman, J. Comput. Phys., 48, 160, 1982.
Akasofu, S.-I., Planet. Space Sci., 29, 495, 1980.
Alfvén, H., and P. Carlquist, Solar Phys., 1, 220, 1967.
Baker, K. D., N. Singh, L.P. Block, R. Kist, W. Kampa, And H. Thiemann, J. Plasma Phys., 1, 1981.
Banks, P.M., C.R. Chappell, and A. F. Nagy, J. Geophys. Res., 79, 1459, 1974.
Baum, G.J., and A. Bratenahl, Magnetic Reconnection experiments, in L. Marton and C. Marton (eds.) Advances in Electrons and Electron Physics, 54, 1, Academic Press, New York, 1980.
Baumjohann, W., Adv. Space Res., 2, 55, 1983.
Baumjohann, W., and K. H. Glaßmeier, Planet. Space Sci., 32, 1984.
Bell, A.R., R.G. Evans, and D.J. Nicholas, Phys. Rev. Lett, 46, 243, 1981.
Birn, J., J. Geophys. Res., 85, 1214, 1980.
Birn, J., and E. W. Hones, J. Geophys. Res., 86, 6802, 1981.
Biskamp, D., Resistive MHD Processes, Physica Scripta, T2/2, 405, 1982.
Biskamp, D., and R. Chodura, Phys. Fluids. 16, 888, 1973.
Bloch, L. P., Cosmic Electrodyn., 3, 349, 1972.
Blottner, F. G., J. Comput. Phys., 48, 246, 1982.
Bodin, H. A. B., A. A. Newton, Nuclear Fusion, 20, 1255, 1980.
Borovsky, J.E., and G. Joyce, Plasma Phys., 29, 45, 1983.
Brackbill, J.U., and D.W.Forslund, J. Comput. Phys., 46, 271, 1982.
Brecht, S. B., J. J. Lyon, J. A. Fedder, and K. Hain, J. Geophys. Res., 87, 6098, 1982.
Brecht, S.B., J. J. Lyon, J. A. Fedder, and K. Hain, Geophys. Res. Lett., 8, 397, 1981.
Burger, P., J. Appl. Phys., 36, 1938, 1965.
Burn, R.D., Plasma Phys., 12, 331, 1970.
Busnardo-Neto, J., P.L. Pritchett, A.T. Lin, and J.M. Dawson, J. Comput. Phys., 23, 300, 1977.
Byers, J.A., B.I. Cohen, W.C. Condit, and J.D. Hansen, J. Comput. Phys., 27, 363, 1978.
Caramana, E.J., R.A. Nebel, and D.D. Schnack, Phys. Fluids, 26, 1305, 1983.
Carrier, G.F., and C.E. Pearson, Partial Differential Equations. Theory and Technique, Academic, New York, 1976.
Chan, C., N. Hershkowitz, and K.E. Lonngren, Phys. Fluids, 26, 1587, 1983.
Chan, C., and N. Hershkowitz, Phys. Fluids, 25, 2135, 1982.
Chiu, Y.T., and M. Schulz, J. Geophys. Res., 83, 629, 1978.
Chodura, R., C.T. Dum, F. Söldner, K.-H. Steuer, Anomalous Transport in Shock Produced Plasmas, Proc. 8th Europ. Conf. Contr. Fusion and Plasma Phys., 1, 93, Prague, 1977.
Chodura, R., C.T. Dum, M. Keilhacker, M. Kornherr, N. Niedermeyer, R. Protz, F. Söldner, K.-H. Steuer, Numerical and Experimental Results on the Production of Weakly Compressed Thermonuclear Plasmas in the High Voltage 5th Conf. Plasma Phys. Cotr. Fusion, Vol.3, 397, IAEA, Vienna 1975.

Chodura, R., Phys. Fluids, 25, 1628, 1982.
Christiansen, P.J., C.T. Dum, and R.L. Lysak, EOS, 63, 1066, 1982.
Cohen, B.I., R.P. Freis, and V. Thomas, J. Comput. Phys., 45, 345, 1982.
Cuperman, S., N. Metzler, and M. Spiegelglass, Astrophys. J., 198, 755, 1975.
Decyk, V.K., Phys. Fluids, 25, 1205, 1982.
Decyk, V.K., and J.M. Dawson, J. Comput. Phys., 30, 407, 1979.
Denavit, J., Phys. Fluids, 22, 1385, 1979.
Denavit, J., J. Comput. Phys., 42, 337, 1981.
Dickman, D.D., R.L. Morse, Phys. Fluids, 12, 1708, 1969.
Dum, C.T., AMPTE: Diamagnetic Effects and Anomalous Transport in Expanding Plasma Clouds, AMPTE Preprint #8 and to be published, 1984.
Dum, C. T., Electrostatic waves and anomalous transport in the solar wind, Proc. Solar Wind V, 1983b, NASA, Conf. Pub. 2280, 369, 1983b.
Dum, C.T., Anomalous Transport Induced by Field Aligned Currents and its Relation to Electromagnetic Coupling, International Astronomical Union Symposium 107, Reidel, Dordrecht, to appear, 1983a.
Dum, C.T., Anomalous Resistivity and Plasma Dynamics, in S.-I. Akasofu and J.R. Kan (eds.) Auroral Arc Formation, 408, AGU, Washington, 1981.
Dum, C. T., Phys. Fluids, 21, 956, 1978
Dum, C.T., and R Chodura, Anomalous Transition from Buneman to Ion Sound Instability, in P.J. Palmadesso, and K. Papadopoulos (eds.), Wave Instabilities in Space Plasmas, 135, Reidel, Dordrecht, 1979.
Dum, C.T., and T.H. Dupree, Phys. Fluids, 13, 2064, 1970.
Evans, D.S., J. Geophys. Res., 79, 2853, 1974.
Evans, D.S., The acceleration of Charged Particles at Low Altitudes, in D.J. Williams (ed.), Physics of Solar Planetary Environments, 730, AGU, Washington, D.C. 1976.
Forbes, T.G., and E.R. Priest, Solar Phys., 84, 169, 1983a.
Forbes, T.G., and E.R. Priest, J. Geophys. Res., 88, 863, 1983b.
Forbes, T.G., and E.R. Priest, Solar Phys., 31, 303, 1982.
Forslund, D.W., K. Quest, J.U. Brackbill, and K. Lee, J. Geophys. Res., 89, 1984.
Galeev, A.A., Magnetospheric Tail Dynamics, in A. Nishida, Magnetospheric Plasma Physics, 143, Center for Academic Publications, Reidel, Tokyo/Dordrecht, 1982.
Galeev, A.A., and L.M. Zelenyi, Sov. Phys., -JETP 43, 1113, 1976 (Zh. Eksp. Teor. Fiz. 70, 2133, 1976).
Glaßmeier, K.-H., J. Geophys. Res., 54, 125, 1984.
Goertz, C.K., and G. Joyce, Plasma Phys., 20, 391, 1975.
Goldstein, H., and K. Schindler, J. Geophys. Res., 83, 2574, 1978.
Guyot, M., and Ch. Hollenstein, Phys. Fluids, 26, 1596, 1983.
Haerendel, G., An Alfvén Wave Model of Auroral Arcs, in B. Hultquist and T. Hagfors (eds.), High Latitude Space Plasma Physics, 515, Plenum, London, 1983.
Haerendel, G., and G. Paschmann, Interaction of the solar wind with

the dayside magnetosphere, in A. Nishida (ed.) Magnetospheric Plasma Physics, Center for Academic Publications, Reidel, Tokyo/Dordrecht, 1982.
Hamilton, J.E.M., and J.W. Eastwood, Planet. Space Sci., 30, 293, 1982.
Harel, M., R.A. Wolf, P.H. Reiff, and R.W. Spiro, J. Geophys. Res., 86, 2217, 1981a.
Harel, M., R.A. Wolf, R.W. Spiro, P.H. Reiff, C.-K. Chen, W.J. Burke, F.J. Rich, and M. Smiddy, J. Geophys. Res., 86, 2242, 1981b.
Harned, D.S., J. Comput. Phys., 47, 452, 1982.
Hashimoto, K., H. Matsumoto, Y. Serizawa, and I. Kimura, J. Geophys. Res., 88, 3072, 1983.
Hershkowitz, N., G.L. Payne, C. Chan, and J.R. DeKock, Plasma Phys., 23, 903, 1981.
Hewett, D.W., J. Comput. Phys., 38, 378, 1980.
Hewett, D.W., and C.E. Seyler, Phys. Rev. Lett, 46, 1519, 1981.
Hockney, R.W., and J. W. Eastwood, Computer Simulation Using Particles, MacGraw-Hill, New York, 1981.
Hoffman, R.A., and C.S. Lin, Study of Inverted-V Auroral Precipitation Events, in S.-I. Akasofu, and J.R. Kan (eds.), Physics of Auroral Arc Formation, 80, AGU, Washington, D.C. 1981.
Hollenstein, Ch., and M. Guyot, Phys. Fluids, 26, 1606, 1983.
Hones, E.W. Jr., J. Birn, S.J. Bame, J.R. Asbridge, G. Paschmann, N. Sckopke, and G. Haerendel, J. Geophys. Res., 86, 814, 1981.
Hones, E.W. Jr., Planet. Space Sci., 23, 393, 1979.
Hoshino, M., and A. Nishida, J. Geophys. Res., 88, 6926, 1983.
Hudson, M.K., W. Lotko, I. Roth, and E. Witt, J. Geophys. Res., 88, 916, 1983.
Hundhausen, A.J., Coronal Expansion and Solar Wind, Springer, New York, 1972.
Iizuka, S., P. Michelson, J.J. Rasmussen, and R. Schrittweiser, Phys. Rev. Lett, 48, 145, 1982.
Israeli M., and S.A. Orszag, J. Comput. Phys., 41, 115, 1981.
Jockers, K., Astron. & Astrophys., 6, 219, 1970.
Joyce, C., and R.F. Hubbard, J. Plasma Phys., 20, 391, 1978.
Junginger, H., The Poynting Vector of Boundary Surface Waves and Resonant Shear Alfvén Waves in a Cold, Inhomogeneous Plasma, submitted to J. Geophys. Res., 1983.
Kalinin, Y., D.N. Lin, L.T. Rudakov, V.D. Ryutov, and V.A. Skoryupin, Sov. Phys., -JEPT 32, 573, 1971 (Zh. Eksp. Teor. Fiz. 59, 1056 1970).
Kamide, Y., Space Sci. Rev., 31, 127, 1982.
Kan, J. R., Space Sci. Rev., 31, 71, 1982.
Kan, J.R., And D.W. Swift, J. Geophys. Res., 88, 69119, 1983.
Kan, J.R., and L.C. Lee, J. Geophys. Res., 85, 788, 1980.
Katanuma, I., and T. Kamimura, Phys. Fluids, 23, 2500, 1980.
Kaufmann, R.L., D.N. Walker, and R.L. Arnoldy, J. Geophys. Res., 81, 1673, 1976.
Kennel, C.F., and Ashour-Abdalla, M., Electrostatic Waves and the Strong Diffusion of Magnetospheric Electrons, in A. Nishida, Magnetospheric Plasma Physics, 245, Center for Acad. Publ., Reidel, Tokyo/Dordrecht, 1982.

Khan, S.A., and T.D. Rognlien, Phys. Fluids, 24, 1442, 1981.
Kindel, J.M., C. Barnes, and D.W. Forslund, Anomalous Dc Resistivity and Double Layers in the Auroral Ionosphere, in S.-I. Akasofu and J.R. Kan (eds.), Physical of Auroral Arc Formation, 296, AGU, Washington, D.C., 1981.
Kruer, W.L., Comments Plasma Phys., 5, 69, 1979.
Langdon, A.B., and B.F. Lashinski, Electromagnetic and Relativistic Plasma Simulation Models, in B. Adler, S. Fernbach, And M. Rotenburg (eds.), Methods in Computational Physics 16, 339, 1976.
Leboeuf, J.N., T. Tajima, C.F. Kennel, and J.M. Dawson, Global Magnetohydrodynamic Simulation of the Two-Dimensional Magnetosphere, in Quantitative Modeling of Magnetospheric Processes, W.P. Olson (ed.), Geophys. Monograph 21, AGU, Washington, D.C. 1979.
Leboeuf, J.N., T. Tajima, C.F. Kennel, and J.M. Dawson, Geophys. Res. Lett., 5, 609, 1978.
Leboeuf, J.N., T. Tajima, C.F. Kennel, and J.M. Dawson, Geophys. Res. Lett., 8, 257, 1981.
Leboeuf, J.N., T. Tajima, and J.M. Dawson, Phys. Fluids, 25, 784, 1982.
Lee, W.W., and J. Okuda, J. Comput. Phys., 26, 139, 1978.
Lemaire, J., and M. Scherer, Planet. Space Sci., 21, 281, 1973.
Lemaire, J., and M. Scherer, Phys. Fluids, 14, 1683, 1971.
Leroy, M.M., D. Winske, C.C. Goodrich, S.S. Wu, and K. Papadopoulos, J. Geophys. Res., 87, 5081, 1982.
Leroy, M.M., and D. Winske, Ann. Geophys., 1, 527, 1983.
Leung, P., A.Y. Wong, B.H. Quon, Phys. Fluids, 23, 992, 1980.
Lin, A.T., Phys. Fluids, 21, 1026, 1978.
Lin, A.T., and J.M. Dawson, Phys. Fluids, 21, 109, 1978.
Lyon, J.J., S.H. Brecht, J.D. Huba, J.A. Fedder, and P.J. Palmadesso, Phys. Rev. Lett, 46, 1038, 1981.
Lyon, J.J., S.H. Brecht, J.A. Fedder, and P. Palmadesso, Geophys. Res. Lett., 7, 721, 1980.
Lysak, R.L., and C.T. Dum, J. Geophys. Res., 88, 365, 1983.
Mallinckrodt, A.J., and C.W. Carlson, J. Geophys. Res., 83, 1426, 1978.
Maltsev, Yu.P., S.S. Leontyev, and W.B. Lyatsky, Planet. Space Sci., 22, 1519, 1974.
Mason, R.J., Phys. Rev. Lett, 47, 652, 1981a.
Mason, R.J., J. Comput. Phys., 41, 233, 1981b.
Mason, R.J., Phys. Fluids, 14, 1943, 1971.
Matte, J.P., and J. Virmont, Phys. Rev. Lett, 49, 1936, 1982.
McPherron, R.L., Rev. Geophys. Space Phys., 17, 657, 1979.
Milroy, R.D., and J.U. Brackbill, Phys. Fluids, 25, 775, 1982.
Miura, A., UCLA preprint, J. Geophys. Res., 89, 801, 1984.
Miura, A., S. Ohtsuka, and T. Tamao, J. Geophys. Res., 87, 843, 1982.
Miura, A., and T. Sato, J. Geophys. Res., 85, 73, 1980.
Mondelli, A., and E. Ott, Plasma Phys., 16, 413, 1974.
Naitou, H., S. Tokuda, and T. Kamimura, J. Comput. Phys., 33, 86, 1979.
Nakagawa, Y., Astrophys. J., 247, 719, 1981.

Nerney, S., and A. Barnes, J. Geophys. Res., 82, 3213, 1977.
Nevins, W.M., Y. Matsuda, and M.J. Gerver, J. Comput. Phys., 34, 226, 1981.
Nielson, C.W., and H.R. Lewis, Particle Code Models in the Nonradiative Limit, in B. Adler, S. Fernbach, and M. Rotenburg (eds.), Methods in Computational Physics 16, 367, 1976.
Okuda, H., C.Z. Cheng, and W.W. Lee, Phys. Fluids, 24, 1060, 1981.
Okuda, H., and M. Ashour-Abdalla, J. Geophys. Res., 88, 890, 1983.
Okuda, H., and M. Ashour-Abdalla, Phys. Fluids, 25, 1564, 1982.
Orlanski, I., J. Comput. Phys., 21, 251, 1983.
Prasad, S.S., D.J. Strickland, and Y.Y. Chiu, J. Geophys. Res., 88, 4123, 1983.
Priest, E.R., Plasma Phys., 25, 161, 1983.
Pritchett, P. L., M. Ashour-Abdalla, and J. M. Dawson, Geophys. Res. Lett., 8, 611, 1981.
Pritchett, P.L., and F.V. Coroniti, J. Geophys. Res., 89, 168, 1984.
Pulliam, D.M., H.R. Anderson, K. Stammes, and M.H. Rees, J. Geophys. Res., 86, 2397, 1981.
Quest, K.B., D.W. Forslund, J.U. Brackbill, and K. Lee, Geophys. Res. Lett., 10, 471, 1983.
Sakanaka, P.H., C.K. Chu, and T.C. Marshall, Phys. Fluids, 14, 611, 1971.
Sato, T. this volume, 1984
Sato, T., Auroral Physics, in A. Nishida (ed.), Magnetospheric Plasma Physics, 197, Center for Acad. Publications, Reidel, Tokyo/Dordrecht, 1982.
Sato. T., J. Geophys. Res., 84, 7177, 1979.
Sato. T., J. Geophys. Res., 83, 1042, 1978.
Sato. T., T. Hayashi, R.J. Walker, and M. Ashour-Abdalla, Geophys. Res. Lett., 10, 221, 1983.
Sato. T., H. Matsumoto, and K. Nagai, J. Geophys. Res., 87, 6089, 1982.
Sato, T., and A. Hasegawa, Geophys. Res. Lett., 9, 52, 1982.
Sato, T., and R.J. Walker, J. Geophys. Res., 87, 7453, 1982.
Sato, T., and H. Okuda, J. Geophys. Res., 86, 3557, 1981.
Sato, T., and T. Hayashi, Phys. Fluids, 22, 1189, 1979.
Sato, T., and T. Iijima, Space Sci. Rev., 24, 347, 1979.
Schindler, K., Macroinstabilities of the Magnetotail, in S.I. Akasfu (ed.) Dynamics of the Magnetosphere, Reidel, Dordrecht, 1980.
Sgro, A.G., and C.W. Nielson, Phys. Fluids, 19, 126, 1976
Silevitch, M.B., J. Geophys. Res., 86, 3573, 1981.
Singh, N., Plasma Phys., 22, 1, 1980.
Singh, N., H. Thiemann, and R.W. Schunk, Geophys. Res. Lett., 10, 745, 1983.
Singh, N., and R. W. Schunk, J. Geophys. Res., 87, 3551, 1982a.
Singh, N., and R.W. Schunk, J. Geophys. Res., 87, 9154, 1982b.
Smith, R.A., Physica Scripta, 25, 413, 1982.
Sojka, J.J., and R.W. Schunk, J. Geophys. Res., 88, 2212, 1983.
Sonnerup, B.U.O., Magnetic Field Reconnection, in L.J. Lanzerotti, C.F. Kennel, and E.N. Parker (eds.), 1 Solar System Plasma Physics 3 46, North Holland Publ., 1979.
Southwood, D.J., and W.J. Hughes, Space Sci. Rev., 35, 301, 1983.

Spicer, D.S., Space Sci. Rev., 31, 351, 1982.
Spicer, D.S., and R.N. Sudan, Beam Return Current System in Solar Flares, International Astronomical Union Symp. 107, 1983, Reidel, Dordrecht, to appear.
Spiro, R.W., M. Harel, R.A. Wolf, and P.H. Reiff, J. Geophys. Res., 86, 2261, 1981.
Steinolfson, R.S., and G. Van Hoven, Phys. Fluids, 26, 117, 1983.
Stenzel, R.L., W. Gekkelman, and N. Wild, J. Geophys. Res., 88, 4793, 1983.
Strickland, D.J., D.L. Book, T.P. Coffey, and J.A. Fedder, J. Geophys. Res., 81, 2755, 1976.
Swift, D., and J.J. Ambrosiano, J. Comput. Phys., 44, 302, 1981.
Swift, D.W., J. Geophys. Res., 88, 125, 1983.
Swift, D.W., J. Geophys. Res., 88, 5685, 1983.
Swift, D.W., and L.C. Lee, J. Geophys. Res., 88, 11, 1983.
Syrovatskii, S.I., Ann. Rev. Astron. Astrophys., 19, 163, 1981.
Tajima, T., and V.C. Lee, J. Comput. Phys., 42, 406, 1981.
Tamao, T., and A. Miura, J. Geophys. Res., 87, 905, 1982.
Tanaka, M., and T. Sato, Phys. Rev. Lett, 47, 714, 1981b.
Tanaka, M., and T. Sato, J. Geophys. Res., 86, 5541, 1981a.
Temerin, M., K. Cerny, W. Lotko, and F.S. Mozer, Phys. Rev. Lett, 48, 1175, 1982.
Terasawa, T., J. Geophys. Res., 86, 9007, 1981.
Thiemann, H., N. Singh, and R.W. Schunk, Formation of V-Shaped Potentials, 6th ESA Symposium on European Rocket & Balloon Programmes, ESA-SP-183, 269, 1983.
Ugai, M., Phys. Fluids, 25, 1027, 1982.
Ugai, M., and T. Tsuda, J. Plasma Phys., 17, 337, 1977.
Vasyliunas, V.N., Mathematical Models of Magnetospheric Convection and its Coupling to the Ionosphere, in B.M. McCormac (ed.), Particles and Fields in the Magnetosphere, 60, Reidel, Dordrecht, 1970.
Wagner, J.S., J.R. Kan, S.-I. Akasofu, T. Tajima, J.N. Lebeouf, and J.M. Dawson, A Simulation Study of V-Potential Double Layers and Auroral Arc Deformations, in S.I. Akasofu and J.R. Kan (eds.), Phys. of Auroral Arc Formation, 304, AGU, Washigton, D.C., 1981.
Winske, D., Phys. Fluids, 24, 1069, 1981.
Winske, D., and M.M. Leroy, Hybrid Simulation Techniques Aplied to the Earth's Bow Shock, this Vol. 1984.
Wu, C.C., R.J. Walker, and J.M. Dawson, Geophys. Res. Lett., 8, 523, 1981.
Zimmerman, G.B., and W.L. Kruer, Comments Plasma Phys., 2, 85, 1975.

SUBJECT INDEX